Atmospheric Effects on Radar Target Identification and Imaging

NATO ADVANCED STUDY INSTITUTES SERIES

Proceedings of the Advanced Study Institute Programme, which aims at the dissemination of advanced knowledge and the formation of contacts among scientists from different countries

The series is published by an international board of publishers in conjunction with NATO Scientific Affairs Division

A	Life Sciences	Plenum Publishing Corporation
B	Physics	London and New York
C	Mathematical and Physical Sciences	D. Reidel Publishing Company Dordrecht and Boston
D	Behavioral and Social Sciences	Sijthoff International Publishing Company Leiden
E	Applied Sciences	Noordhoff International Publishing Leiden

Series C – Mathematical and Physical Sciences

Volume 27 – Atmospheric Effects on Radar Target Identification and Imaging

ISBN-13: 978-94-010-1533-2   e-ISBN-13: 978-94-010-1531-8
DOI: 10.1007/978-94-010-1531-8

Published by D. Reidel Publishing Company
P.O. Box 17, Dordrecht, Holland

Sold and distributed in the U.S.A., Canada, and Mexico
by D. Reidel Publishing Company, Inc.
Lincoln Building, 160 Old Derby Street, Hingham, Mass. 02043, U.S.A.

Atmospheric Effects on Radar Target Identification and Imaging

Propagation Effects on the Non-Ionized Atmosphere on the Presentation and Analysis of Radar Targets, Especially in the mm- to m-Range of the Electromagnetic Spectrum

Proceedings of the NATO Advanced Study Institute held in Goslar/Harz, F.R.G., September 22 – October 3, 1975

edited by

H. E. G. JESKE

Meteorologisches Institut der Universität Hamburg, F.R.G.

D. Reidel Publishing Company

Dordrecht-Holland / Boston-U.S.A.

Published in cooperation with NATO Scientific Affairs Division

CONTENTS

PART III
WORKSHOP REPORTS

PREFACE

The Advanced Study Institute (ASI) under discussion was
initiated by the "Special Programme Panel on Radio-
meteorology" of the Scientific Affairs Division of
NATO. The domain of this panel - and consequently the
topics of their former ASI-programmes - is the influence of the non-ionized atmosphere on electromagnetic
wave propagation, its prediction and its use as a remote sensing technique. It is the final goal to inform
radio and radar engineers about the various defects
caused by the propagation medium atmosphere. Today
there exist high-sensitive radar systems which can provide identification and produce images of distant objects very accurately by measuring a) the effect of the
target on the shape of a short radar pulse, or b) the
wave front (phase and amplitude distribution) and its
orientation in space. But usually the radar-to-target
path is through the inhomogeneous and turbulent atmosphere and so the absolut limits of the system are very
often determined by this atmosphere. It was the plan of
this ASI to arrange an interdisciplinary information
exchange between radar experts and propagation specia-
lists in order to get a better understanding of the
susceptibility to atmospheric effects and to develope
new methods that will reduce or correct these errors.
The lectures given and especially the intensive discussions during the workshop sessions contributed to
this aim. Special points of clarification were:

a) to identify the signal properties that are used in
   the present techniques for target identification
   and imaging,

b) to consider the characteristics of various types of
   objects in terms of their effects on these signal
   properties,

c) to identify how errors in the signal measurements
   are translated into errors in target properties, and

d) to examine known and expected effects from transmission through the atmosphere and interpret these
   in terms of fundamental limitations in the performance of systems based on the present techniques.

This volume includes most of the contributions and
workshop reports given at the ASI. At the meeting the
programme was divided into three sections:
a) Fundamental problems, b) Atmospheric influences,
and c) Specific techniques and their application.
Part a) and c) are combined here in chapter I. Chapter
II relies mainly on the original programme. Under
chapter III the workshop reports are collected.

Both groups assembled at the ASI had fruitful and
stimulating discussions when confronted with the
problems of the other side.

The pleasant setup at the hotel "Der Achtermann" at
Goslar/Harz contributed to a successful meeting.

The scientific part of the ASI could not have come to
pass without the intensive and indefatigable help of
the members of the programme committee of this ASI:
D.T. Gjessing, E.M. Kennaugh, and M.C. Thompson. I wish
to express my sincere thanks to them.

I may also thank cordially all participants for a good
cooperation and vivid engagement in their different
functions as lecturer, session chairman, working group
chairman or student. Last not least I want to thank the
ASI-staff for effective cooperation in a delightful
atmosphere.

For financial and personal support I would like to
thank the "Forschungsgruppe für Radiometeorologie an
der Universität Hamburg - Institution der Fraunhofer-
Gesellschaft".

Summarizing I would say the ASI-Goslar was worthwile,
indeed.

H. Jeske

LIST OF PARTICIPANTS

A.V. Alongi, Calspan Corporation, Buffalo, New York,
USA   +), ++)
M.E. Bechtel, Calspan Corporation, Buffalo, New York,
USA   +)
K.D. Becker, Universität des Saarlandes, Fachbereich
Elektrotechnik, Saarbrücken, FRG   ++)
M. L. Boithias, Centre National d'Etudes des Télé-
communications (CNET), Issy les Moulineaux, France
(temporary)
S.R. Brooks, Marconi Research Laboratories,
Chelmsford, Essex, UK   +)
P.J. Cabion, National Institute for Communication
Research, Johannesburg, South Africa
K.G. Corless, Royal Radar Establishment,
Great Malvern Worcs, UK   +), +++)
C. Fengler, McGill University, Dep. of Electrical
Engineering, Montreal, Que., Canada   +) (temporary)
W.L. Flock, University of Colorado, Boulder, Colorado,
USA   +), ++)
W. Gabsdil, Fa. Telefunken, Ulm, FRG
D.T. Gjessing, Norwegian Defence Research Establish-
ment, Division for Electronics, Kjeller, Norway +), ++)
O. Godart, Université de Louvain, Inst. d'Astronomie
et de Géophysique, Louvain-la-Neuve, Belgium   ++)
(temporary)
G. Graf, Deutsche Forschungs- und Versuchsanstalt für
Luft- und Raumfahrt e.V., Oberpfaffenhofen/Obb.,
FRG   +)
J.T. Hall, Armament Development Test Center, Florida,
USA   ++)
M.P.M. Hall, Appleton Laboratory, Science Research
Council, Slough, Bucks, UK (temporary)
H.S. Hayre, University of Houston, Houston, Texas,
USA   +)

+)   Lecturer
++)   Session chairman
+++) Working group chairman

R.A. Helvey, Pacific Missile Test Center, Point Mugu,
California, USA

D.B. Hodge, The Ohio State University, Dep. of Electrical
Engineering, Columbus, Ohio, USA   +)

H. Jeske, Meteorologisches Institut der Universität
Hamburg, FRG

E.M. Kennaugh, The Ohio State University, Dep. of
Electrical Engineering, Columbus, Ohio, USA   +), ++)

T. Kester, NATO Scientific Affairs Division, Brussels,
Belgium (temporary)

A.G. Kjelaas, Norwegian Defence Research Establishment, Kjeller, Norway

N. Klint Hansen, Danish Defence Research Establishment, Copenhagen, Denmark

E. Lakatsch, Forschungsgruppe für Radiometeorologie
an der Universität Hamburg, Hamburg, FRG

K.-J. Langenberg, Universität des Saarlandes, Fachbereich Elektrotechnik, Saarbrücken, FRG,   +)

O. Loevhaugen, Central Institute for Industrial
Research, Blindern-Oslo, Norway

K. Magura, Forschungsinstitut für Hochfrequenzphysik,
Wachtberg-Werthhoven, FRG

E. Mehlum, Central Institute for Industrial Research,
Blindern-Oslo, Norway   +)   (temporary)

A. Nania, Richerche Operative Meteorologia
Aeronautica, Brindisi, Italy

J. Nilsson, Research Institute of National Defence,
Stockholm, Sweden

R.L. Olsen, Government of Canada, Dep. of Communications, Ottawa, Ontario, Canada

A. Plaisant, NATO-SACLANT ASW Research Centre,
La Spezia, Italy

S. Marder, Environmental Research Institute of
Michigan, Ann Arbor, Michigan, USA   +), ++)
(temporary)

D. Rauch, NATO-SACLANT ASW Research Centre,
La Spezia, Italy

J.H. Richter, Naval Electronic Laboratory Center,
San Diego, California, USA   +), ++)   (temporary)

R.C. Runnels, Texas A+M University, Dep. of Meteorology,
College Station, Texas, USA

K. v.Schlachta, Forschungsinstitut für Funk und
Mathematik, Wachtberg-Werthhoven, FRG   +), +++)

H.E. Speckter, Seezeichenversuchsfeld, Koblenz-Lützel,
FRG

G. Tacconi, Istituto di Elettrotecnica, Genova, Italy +)

R.G. Taylor, ASWE, Portsmouth, UK   +++)

M.C. Thompson, Institute for Telecommunication
Sciences, Boulder, Colorado, USA   +)

D. W. Thomson, Dep. of Meteorology, The Pennsylvania
State University, University Park, Pennsylvania,
USA   +)
E. Topuz, I.T.U. Elec. Eng. Fac., Istanbul, Turkey
B.L. Trotter, NOAA, Boulder, Colorado, USA
M. Vogel, DFVLR, Institut für Flugfunk und Mikrowellen,
Oberpfaffenhofen bei München, FRG   +++)
W. Vogel, University of Texas, Electrical Engineering
Research Laboratory, Austin, Texas, USA   +)
A.T. Waterman, Stanford Electronics Laboratories,
Radioscience Laboratory, Stanford, California,
USA   +), ++)
B. Yazgan, Teknik Universite, Istanbul, Turkey

Staff:

H. Cassebaum
R. Eissing   } Forschungsgruppe f. Radio-
Y. Jeske (temporary)   } meteorologie an der Universität
I. Rapson   } Hamburg (Institution der
G. Weber   Fraunhofer-Gesellschaft)

INTRODUCTION

M. C. Thompson, Jr.

U. S. Department of Commerce,
Office of Telecommunications,
Institute for Telecommunication Sciences

The potential of radio waves to detect the presence and certain properties of distant objects was demonstrated in the original experiments by Hertz. The use of Lecher wires in which the standing waves on a transmission line are used to measure the wavelength of oscillators was an early practical application. Refined versions of this method are used to determine complex impedance of various radio frequency devices such as antennas, attenuators, etc.

As a result of many advances in technology and understanding of electromagnetic wave behavior, modern radars are capable of detecting and determining the characteristics of complex objects at distances of hundreds of kilometers. To do this has required advances in many areas such as the development of high power microwave sources, sensitive low-noise receivers, sophisticated filtering techniques and complex methods of data analysis and interpretation.

One class of these microwave systems can be used to obtain relief maps of the earth's surface similar to those made by aerial photography, but not limited to good seeing conditions. The example in figure 1 shows such a map of a mountainous tropical region nearly always covered with clouds. Until the radar imaging technique became available, there were no maps of this region.

In addition to the generation of terrain maps for geographic purposes such imaging techniques may be used for other purposes such as to record and describe the topographical development of cities, or to monitor surface changes from natural causes like floods, erosion, or earthquakes.

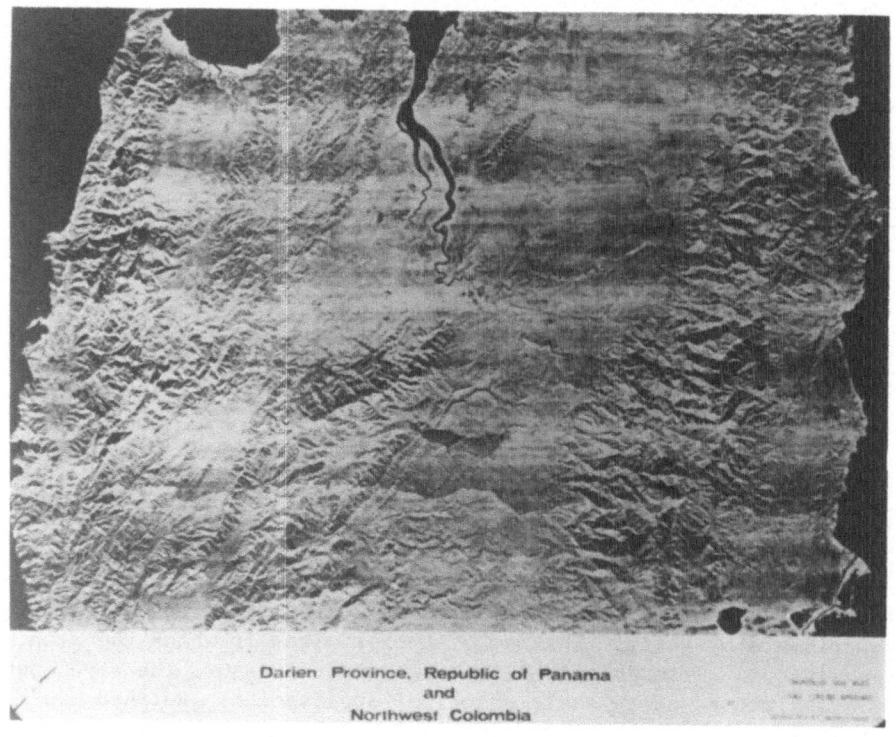

Figure 1. Map of Darien Province, Panama, made by side-looking
airborne radar (SLAR). U. S. Army Photograph

Another technique being developed makes use of the fact that
most natural and man-made objects are extended scatterers. Thus,
if any airplane is illuminated with a single pulse from a radar
transmitter, each element of the aircraft will reflect energy
depending on its electrical properties, its angular orientation,
and its range from the radar. The received pulse or echo is a
combination of these contributions from all elements of the target,
and, thus, in principle, contains information on the target characteristics. By the use of sufficiently short duration pulses,
it becomes practical to resolve details from the echo and, under
certain circumstances, to identify the object and deduce its
attitude in addition to its position in space.

In both techniques, the performance of the radar system is
influenced by the medium through which the electromagnetic waves
are propagated.

Various meteorological phenomena cause the signal velocity
or refractive index in the troposphere to vary randomly in space

and time. The resulting variations in delay time and wave front characteristics of the radar echo constitute a source of error in both synthetic aperture and short pulse radars. In the first type, distortion of the phase front is the principal effect, and in the second, pulse-broadening and variation in the propagation delay time can introduce uncertainties in the estimates of target characteristics.

It is important to understand these atmospheric effects for two reasons:

1)   to avoid expensive overdesign of systems whose performance will be limited by the atmosphere, and

2)   to help in developing new techniques that will minimize these errors.

The principal objectives of the lectures which follow are to:

1)   identify signal properties used in target identification and imaging,

2)   consider the characteristic signatures of various types of objects,

3)   show how errors in signal measurements translate into errors in target properties, and

4)   examine atmospheric properties in terms of their limitations on system performance.

Our lecturers represent several areas of research and we are confident that our discussions in this ASI will provide a valuable interdisciplinary exchange of information that will help in defining and solving the problems of "Atmospheric Effects on Radar Target Identification and Imaging."

P A R T I

TARGET CHARACTERISTICS AND THEIR DETERMINATION
BY MODERN RADAR TECHNIQUES

SHORT-PULSE TARGET CHARACTERISTICS

Marley E. Bechtel

Calspan Corporation
P. O. Box 235
Buffalo, New York 14221
U. S. A.

ABSTRACT. A survey of radar scattering principles related to the estimation of short-pulse (or, more generally, high-resolution) characteristics of radar targets is presented. The discussion is presented in the following sections: 1. Introduction; 2. Scattering centers on conducting targets; 3. Techniques for approximating scattering-center contributions; 4. Computation of short-pulse responses of radar targets; 5. Analytical approximations for specific scattering centers; 6. Examples of short-pulse target responses; 7. Relationships between FM/CW, chirp, and short-pulse radars; 8. Special considerations for short-pulse imagery; 9. Summary and conclusions; 10. Acknowledgments; 11. References.

1. INTRODUCTION

1.1 Historical background

The earliest use of the electromagnetic scattering characteristics of an unreachable object to permit inference of its physical nature probably occurred about 2500 years ago in Greece when various philosophers were considering the nature of the moon. The rough, solid surface of the moon was correctly deduced by some of these ancients on the basis of their observations of its reflected light. The fact that the moon does not produce a reflected image of the sun was interpreted, at least as early as Plutarch (c. 100 AD), as proof that the surface of the moon is not polished but rough like the surface of the earth.

An essentially correct explanation of the necessity that the surface of the moon be rough for it to appear as it does is

Jeske (ed.), Atmospheric Effects on Radar Target Identification and Imaging, 3–53.

contained in the following lines: "...Now let us consider what
would happen were the glass a sphere, or the lunar orb polished
like the glass... Only that small part [of the Moon] whose inclination is just right to reflect the light to the place of the
watching eye could ever be seen shining; all the rest of the reflecting surface would be obscure. Hence, were the Moon as smooth
as a looking glass, only a very small part would be seen to be
illuminated by the Sun, although the whole hemisphere is exposed
to the Sun's rays, and the rest would be invisible..." This description of specular reflection, given by Galileo in his "Dia-
logue on the Great World Systems" in 1632, can hardly be improved
upon in 1975.

Radar-wave scattering, the topic of interest here, is entirely a subject of the Twentieth Century. Although the use of artificially generated radio waves for the detection of remote objects
was proposed as early as 1900, it was not until the 1930's that
useful radar systems were developed. From then until after the
Second World War, the primary goal of any radar was the detection
and the determination of the location and velocity of a target.
Aside from a few special cases, such as the use of propeller
modulation for aircraft identification, little information beyond
signal amplitude and Doppler shift were available from which to
deduce target characteristics. This limitation on radar capability was imposed by several factors: peak power, bandwidth, and
receiver sensitivity and noise factor were not good enough to
permit the determination of target properties from the received
pulse shape.

Only with the introduction of chirp radars and improved
receivers in the 1950's and 1960's did it become practical to
obtain sufficiently detailed target-scattering-property estimates
to permit some degree of target identification. Since that time,
interest has grown in the use of radar for target identification,
and analytical target modeling techniques have also been developed
because of their value in establishing limitations on radar performance and in establishing suitable data-processing techniques.
The remainder of this presentation will concentrate on target-
modeling techniques, various types of high-resolution radars
being considered by other lecturers.

Electromagnetic scattering analysis began about a century
ago as a part of the study of light. Lord Rayleigh, for example,
computed the diffraction of light by small spheres, and Mie, in
the early 1900's, developed the complete formulation for scattering from spheres. During the Second World War, some progress
in radar scattering analysis, primarily based upon geometrical
and physical optics, was made. Scattering from the surface of
the sea and from precipitation were also investigated during that
period.

Modern radar scattering analysis dates from the introduction, about 25 years ago, of the geometrical theory of diffraction (GTD or GDT) by J. B. Keller at New York University. This technique, still under active development, is based upon asymptotic expansion techniques permitting the computation of the contributions from individual scattering centers and relies upon the essentially local nature of the scattering phenomenon.

Computation of high-resolution radar responses of targets by a variety of techniques has been a matter of interest for the past 10 or 15 years. Early investigations of such techniques were made by the author and others at Calspan Corporation (formerly the Cornell Aeronautical Laboratory, Inc.) and by E. M. Kennaugh and others at The Ohio State University. During this period, high-resolution measurement facilities were also developed at several organizations including Convair, Micronetics, Calspan, and Sperry-Rand.

The preceding historical survey is obviously very limited in scope, leaving out a great many individuals and organizations, but it should help to put in perspective some of the principal developments leading up to our present position. In the following sections we discuss some of the basic considerations involved in short-pulse scattering from radar targets.

1.2 Basic types of electromagnetic scattering

At least four basic types of electromagnetic scattering can be considered, and each of them in turn can take various forms. Some will be discussed in greater detail in subsequent sections.

Specular reflection, when it occurs, usually produces a larger amount of scattering than other forms. The requirement for specular scattering to occur is that the target be smooth relative to the illuminating wavelength, that the surface normal be parallel to the incident ray (or, in a bistatic system, lie on the bisector of the lines from the reflection point to the source and to the observer), and that the radii of curvature at the point of reflection be comparable with, or larger than, the wavelength, although in some cases this latter requirement can be significantly relaxed. Note that for most bodies specular reflection is strongly dependent on orientation as well as on shape; for many radar targets specular reflections occur only for very limited ranges of aspect angles.

Diffuse scattering is produced when the surface is rough with a height variation larger than a fraction of a wavelength, the limiting height variation depending upon the angle of incidence as well as on the wavelength. The most significant feature of diffuse scattering is that it is not so sensitive to aspect angle as specular scattering, in general. For many targets such as aircraft and missiles, rough-surface scattering is not significant at conventional radar frequencies, although it may become significant for millimeter waves and it is certainly important at laser frequencies. Even for more-conventional radar frequencies, rough-surface effects can be important for observation of the sea or of ground returns and for planetary radar astronomy.

Internal reflection and refraction are important contributors to radar returns from dielectric bodies. For most radar applications, the most important such scatterers are hydrometeors, i.e., raindrops, hailstones, snowflakes, etc. The Luneberg lens, a graded-dielectric sphere that can be made reflective on one side and used as a retrodirective scatterer, also should be mentioned as an important radar scatterer whose characteristics depend strongly on internal refraction effects. Note that this device is, in effect, a specular reflector, but that it is a specular reflector over a wide range of aspect angles, much like a corner reflector. Analyses of scattering from dielectric structures more complex than the sphere and spheroid have been made. For example, over a limited range of aspect angles an analysis based on geometrical and physical optics has yielded good results for a dielectric rectangular parallelepiped [1]. Analyses of effects of a plasma sheath on backscattering from an edge of a reentry vehicle have also been performed [2], as have analyses of the backscattering properties of a proposed dielectric aircraft wing [3]. Here we will concentrate almost entirely on scattering from conducting targets, because the inclusion of dielectric targets would greatly lengthen the necessary discussions.

For many types of radar targets the most important sources of backscattered energy are sharp discontinuities in metallic surfaces. Such discontinuities give rise to scattering contributions that tend to be much smaller than those produced by specular reflection but that are present over a wide range of aspect angles. For many radar targets, diffraction effects are the entire source of the radar return at most aspect angles. Other types of diffraction such as creeping waves and reentrant-cavity scattering also can be important (see Section 2.3).

2. SCATTERING CENTERS ON CONDUCTING TARGETS

2.1 Reality and significance of scattering centers

When an oscillatory waveform excites a conducting body with dimensions large relative to the wavelength, scattering arises at discontinuities in the contour of the body, for example, at specular-reflection points and at changes in curvature (or in spatial

derivatives of such curvatures) of the body. When the contour of
the body is smoothly varying, and when it is so oriented that
specular reflection does not occur in the direction of the observer,
cancellation of the scattering from nearby elements of the surface
results in almost no scattered signal at the observer. The effect
is analogous to that encountered in integration of rapidly oscil-
lating functions in which the principal contributions to the integral arise at stationary-phase points and at the end points of
the range of integration. Scattering can thus be considered to
arise from a discrete, usually small number of these so-called
scattering centers on the body. Principal classes of scattering
centers are described in Section 2.3, and analytical approximations
to their scattering amplitudes are given in Section 5.2. More
detailed discussions of some types of scattering centers have been
given by Bechtel and Ross [4].

Although the scattering-center concept arises from the mathematical analysis of the scattering process, it should not be regarded as only an artifice. Observation of radar targets with
ultra-high-resolution radars produces returns in accord with the
predictions of scattering-center theory. In some cases, analytical
formulations may not yet be sufficiently developed to yield accurate amplitude estimates, but the radar returns always arise
from scattering centers rather than from the surface as a whole,
provided the illuminating waveform is oscillatory and the target
is at least several wavelengths in extent, a common situation
with many operational radars.

The scattering-center concept is highly useful for analysis
of radar scattering because consideration of a small number of
types of scattering centers is sufficient to permit estimation of
the scattering from many types of relatively complex radar targets.
In addition, the amplitudes of returns from appropriately located
scattering centers represent the limits to the performance of a
short-pulse radar observing the target, because, in general, the
best the radar could do is to resolve the scattering centers in
range and azimuth, the latter obviously requiring either a very
high angular resolution or some form of Doppler processing and
therefore a more elaborate system than one simply providing very
high range resolution. Here we will restrict the discussion to
high resolution in range only, except for a very brief mention of
angular glint in Section 8.4. As will be indicated shortly, an
operational radar will generally have insufficient bandwidth to
permit resolving all scattering centers on a typical target; the
received signal will arise from combinations of scattering centers
as well as from some individual scattering centers.

2.2 Bistatic vs. monostatic radar

Although most of the discussion given here is for monostatic radars, i.e., radars having transmitting and receiving antennas in the same (or very nearly the same) location, similar considerations apply for bistatic radars, i.e., radars having separated transmitting and receiving antennas. In general, provided the angular separation between antennas (as viewed from the target) is not too large, the bistatic return is the same as the monostatic return along the bisector of the transmitter-target-receiver (bistatic) angle with the radar frequency reduced by a factor of $\cos(\beta/2)$, where β is the bistatic angle [4(p.41),5]. The main limitations on this approximation are that it does not work well for reflex (reentrant) scatterers or other scattering centers having narrow lobe structures of their own. For many typical scattering centers, such as sharp edges at the ends of cylinders, for example, the simple technique for relating bistatic to monostatic scattering is valid for bistatic angles of 10 degrees or more. At wide bistatic angles, geometric shadowing effects dominate the bistatic scattering process and must be accounted for in each particular case.

2.3   Scattering-center classes

Scattering centers on conducting bodies are of several types. Qualitative descriptions of the principal types, together with a few pertinent references, are given in this section; specific formulations are given in Section 5. Much more extensive bibliographies are available elsewhere; see, for example, Reference 4 and the special issues of the Proceedings of the IEEE for August 1965 and for November 1974.

Specular scattering, as described earlier, arises when a surface on a target has a normal in the same direction as the radar (assuming, from here on, monostatic radars). Such a surface area may be flat, as for a flat plate, singly curved, as for a cylinder viewed normally to its axis or a cone viewed normally to its generator, or doubly curved, as for a spheroidal scatterer. Conventional geometrical or physical optics provide quite accurate estimates for scattering contributions from specular points, often even if the radii of curvature are not much greater than the wavelength. Specular scattering, when it occurs, usually is a dominant contributor to the radar return from a body.

A second class of scattering centers of great practical importance arises at sharp edges such as those at the ends of a right circular cylinder or right circular cone or on a metallic prism or flat plate. Such edges produce rapid changes in surface currents and a consequent radiation from the neighborhood of the edge. If the edge lies in a constant-phase plane, the entire edge becomes a ring discontinuity and contributes to the return as a

whole. If the edge is viewed obliquely, only the neighborhoods of
the nearest and farthest points (strictly speaking, stationary·
points of the path to the edge) contribute. Thus a ring discontinuity acts as a single scattering center if observed from one
aspect angle and becomes two scattering centers at other aspect
angles, although one of these scattering centers may not be observed because of shadowing effects. Sharp-edge diffraction is
important because for many simple target classes it represents
the principal source of radar returns at all angles other than
those producing specular returns. The geometrical theory of diffraction (GTD) has proved very useful for estimating the returns
from sharp edges on bodies of revolution [4, 6, 7], on rectangular
cylinders (prisms)[8], and on flat plates [9]. Although single
diffraction is adequate in most cases for solid bodies, multiple
diffraction, in which rays diffracted from one scattering center
are then diffracted from another scattering center, must be included for flat-plate scattering. In some cases, such as diffraction from the discontinuity at the rear of a right circular cone,
the basic GTD is inadequate for one polarization and more elaborate
analyses are required [10, 11].

On some bodies, edges may be rounded off rather than being
sharp. If the radius of curvature is much less than a wavelength,
the edge behaves as if it were sharp. If the radius of curvature
is greater than a wavelength or two, the scattering process is
essentially specular for suitable aspect angles and, for nonspecular angles, is edge diffraction from a second-order edge (i.e.,
one with continuous first derivatives but discontinuous second
derivatives), the latter case producing a very small backscattered
field. For the large-radius case, asymptotic developments based
on physical optics or the GTD can be used. Ross [12] and Ross
and Hamid [13] have produced numerical solutions to the problem
of the wedge with a small-radius rounded edge which can be used
to estimate the effects of edge rounding on body scattering
characteristics (see also Section 5.3).

Ducts, reentrant cavities, and so-called reflex scattering
centers such as corner reflectors provide a set of very difficult
analytical problems to which satisfactory solutions are not yet
available except in special cases. Cavities and ducts such as
jet-engine intakes and exhausts are hard to analyze because of
the complex propagation processes that can exist, especially if
the duct dimensions are not much greater than the wavelength so
that the duct behaves like a waveguide. For simple, terminated
rectangular or circular ducts solutions are available [14], but
these results are difficult to compute and cannot be generalized
to more complex shapes such as those usually present in jet engines.

Pointed and rounded tips can be approximated well by the

expressions developed from physical optics. A rounded tip produces a specular return (if viewed from an appropriate specular angle) that is essentially equal to the specular return (not the total return) from a sphere of the same radius even for very small radii [15(pp. 134-136)] (see also Section 5.5).

Finally, there is a form of scattering center produced by creeping waves, i.e., waves that propagate around a singly or doubly curved surface and then are launched back towards the radar. Such waves continue to radiate as they propagate on a curved surface, shedding their energy as they go, thus producing an exponential rate of decay with propagation distance on the surface. Analysis of creeping-wave contributions is possible using the GTD [16], but for scattering from spheres there is some question as to the validity of the higher-order contributions as derived from the basic GTD [17], although Senior and Goodrich [18] have developed a valid creeping-wave formulation for the sphere starting from the exact solution.

3.   TECHNIQUES FOR APPROXIMATING SCATTERING-CENTER CONTRIBUTIONS

Analytical techniques for the computation of scattered fields range from exact methods to very crude approximations useful only for order-of-magnitude estimates of scattered-field levels. The literature in this field is vast and only a very limited number of references can be cited here. Two of the best sources of formulations are the books by Bowman, Senior, and Uslenghi [19] and by Ruck, Barrick, Stuart, and Krichbaum [20].

3.1 Exact methods using boundary-value solutions

Exact methods based on evaluations of boundary-value solutions are limited to simple bodies such as the sphere or spheroid, infinitely long cones or paraboloids of revolution, etc. for which the surface of the body coincides with a coordinate surface in a coordinate system in which the wave equation is separable. Solutions of such problems have proved valuable in the development of asymptotic techniques but are themselves of limited usefulness because of their limited applicability to common radar targets and the complex series of computations usually needed to obtain numerical results. Exact solutions, where they exist, are given by Bowman, Senior, and Uslenghi [19] and in other references.

The principal class of exact solutions that has found use is that involving two-dimensional scattering problems whose asymptotic solutions can be used in the GTD formulations of scattering from three-dimensional bodies (Section 3.4). Other applications of exact solutions have involved quasi-exact techniques discussed in Section 3.5.

3.2 Geometrical optics

Specular scattering from doubly curved surfaces can be analyzed using simple geometrical optics which basically assumes
that the wavelength is vanishingly small. The method is based
upon Fermat's principle and simply accounts for the divergence
of the reflected ray paths produced by the local curvature of the
body near the reflection point. If one or both radii of curvature
are infinite, as, for example, in the case of a cylinder or a flat
plate, geometrical optics, being unable to account for diffraction
effects which depend on a nonzero wavelength, cannot be used.

Geometrical optics can also sometimes be used to estimate
refraction by dielectric bodies [1], but such applications are
outside the scope of the present discussion.

3.3 Physical optics

Physical optics is a technique in which the Chu-Stratton
radiation integral, which is itself exact when the surface currents are known, is evaluated on the assumption that the surface
current at a point in the directly illuminated region is equal to
the surface current that would flow on an infinite plane tangent
to the surface at that point [20(pp. 50-63), 21(p. 27-20)]. It
is assumed that the surface current in geometrically shadowed
regions is zero. A basic assumption for physical optics to apply
is that body dimensions are much greater than the wavelength and
that the source is very far from the body. Evaluation of the
scattered field using physical optics is straightforward, as only
a single integration across the body is usually required.

By means of asymptotic expansions of the physical-optics integral, one can show that the principal contributors to the scattered field are specular reflection points and edges or other
discontinuities in the surface of the body. These regions of
principal contribution are the scattering centers discussed above.

Because of the assumption regarding surface currents, physical optics must be used with considerable caution, incorrect predictions occurring in many situations. For example, the assumption of zero surface current in a shadowed region often produces
a spurious "shadow-boundary" return that in fact does not exist.
(Although there may be a diffracted component of the scattered
field that appears to arise at the shadow boundary, it arises
from diffraction around the rear of the body produced by creeping
waves or multiple diffraction, not from a discontinuity in surface

current at the shadow boundary.) Another limitation of physical optics is that it predicts no depolarization of the backscattered field except in special cases involving multiple reflections. Physical optics does predict depolarization for bistatic scattering, but the results obtained there do not obey reciprocity and therefore must always be considered as a questionable approximation at best. Finally, it must be pointed out that because diffracted rays are not considered this method cannot account for modifications in the diffracted field produced by perturbations in the geometrically shadowed portion of the body.

Although, as we have mentioned, physical optics is a technique that must be used with considerable caution because of its use of incorrect surface-current estimates, the method has served well in approximating returns from certain types of scattering centers. Either monostatic or bistatic reflection from specular points is quite accurately predicted by physical optics, which, unlike geometrical optics, can be used for surfaces having infinite radii of curvature. Another useful application of physical optics is to the approximation of scattering from a second-order edge, as, for example, a smooth join of a spherical nose to a cone or a cylindrical edge to a wedge. Physical optics also gives the correct result for scattering from the tip of a circular cone, even though the radii of curvature become much less than a wavelength, but this result appears to be a coincidence and cannot be used to indicate general validity of physical optics when radii of curvature are small.

Therefore, although physical optics is subject to serious limitations it does form an essential tool for practical scattering analysis, provided the user is sufficiently aware of its limitations.

3.4 Geometrical theory of diffraction

The geometrical theory of diffraction (GTD or GDT in the literature) was originated by J. B. Keller in the early 1950's and has led to great advances in diffraction theory. The literature in this area is vast and no attempt will be made here to give a complete bibliography; extensive lists of references are given elsewhere [4, 12, 16, 20, 22-26], and we cite here only a few pertinent references. The best introduction to the principles of the GTD is that of Keller [22]. The most complete discussion of the underlying mathematics is probably Lewis and Keller [25], although that report preceded the more recent developments in uniform asymptotic expansions. A more recent discussion of the GTD and related theories and an extensive, up-to-date bibliography are given by Borovikov and Kinber [26]. Very extensive discussions of the GTD and related asymptotic analyses of diffraction problems are given by Felsen and Marcuvitz [27].

The basic premise of the GTD is that scattering of an electromagnetic (or acoustic) wave at high frequencies is largely a local
process, the scattering from a given point on a body depending on
its geometry in the neighborhood of that point. Fermat's principle is extended to include diffracted, as well as reflected or
refracted, rays. These diffracted rays are assumed to behave like
ordinary rays away from the diffraction region on the body, the
initial value of the field on a diffracted ray being obtained by
multiplying the field on an incident ray by a diffraction coefficient. This diffraction coefficient for a three-dimensional
body is obtained from a transformation involving a geometrically
determined divergence factor and from the asymptotic solution of
a two-dimensional canonical problem [22, 4]. The canonical problem that has served as the basis for many practical GTD solutions
to scattering problems such as cylinders, frustums, and cones is
that of diffraction from a wedge or a half-plane.

When the canonical problem can be solved rigorously and an
asymptotic expansion developed, as in the case of diffraction by
a wedge, the solution to a related three-dimensional problem can
usually be found with much greater accuracy than is the case when
physical optics is used, because the surface currents are much
more closely approximated. Depolarization effects are usually
predicted by the GTD for monostatic scattering, contrary to the
case when physical optics is used. For example, cross-polarized
backscattering from a right-circular cylinder, for some incident
polarizations, is quite accurately predicted using the GTD but is
predicted as zero when physical optics is used [21(p. 27-25),
4(p. 69)]. The GTD, because it permits inclusion of diffracted
rays, also allows inclusion of multiple-diffraction effects and
shadow-region contributions to the diffracted field, although such
solutions tend to become very complex and cannot be included here.

The original, and still very useful, form of the GTD has
singular regions in which the solution fails; such regions include
reflection and shadow boundaries and caustics. In the past few
years, uniform asymptotic expansion techniques have been developed
which do not fail in one or more of these regions [23, 26]. The
uniform asymptotic expansion techniques tend to be much more complicated than the original form and have not yet found wide use,
but they will no doubt find increasing applications in the future.
Recently, the GTD has also been applied to diffraction by an edge
or a vertex of an interface separating two media of different refractive indices, including diffraction by dielectric wedges or
cones [28].

Although the GTD is the most widely used asymptotic approach
to scattering analysis, it should be noted that other, closely

related techniques exist. The most used of these alternative techniques is the physical theory of diffraction developed by Ufimtsev in the U.S.S.R. [24, 26]. This method involves the asymptotic expansion of a "uniform" and a "nonuniform" component of the scattered field, the first being that predicted by physical optics, and the second, that produced by departures of actual surface currents from those predicted using the physical-optics assumption. The main difference in concept is that the GTD considers diffracted fields directly whereas the physical theory of diffraction works from equivalent surface currents. Knott and Senior [24] give a detailed comparison of these two techniques as well as a third, related technique called the method of equivalent currents.

Finally, we note that the GTD and related asymptotic techniques, although of interest here as methods of predicting diffraction from radar targets, also have been applied to such problems as antenna analysis, diffraction from discontinuities in waveguides, and radiation from leaky-wave structures [26, 27, 29, 30].

3.5 Numerical integration (quasi-exact) techniques

It is possible to solve Maxwell's equations through the use of numerical solutions to the appropriate integral equations for any target configuration. Because the solution is rigorous, there is no fundamental limitation that the body be very large, or very small, relative to the wavelength. The fundamental restriction on accuracy is imposed by the need to truncate an infinite set of simultaneous integral equations to obtain a finite-size matrix which can then be inverted using a digital computer. As more and more equations are retained, the solution becomes, in principle, more accurate [20(pp. 100-105)].

Numerical solutions are frequently referred to as the method of moments [31], and considerable work has been done using these techniques. Although it was hoped, when numerical-approximation techniques were first being developed about 15 years ago, that improvements in digital-computer capability would lead to solutions for all scattering problems, it now appears clear that even with very-high-speed computers the matrix-analysis problem will be of unacceptably high order for bodies many wavelengths in size. Of even more fundamental concern is the fact that the matrices obtained tend to be ill-conditioned, i.e., that the solution to the scattering problem tends to exhibit large changes following small changes in matrix elements. Thus, even though the solution may converge mathematically, it may not converge when the mathematical operations must be effected on a computer having a finite word length. Klein and Mittra [32] have shown that it is possible to obtain a condition number, or stability parameter, that is useful

in indicating whether a solution is ill conditioned. A third
disadvantage to the method of moments is its inability to provide
physical insight into the locations and scattering characteristics
of individual scattering centers or to permit easy approximations
to returns from scattering centers.

Because moment-method solutions are generally only useful
for small bodies, typically those with major dimensions no greater
than a few wavelengths, and because the GTD and other asymptotic
techniques are useful for bodies larger than a few wavelengths,
it is natural to attempt to combine the two forms of approximation
using the method of moments to obtain the scattering characteristics of small but complex scatterers, e.g., antennas, on the body
and the GTD method for other scatterers on the body. Work is
being conducted in this area, but as yet no widespread use has
been made of such procedures. In another approach to the extension of the capability of the method of moments, Burnside, Yu,
and Marhefka [33] have shown that by subtracting out a diffraction current based on the GTD solution one can greatly reduce the
required number of samples in the method of moments solution.

A quite separate form of numerical solution has been developed by C. L. Bennett. This method [34] involves an iterative
solution in the time domain taking advantage of the finite propagation velocity to simplify the solution. With modifications and
extensions, this method is claimed to be valid for bodies much
larger than can be treated using the more common method of moments.
Although this technique offers considerable promise in radar scattering analysis, it is outside the scope of the present discussion.

3.6 Measurement

The importance of measured scattering-center amplitudes must
not be forgotten. In many cases, particularly those involving
scattering from jet-engine ducts or aircraft cockpits, analytical
techniques are insufficiently advanced to permit valid estimates
of scattering-center contributions to be made. Scattering measurements of such contributions are of great use in establishing the
characteristics of these scattering centers. High-resolution
radar measurements are particularly valuable in permitting isolation of returns from specific scattering centers [21(pp. 27-16
to 27-18), 35]. Measurements are also important for the verifi-
cation of analytical approximations to scattering-center contributions, a matter to be discussed further in Section 8.1.

4. COMPUTATION OF SHORT-PULSE RESPONSES OF RADAR TARGETS

Several equivalent methods are used for the analytical

estimation of short-pulse (or, more generally, high-resolution) radar responses of targets. Because FM/CW and chirp radars give high-resolution responses that are essentially the same as a related short-pulse response (see Section 7), in the remainder of this section we consider only a short-pulse radar that transmits either a sinusoid with a specified envelope or a true short pulse.

4.1 Transient response from CW response using Fourier transform

If the CW scattering response of the target is known as some transfer function $G(f)$, the transient response is given by [36]

$$f_o(t) = \int_{-\infty}^{\infty} F(f)H(f)G(f)\exp(i2\pi ft)df \qquad (1)$$

where

$F(f) = \int_{-\infty}^{\infty} f_i(t)\exp(-i2\pi ft)dt$ is a Fourier transform,

$H(f)$ = bandpass characteristic of the system, and

$f_i(t)$ = input waveform.

The CW scattering characteristics of the target must be known (or well approximated) over a range of frequencies for which $F(f)H(f)$ is significantly different from zero, here defined as BW.

Two cases can be considered. First, suppose the transmitted signal is a true short pulse, as contrasted with the more common pulse-modulated sinusoid discussed below. The signal spectrum is in this case centered on zero, and significant components will exist for frequencies from some lower limit F_1 to an upper limit BW. F_1 is greater than 0 because frequencies approaching 0 cannot be radiated. The output of this system is then given by

$$f_o(t) = \text{Re}\{ \int_{F_1}^{BW} 2F(f)H(f)G(f)\exp(i2\pi ft)df \} \qquad (2)$$

Of more common interest is the pulse-modulated sinusoid

$$f_i(t) = f_1(t)\sin(2\pi f_o t + \phi) \qquad (3)$$

where

$f_1(t)$ is an envelope function,

f_o is the carrier frequency, and

ϕ is the phase reference for the sinusoid.

In a high-resolution radar, the bandwidth of the system may be very large and $f_1(t)$ may be nonzero for only a single cycle of the carrier, although usually several cycles of the carrier will be present. The bandwidth BW will now be roughly centered around

f_o, extending from $f_o - BW/2$ to $f_o + BW/2$.

It is convenient to formulate the problem in terms of the Fourier transform of the envelope, i.e.,

$$F_1(f) = \int_{-\infty}^{\infty} f_1(t) \exp(-i2\pi ft) dt \qquad (4)$$

The Fourier transform of $f_1(t)$ is then given by

$$F(f) = -0.5i \left[\exp(i\phi)F_1(f-f_o) - \exp(-i\phi)F_1(f+f_o) \right] \qquad (5)$$

The second term, the folded-spectrum contribution, is frequently nonnegligible for high-resolution applications and must usually be retained. We can now write equation 1 in the form

$$f_o(t) = \text{Re}\{ -i\exp(i\phi)\exp(i2\pi f_o t)A\exp(iv) \} \qquad (6)$$

where $A\exp(iv)$ is a complex envelope given by

$$A\exp(iv) = \int_{-f_o}^{\infty} \left[F_1(f) - \exp(-i2\phi)F_1(f+2f_o) \right] H(f+f_o) \\ G(f+f_o)\exp(i2\pi ft) df \qquad (7)$$

and this function can be evaluated using the fast-Fourier-transform (FFT) algorithm on a digital computer. In practice, it can be assumed that $H(f+f_o)$ is zero except for a range of frequencies from $f_o - BW/2$ to $f_o + BW/2$, and these limits become, in effect, the limits of the integration. Equation 7 is exact, the only approximations involved being those used for the F_1, H, and G functions. $A\exp(iv)$ can be seen to give in-phase and quadrature components of the signal if equation 6 is rewritten in the form

$$f_o(t) = (A\cos v)\sin(2\pi f_o t + \phi) + (A\sin v)\cos(2\pi f_o t + \phi) \qquad (8)$$

which shows that $A\cos(v)$ is the envelope of the in-phase component of the waveform and that $A\sin(v)$ is the envelope of the quadrature component of the waveform. For many applications only this envelope will be of interest, and it is obvious that once the complex envelope is available the complete waveform can easily be reconstructed using equation 6 or equation 8.

4.2 Transient response from impulse, step, or ramp response

If the response of a radar target to an impulse (delta function), step, or ramp excitation were known, simple Fourier analysis or convolution techniques would permit the computation of the transient response for any excitation [20(pp. 63-67), 37]. As this technique is the subject of a separate presentation by Dr. Kennaugh, we only give a brief summary here. Approximation to the impulse response is difficult, involving the same sorts

of problems encountered in any diffraction analysis. Kennaugh
and Moffatt [37] have shown that the use of physical optics for
the high-frequency response combined with moment conditions (not
to be confused with the method of moments discussed above) imposed
by the low-frequency (Rayleigh-region) scattering characteristics
of the body can lead to an impulse response from which good results
can be obtained even in the resonance region. As with conventional
physical optics, however, polarization information and shadow-
region contributions are lost, although it may be possible to
modify the impulse response to reduce some of the errors inherent
in physical optics; for example, a shadow-boundary contribution
can artificially be removed from the impulse response.

Attempts to base the impulse response of a target on the GTD
have also been made [38]. Here one would use the GTD expression
for the diffraction from each scattering center to compute an impulse response and then add these responses to obtain the impulse
response of the entire body. In this way, one would gain the advantage of the GTD over physical optics including polarization
dependence and inclusion of diffracted-ray and shadow-region contributions.

Unfortunately, it turns out that for some scattering centers
the GTD-derived impulse response is noncausal, apparently as a result of the specification of the transfer function in amplitude
and phase over an infinite range of frequencies [38(pp. 17-22)].
The analysis must be modified in some way so that at low frequencies the body as a whole is considered, not individual scattering
centers on the body. In addition, some modification to the GTD-
prescribed amplitude or phase functions may be necessary at high
frequencies. This problem remains open for further investigation.
One promising route might be to seek a solution to the problem
of diffraction of a pulse by a wedge and to use this result as
the canonical solution on which to base a time-domain form of the
GTD [39]. For a discussion of time-domain analysis of the scattering of pulses by wedges and half planes see, for example,
Friedlander's book [40].

4.3 Transient response from summation of transient responses of individual scattering centers

An alternative form of transient analysis, and one equally
as valid as those described above, involves the computation of
the transient responses, for the excitational waveform of interest,
of each scattering center and the subsequent summation of these
responses. This technique has been applied for scattering from
circular cylinders [38] with good results. Computations of this
type can be facilitated using a generalized form of paired-echo
analysis [38(Appendix A), 41]. Although this computational

approach may be advantageous in some applications, it can become cumbersome to execute and has been largely abandoned in favor of the method described in Section 4.1.

5.   ANALYTICAL APPROXIMATIONS FOR SPECIFIC SCATTERING CENTERS

In the preceding section we indicated how the transient response of a radar target can be computed from its CW response. Although direct computation of the transient responses of scattering centers is a promising area for future investigations, most available results are for CW scattering and we restrict the discussion to that case. Only a few basic results can be given here because of the great complexity and extent of radar scattering analysis. We restrict our discussion to backscattering from smooth metallic bodies, assume all relevant body dimensions to be greater than about 2 wavelengths, and assume the target is far from the radar. Generalizations of the results to permit rough-surface scattering or finite conductivity would be feasible (see, e.g., [42]), as would relaxation of the other restrictions cited here, but time and space do not permit such a discussion here.

One problem in approximating CW radar responses of scattering centers for subsequent transient analysis is the lack of phase information: many sources are concerned with RCS rather than with field strength and therefore do not preserve all necessary phase information. This problem becomes particularly serious if approximations to the scattering from different scattering centers must be obtained from different sources. One common source of difficulty is the use of scattering coefficients for H fields in some cases and for E fields in other cases, a factor of -1 being needed to convert from one to the other. Also very important is that the scattering contributions from all scattering centers be computed for $\exp(i2\pi ft)$ time dependence if $\exp(i2\pi ft)$ time dependence is used in the Fourier analysis (as is the case in Section 4.1). The use of $\exp(-i2\pi ft)$ time dependence in a transfer function will cause a reversal of the time scale. Throughout this section we assume E fields with an $\exp(i2\pi ft)$ time dependence.

5.1 Specular reflection

Specular reflection occurs whenever the normal to the surface of the target at the reflection point coincides with the radar line of sight. The discussion here is limited to the case in which the principal radii of curvature are of the order of a wavelength or greater and the first Fresnel zone occupies more than about one square wavelength. Specular reflection is very important for when it occurs it usually dominates other sources of scattering on the body.

Three principal types of specular reflection are considered here: doubly curved surfaces, singly curved surfaces, and flat surfaces. All of these types of reflection share one feature: they do not depolarize linear polarizations, although they will reverse the sense of circular polarization for backscattering; at some wide bistatic angles, complete depolarization of linear polarization can occur, but we do not consider that case here. Because there is little if any depolarization of the backscattered field (for linear polarization), the physical optics approximation should give good results, and experience indicates that such is the case.

In the remainder of this section, we will define the target transfer function by $G(k)$, defined as

$$\underline{E}_s = \underline{E}_i \frac{\exp(i2\pi ft - i2kr)}{r} G(k) \qquad (9)$$

where

$\underline{E}_s$ = scattered electric field,

$\underline{E}_i$ = incident (plane-wave) electric field,

r = distance from radar to reference point on target,

f = frequency,

t = time,

k = wave number = $2\pi f/c$, c = 2.997925×10^8 mps.

The $G(k)$ function serves as the transfer function of the target in the transient-response computation outlined in Section 4.1; note that the wave number k must be written as frequency when the Fourier analysis is performed. The radar cross section is related to $G(k)$ by

$$RCS = 4\pi |G(k)|^2 \qquad (10)$$

If an asymptotic expansion of the physical-optics integral is used, the transfer function for a doubly curved surface is found to be

$$G(k) = -0.5(a_1 a_2)^{0.5} \qquad (11)$$

where a_1 and a_2 are the principal radii of curvature at the specular point. Note that this result is independent of frequency. For a sphere, $a_1 = a_2 = a$ leads to the usual result for a sphere,

$$G(k) = -0.5a \qquad (12)$$

For the special case of a sphere, much more accurate expressions for $G(k)$ are available, these results arising from an asymptotic expansion of the exact solution [18, 43]. We have

$$G(k) = -0.5a\left[1 + 0.5(ka)^{-4} + i(2ka)^{-1}\right] \qquad (13)$$

For a greater than the wavelength, the RCS is changed by less than 0.03 dB by inclusion of the correction terms.

Singly curved surfaces occur on cylinders and cones; on such a body, the specular reflection arises on a line on the surface along which the surface normal is parallel to the radar line of sight. From an asymptotic expansion of the physical-optics integral, we can show that

$$G(k) = 0.5(k/\pi)^{0.5}\exp(-i3\pi/4) \int_0^L R^{0.5}dz \qquad (14)$$

where
   z is a coordinate along the specularly reflecting line,
   L is the length of the line, and
   R is the radius of curvature of the surface at the line in a plane normal to the line. Note that the specularly reflecting line must lie in a constant-phase plane but need not be a straight line in that plane, although it will be in the examples discussed here. Figure 1 shows the coordinates used. For a right-circular cylinder, $r_1=r_2=a$ and $\alpha=0$; for this case

$$G(k) = 0.5(ka/\pi)^{0.5}L\exp(-i3\pi/4) \qquad (15)$$

and the RCS is thus

$$RCS = kaL^2 , \qquad (16)$$

the usually quoted value. For the specular reflection from a

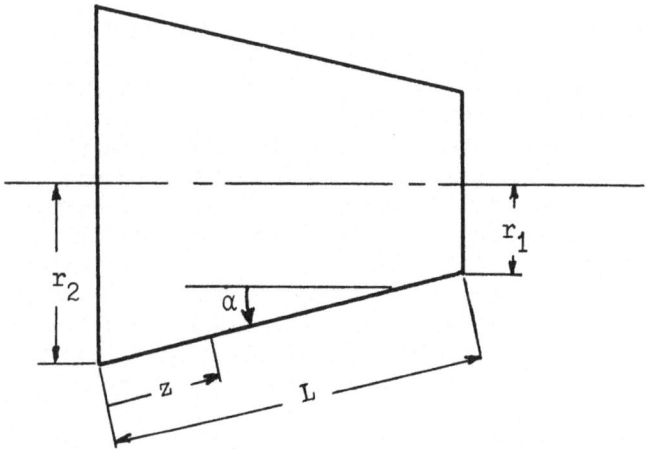

Fig. 1. Coordinates for specular reflection from a line

generator of a frustum of a right-circular cone (or a pointed right-circular cone if $r_1=0$), we have $R=(r_2-z\sin\alpha)/\cos\alpha$ and

$$G(k) = [k/\pi\cos\alpha]^{0.5}[r_2{}^{1.5}-r_1{}^{1.5}][3\sin\alpha]^{-1}\exp(-i3\pi/4) \quad (17)$$

and the RCS is

$$RCS = 4k[r_2{}^{1.5}-r_1{}^{1.5}]^2[9\cos\alpha\sin^2\alpha]^{-1} \quad (18)$$

which is, for $r_1=0$, the standard value for a pointed right-circular cone [6]. Although the specular return from a cone is occasionally approximated by that from a circular cylinder having a radius equal to some mean radius of the cone, equations 17 and 18 are more accurate and nearly as easy to use. Also, it is easy to show that equation 17 approaches equation 15 as $\alpha\rightarrow0$.

Finally, we consider the case of a flat surface normal to the line of sight. From asymptotic physical optics we obtain

$$G(k) = -ikA/(2\pi) \quad (19)$$

and

$$RCS = 4\pi A^2/\lambda^2 \quad (20)$$

where A is the area of the specularly reflecting region and λ is the wavelength. The shape of the area is not important provided no dimension across the area is much less than the wavelength.

5.2 Diffraction by sharp edges

Although specular scattering can be a dominant contributor to scattering when it occurs, at many observation angles specular scattering does not occur. Further, if a high-resolution radar is used, a nonspecular return may occur at a time different from that of the specular return so that it is a significant part of the output of the radar. Sharp edges are common contributors to the radar return from a target. Examples are the ends of cylinders, frustums, and combinations of these shapes; such axially symmetric bodies are easily treated using the GTD. We will also indicate how the GTD can be applied to straight, rather than curved edges, for example, to backscattering from a rectangular cylinder or prism. Some of the basic formulations are given here, but it is not possible to include more than a few practical examples.

The canonical problem on which the GTD relies for edge scattering is the diffraction of a plane wave by a perfectly conducting wedge, a two-dimensional problem that is used, with geometrical corrections, in the solution of a three-dimensional problem. The

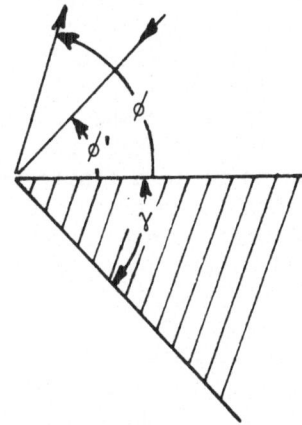

Fig. 2. Geometry of wedge

geometry of the wedge is shown in Figure 2. We assume here that
the rays from the transmitter and the receiver both are normal
to the edge, and soon we will let $\phi = \phi'$ to get the backscattering
result.

For plane-wave incidence, Kouyoumjian and Pathak [23(eq.20)]
show that

$$
\begin{bmatrix} E_\parallel^s \\ E_\perp^s \end{bmatrix} = \begin{bmatrix} E_\parallel^i D_s \\ -E_\perp^i D_h \end{bmatrix} \frac{\exp(i2\pi ft - i2kr)}{\sqrt{r}} \tag{21}
$$

where $\parallel$ indicates the electric vector is parallel to the edge and
$\perp$ indicates it is perpendicular to the edge. Superscript i indicates incident fields, and superscript s, scattered (diffracted)
fields. We reverse the sign of the D_s factor from that used in
Kouyoumjian and Pathak's ray-oriented coordinates, because their
$\hat\beta_0'$ and $\hat\beta_0$ are oppositely directed. The diffraction coefficients
are given, for most aspect angles ϕ' and ϕ, by [23(eq. 24)]

$$
D_{\substack{s \\ h}} = \frac{\sin(\pi/n)\exp(-i\pi/4)}{n(2\pi k)^{0.5}}\{[\cos(\pi/n) - \cos((\phi-\phi')/n)]^{-1}
$$

$$
\mp [\cos(\pi/n) - \cos((\phi+\phi')/n)]^{-1}\} \tag{22}
$$

where
$$n = 2 - \gamma/\pi$$
and the upper sign gives D_s and the lower sign gives D_h.

Most applications of the GTD have used the diffraction coefficient given by equation 22, and this expression is used in the

remainder of this section. The main disadvantage of equation 22 is that it yields infinite values for observation angles along reflection or shadow boundaries. For backscattering, the shadow boundary poses no problem, but if $\phi+\phi' = 2\phi = \pi$ or if $2\phi = (2n-1)\pi$ the second term in square brackets in equation 22 becomes infinite. In some cases, a pair of scattering centers will interact for specular-reflection angles in such a way as to yield a finite result which agrees with the specular-reflection value predicted by physical optics. In other cases, the singularity can be removed using a caustic-correction factor. In still other cases, the singularity cannot be removed.

The blowup of equation 22 at reflection boundaries can be avoided by using a uniform asymptotic expansion of the diffraction coefficient. Kouyoumjian and Pathak [23(eq. 25)] give such an expansion, but its application is much more complicated than that of equation 22. At the reflection boundary, the amplitude of the diffracted field is finite, but its sign undergoes a reversal so the sum of reflected and diffracted fields is continuous, the reflected field also being discontinuous in this region. Details of these applications have not yet been worked out, but it is obvious that the uniform-expansion techniques offer much promise.

First, we will consider the application of equations 21 and 22 to backscattering from axially symmetric bodies such as cylinders and frustums of cones. The divergence factor required to convert from two-dimensional to three-dimensional scattering is [22(section 4)]

$$DF = [r(1+r/y_1)]^{-0.5} \qquad (23)$$

where
    r = distance from edge to observation point, and
    y_1 = distance from edge to the caustic of diffracted rays. This factor is derived from the geometry of the ray paths, accounting for the fact that as the rays from a curved edge diverge their energy density must decrease. When the edge is a plane curve, y_1 can be found from [22(eq.21)]

$$y_1^{-1} = -\dot{\beta}/\sin(\beta) - \cos(\delta)/(y\sin^2(\beta)) \qquad (24)$$

where
    y = radius of curvature of edge,
    β = angle between incident ray and positive tangent to edge,
    $\dot{\beta}$ = $d\beta/ds$
    s = arc length along edge, and
    δ = angle between diffracted ray and normal to the edge lying in the plane of the edge and directed towards its center of curvature.

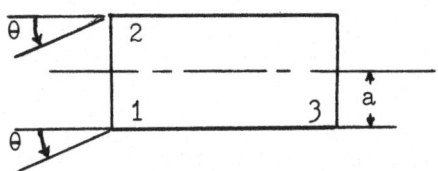

Fig. 3. Divergence-coefficient geometry

The only case to be considered here is that of a circular edge such as that at the end of a circular cylinder or frustum, the geometry being shown in Figure 3. Only the geometry of the curved edge is important, the wedge angle only entering the diffraction coefficients D_s and D_h and not the divergence coefficient DF. Two points on the circular edge contribute (for θ not too near 0), the point nearest the radar, shown as 1 in the figure, and the point furthest from the radar, shown as 2. At the near point we have $\beta = \pi/2$, $\dot{\beta} = -\sin\theta/a$, $\delta = \pi/2 + \theta$, and

$$DF_1 = r^{-1}[a/(2\sin\theta)]^{0.5} \tag{25}$$

The same form of coefficient is obtained at scattering center 3, although a different radius a may occur there. At the point 2, $\dot{\beta} = \sin\theta/a$, $\delta = \pi/2-\theta$, and

$$DF_2 = r^{-1}[a/(-2\sin\theta)]^{0.5} = (i/r)[a/(2\sin\theta)]^{0.5} \tag{26}$$

where this branch of the square root is the correct one for $\exp(i2\pi ft)$ time dependence. When these values of DF are used in conjunction with equations 21 and 22, the $1/r$ in DF_1 and DF_2 becomes $1/\sqrt{r}$ because a $1/\sqrt{r}$ factor has already been incorporated in equation 21.

Suppose the body of interest is a frustum of a right-circular cone, as shown in Figure 4. Scattering from each scattering center on the frustum is given by equations 21, 22, 25, and 26 by multiplication of the right side of equation 21 by $DF/\sqrt{r}$ to account for divergence of the ray bundles in three-dimensional scattering. At scattering centers 1 and 3 we use equation 25 for DF, and for scattering centers 2 and 4, equation 26. Scattering center 4 is directly observed only for $\theta < \alpha$. (We will not consider here the effects of multiple diffraction.) Two target transfer functions result, one, $G_\parallel(k)$, when the E vector is parallel to the tangent to the edge at the scattering center, and the other, $G_\perp(k)$, when the E vector is perpendicular to the tangent to the edge at the scattering center. We have, for each scattering center,

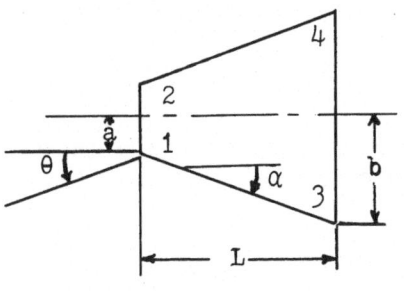

Fig. 4. Geometry for scattering from frustum

$$G_\parallel(k) = rD_s(DF)\exp(i2kd) \text{ , and} \tag{27}$$

$$G_\perp(k) = -rD_h(DF)\exp(i2kd) \tag{28}$$

where d is the distance, projected onto the line of sight, from the r=0 point to the scattering center and is positive if towards the radar. The target transfer function is the sum of the scattering-center transfer functions. In evaluating equation 22 the correct value of n must be used for each scattering center. Note that the r which appears as an amplitude factor in equations 27 and 28 can be replaced by the r to the center (or other reference point) of the target as we assume far-field conditions; d is used to account for phase shifts corresponding to scattering-center locations. In this discussion we assume r=0 to correspond to a point on the axis of the frustum midway between the bases. The transfer function for scattering center m is given by

$$_mG_\perp^\parallel(k) = \frac{\sin(\pi/n_m)\, A_m}{2n_m(\pi k\sin\theta)^{0.5}} \left[\mp B_m + C_m\right]\exp(iD_m) \tag{29}$$

where

$$n_m = 1.5 - \alpha/\pi \quad , \quad m = 1,2$$
$$ = 1.5 + \alpha/\pi \quad , \quad m = 3,4$$

$$A_m = \sqrt{a} \quad , \quad m = 1,2$$
$$ = \sqrt{b} \quad , \quad m = 3,4$$

$$B_m = \left[\cos(\pi/n_m) - 1\right]^{-1}$$

$$C_1 = \left[\cos(\pi/n_1) - \cos((\pi+2\theta)/n_1)\right]^{-1}$$

$$C_2 = \left[\cos(\pi/n_2) - \cos((\pi-2\theta)/n_2)\right]^{-1}$$

$$C_3 = \left[\cos(\pi/n_3) - \cos(2(\theta+\alpha)/n_3)\right]^{-1}$$

$$C_4 = \left[\cos(\pi/n_4) - \cos(2(\theta-\alpha)/n_4)\right]^{-1}$$

$$D_1 = 3\pi/4 + kL\cos\theta + 2ka\sin\theta$$

$$D_2 = 5\pi/4 + kL\cos\theta - 2ka\sin\theta$$

$$D_3 = 3\pi/4 - kL\cos\theta + 2kb\sin\theta$$

$$D_4 = 5\pi/4 - kL\cos\theta - 2kb\sin\theta$$

and where the notation $G_\parallel(k)$ indicates that the $G_\parallel$ result is obtained if the upper sign is used, and $G_\perp$, when the lower sign is used.

The transfer function for the frustum is the sum of the transfer functions for m = 1, 2, and 3, and, if $\theta<\alpha$, 4. For $\theta>\pi/2$, scattering center 4 will again become visible and scattering center 2 will be shadowed; the expression for the transfer

function for scattering center 4 must also be modified,
$\theta-\alpha$ in C_4 being replaced by $\theta-\pi/2$.

As mentioned earlier, at certain angles the value of D_s or
D_h becomes infinite. In some cases, the GTD singularities can be
combined to yield a finite (and sometimes correct) result even if
D_s and D_h of the individual scattering centers become infinite.

The singularity obtained near $\theta=0$ (axial incidence) is a
real one in the sense that even the combined returns from scattering centers 1 and 2 become infinite. It is possible, however,
to recognize in an expansion for small θ of the transfer function
for scattering centers 1 and 2 the asymptotic expansion of
$J_1(2ka\sin\theta)$ [7]. We can then use this function near $\theta=0$, getting

$$_1G(k) + {}_2G(k) = -ika^2 \left[\frac{J_1(2ka\sin\theta)}{2ka\sin\theta} \right]\exp(ikL\cos\theta) \qquad (30)$$

This result is independent of polarization, so we omit the
subscripts on $G(k)$. This combined transfer function is well-
behaved at $\theta=0$ since $J_1(z)/z\rightarrow0.5$ as $z\rightarrow0$. Also, note that at
$\theta=0$ equation 30 becomes equivalent to equation 19 which represents
the physical-optics value for specular returns from a flat plate
(in this case, a disk). In deriving equation 30 we have omitted
a small, polarization-dependent term which may well be spurious;
in any case, it vanishes at $\theta=0$. This result is independent of
cone angle α, thus applying to frustums, cylinders, or disks.
In a way, this result is surprising because it leads to a result
depending on a surface area from the behavior of effects occurring
at only two points. Although the returns from scattering centers
3 and 4 also become infinite at $\theta=0$, they can be shown to produce
a finite return as well.

Specular incidence on the side of the frustum causes the returns from scattering centers 1 and 3 to become infinite. For a
frustum, there appears to be no way to combine these returns to
yield a finite value, so at the specular angle we must use equation 17. If $\alpha=0$, the frustum becomes a right-circular cylinder.
For this special case, we can combine the returns from scattering
centers 1 and 3 to obtain, for angles near $\theta=\pi/2$,

$$_1G(k) + {}_3G(k) = \frac{(L/2)\exp(-i3\pi/4)(ka)^{0.5}}{(\pi\sin\theta)^{0.5}} \left[\frac{\sin(kL\cos\theta)}{kL\cos\theta}\right]$$

$$\times \exp(i2ka\sin\theta) \qquad (31)$$

which, at $\theta=\pi/2$, agrees with the physical-optics result, equation
18. For the specular reflection from a frustum, we might use the
angular dependence in the $\sin(z)/z$ function in equation 31 with
the angle θ changed to $\theta+\alpha$ in the $\sin(z)/z$ function but left
unchanged in the $1/\sqrt{\sin\theta}$ factor.

Although we assumed a frustum of a right-circular cone in the preceding example, the results given are more general. If α=0, the results apply for a right-circular cylinder. For a pointed cone, use scattering centers 3 and 4 plus some approximation for scattering from the tip of a cone (Section 5.5). For a compound body such as a cone-cylinder or cylinder-flare, the same expressions apply provided appropriate values of n are used and all of the unshadowed scattering centers are considered. Examples of results obtained from analysis and measurement are given, for example, by Kell and Ross [21(Figs. 38 and 39)].

The values of $G_{\parallel}$ for scattering centers 3 and 4 are not given very accurately by the GTD for $\theta < 40°$. This phenomenon has been investigated in considerable detail [36, 38] and some of the improved approximations obtained have been published by Ross [10] and by Bechtel [11]. The formulations obtained for cones at angles small enough so the base discontinuity is entirely illuminated give very good results [10]. Attempts to extend these results have led to improved accuracy but still do not predict the scattering characteristics as well as might be desired [11], although inclusion of a higher-order correction term with the GTD solution does help significantly [36]. In the special case of scattering from the rear of a cylinder, the modified solution [11] does give quite good results. Because of the complexity of the modified solutions, they are not included here, but the user of the GTD results given here should be aware of the possibility of errors in some applications and of the availability of more accurate approximations.

So far, our discussion has been limited to bodies of revolution. Another situation of interest is a finite-length edge such as the edge of a cube. For such an edge, the divergence coefficient described above cannot be applied, as it becomes infinite. Instead, we can assume the edge to radiate in an azimuthal pattern given by D_s and D_h and in a longitudinal pattern determined by the path lengths from the radar to scattering elements along the edge. This procedure is a reasonable approximation for a line of sight not very far from normal to the edge; as the angle between these lines becomes small, the backscattering from the edge also tends to become small and therefore of less concern. More accurate approximations are no doubt possible but are not considered here.

Conversion from two-dimensional to three-dimensional diffraction has been described previously [45] using an extension of a method derived by van de Hulst [46]. The assumption we make is that the conversion from cylindrical-wave scattering, i.e., the two-dimensional case, to three-dimensional scattering will be the same for an edge as it is for a long cylinder, the case van de Hulst treats. For a two-dimensional case, he has [46(p. 302)]

$$u_2 = [2/(\pi kr)]^{0.5} T(\theta) \exp(i(2\pi ft-kr)-i3\pi/4) \qquad (32)$$

where

u_2 = axial component of field (either E or H),

r = distance to observer,

$T(\theta)$ = azimuthal scattering pattern of cylinder,

and plane-wave incidence is assumed. For a finite cylinder of length L, viewed at a (small) angle ϕ off broadside, he gives

$$u_3 = \frac{L}{i\pi r} \left[\frac{\sin(kL\phi/2)}{kL\phi/2}\right] T(\theta) \exp(i(2\pi ft-kr)) \qquad (33)$$

as the far-zone backscattered field. We can thus get u_3, given u_2, from

$$u_3 = u_2 \, L\left[\frac{k}{2\pi r}\right]^{0.5} \exp(i\pi/4) \left[\frac{\sin(kL\phi/2)}{kL\phi/2}\right] \qquad (34)$$

For large values of ϕ, the relationship is questionable, but if $L \gg \lambda$ the field contribution becomes small. It is probable that a $\cos^2\phi$ multiplier should appear on the right-hand side of equation 34, but for small ϕ the correction is not significant; note that for $\phi \neq 0$ the D_s and D_h expression in equation 22 actually has a $|\cos\phi|$ in the denominator (omitted in our expression because we generally ignore off-normal incidence here) so the correction is really by $|\cos\phi|$ rather than by $\cos^2\phi$. An expression of the form of equation 34 can also be established for applications with curved wavefronts (Fresnel-region diffraction)[45] and has recently been extended to allow inclusion of arbitrary antenna patterns (to be published), although these formulations are very complex and require a large digital computer for evaluation.

Note that equation 34 can also be used to derive the ratio of RCS to two-dimensional RCS, i.e.,

$$\frac{RCS_3}{RCS_2} = \frac{4\pi r^2 |u_3|^2}{2\pi r \, |u_2|^2} = \frac{2L^2}{\lambda} \left[\frac{\sin(kL\phi/2)}{kL\phi/2}\right]^2 \qquad (35)$$

which is, for $\phi=0$, the conversion factor used by Ross [9].

Assume a rectangular cylinder such as that shown in Figure 5, and assume the line of sight is perpendicular to edges 1, 2, and 3. These edges, each of length 2h, will then form the primary sources of scattering from the cylinder. Using equations 9, 21, 22, and 34 we find the GTD expression for the transfer function corresponding to the field from scattering center 1 (i.e., edge 1) to be

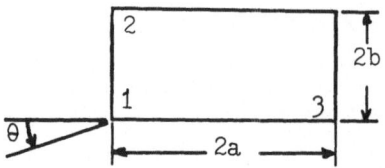

Fig. 5. Rectangular cylinder (prism)

$$_1G_{\parallel}^{\perp}(k) = \frac{h\sin(\pi/n)}{\pi n} \left\{ \pm \left[\cos(\pi/n)-1\right]^{-1} - \left[\cos(\pi/n)- \right.\right.$$
$$\left.\left. \cos((\pi+2\theta)/n)\right]^{-1} \right\} \exp\left[ik(a\cos\theta+b\sin\theta)\right] \qquad (36)$$

where r=0 corresponds to the center of the rectangular cylinder.
For edge 2, replace $(\pi+2\theta)/n$ by $(\pi-2\theta)/n$ and b by -b. For edge
3, replace $(\pi+2\theta)/n$ by $2\theta/n$ and a by -a. In a paper on scattering
from rectangular cylinders [8], two corrections should be noted:
the factors of 2b in the numerators of equations 2, 3, and 4 should
be 4b, and the $\pi/4$ phase shifts in equations 5, 6, and 7 do not
belong--conversions from two-dimensional to three-dimensional
scattering there used the square root of our equation 35 rather
than using our equation 34, thus losing a factor of $\exp(-i\pi/4)$.
Note further that we have the negative of the complex conjugate
of the results given previously because of our use of E fields
rather than H fields and of $\exp(i2\pi ft)$ rather than $\exp(-i2\pi ft)$.

Equation 36 is easily generalized for corners that are not
right angles by using the correct n for each corner and by adjus-
ting the θ-parameter expression to give $\phi+\phi'$ in equation 22. The
phase terms must also be adjusted for the appropriate geometry,
and shadowing of the edges must be accounted for.

For specular incidence on the flat faces of the prism, the
GTD expressions become infinite as they did for the circular cylinder. For a rectangular cylinder, the two scattering centers (edges)
bounding the specularly reflecting surface combine to give the
specular reflection from a flat plate (equation 19), just as the
scattering centers led to the correct broadside response from the
circular cylinder. Here the result is surprising because at broadside incidence four, rather than two, edges are contributing.

For generalized angles of incidence for which the line of
sight is not perpendicular to any edge, a similar procedure to
that of equation 36 but retaining the angle dependence of equation
34 might lead to useful results, at least in regions of relatively
strong backscattering, but this approach has apparently not yet
been tried. The problem cited above regarding the ability of two
of the four scattering centers to represent the specular return
indicates that difficulty will be encountered in formulating

general-incidence backscatter near specular incidence for generalized angles.

A related problem is that of backscattering from rectangular
flat plates. Although in principle this problem can be treated in
a manner analogous to that used for the rectangular cylinder, the
problem is far more complex because of the need to include multiple-diffraction effects [9]. Further work, particularly for near-
grazing angles of incidence, in needed. Because of the great complexity of the expressions needed for flat plates, the reader is
referred to Ross's paper [9] and to the discussion by Ruck, et.
al. [20(pp. 508-535)].

5.3 Rounded edges

In the preceding section, we gave some examples of methods
for approximating backscattering contributions from sharp edges
or discontinuities on radar targets. Many radar targets of potential interest have edges that are slightly rounded rather than
being sharp. Because the radius for which the rounded edge ceases
to produce the same scattered field as a sharp edge depends on
wavelength, edge rounding becomes a more significant factor at
short wavelengths such as are likely to be used for radar imaging
purposes. Ross [12] and Ross and Hamid [13] have developed an
exact formulation for the rounded edge which can, for small radii,
be treated as the sharp-edge result plus a perturbation term.
Unfortunately, the perturbation term cannot be expressed in any
simple form but must be computed numerically. This term depends
on the wedge angle, the radius of the rounding, polarization, and
viewing angle. Numerous curves showing these perturbation contributions are given in the references [12, 13], but these results
are too complex to include here. We note simply that a rounding
of radius a will increase the sharp-edge RCS by 6 dB if ka=2 (for
E parallel to edge) or if ka=0.8 (for E normal to edge) when the
wedge angle is 60°; for a 90° wedge, the corresponding radii give
ka=3 and ka=1.5; for a 120° wedge, they are >4 and 4, respectively.
In his dissertation, Ross [12] also gives numerous examples of the
effects of edge rounding on scattering from simple bodies.

The method of solution used by Ross cannot be applied if the
rounding of the edge becomes too large, because although the formulation is exact the matrices that are involved in the determination
of certain coefficients are ill conditioned, limiting solution
capability to ka values less than 4 in general but up to 7 in some
cases. For larger radii, physical optics can be used to determine
the return from the cylindrical part of the rounded wedge and also
the (relatively small) return from the cylinder-wedge junction, an
edge of second order [12(Section 3)].

For a second-order edge formed by the junction of two parabolic cylinders having radii of curvature $1/a_1$ and $1/a_2$ at their
join (Figure 6), Senior [47] has derived diffracted fields for the
second-order edge from which one can obtain diffraction coefficients
analogous to those given earlier for first-order edges (equation
22). When Senior's result is specialized for backscattering, the
diffraction coefficient equivalent to the earlier one for sharp
edges is

$$D_s \atop h = \Big[\frac{2}{\pi k} \Big]^{0.5} \exp(i\pi/4) \Big[\frac{a_2 - a_1}{2k} \Big] \Big[\frac{\mp(1-\cos(2\alpha))-2}{(2\sin\alpha)^3} \Big] \qquad (37)$$

and this diffraction coefficient can be used in the same manner as
that in equation 22. Note that this expression becomes infinite
for $\alpha = 0$ just as the other edge results do when a specular reflection occurs. Also note that a flat surface can occur on one side
of the junction here if either a_1 or a_2 is zero. If $a_1 = a_2$, the
discontinuity becomes zero and no diffraction occurs. Diffraction
by a second-order edge varies as $1/(k\sqrt{k})$ whereas that by a first-
order edge varies as $1/\sqrt{k}$. Much more extensive results, applicable
for edges and vertices of all orders, have been given by Kaminetzky
and Keller [48] who also consider the aspect angles for which equation 37 becomes infinite. Their method can even be applied to
dielectric bodies in some cases [28, 48].

5.4 Ducts and reentrant structures (reflex scatterers)

Two common types of reentrant or reflex scatterers are jet-
engine ducts and corner reflectors. Such scatterers as ducts, although in principle amenable to optics-type solutions (if wavelength
becomes comparable to duct diameter, much more complex processes
occur; see, e.g., Witt and Price [14]), generally cannot be analyzed
very accurately. In jet-engine ducts, for example, the interior
contour of the duct tends to be
complex and the termination is a
set of turbine or compressor
blades having complicated scattering properties of their own.
Application of simple models involving flat-plate equivalent
areas and empirical factors may
work well, but in some cases these
results have been found to be in
error by one to two orders of
magnitude in amplitude and to give
incorrect angular dependence as
well. To date, no reliable analytical methods for estimating

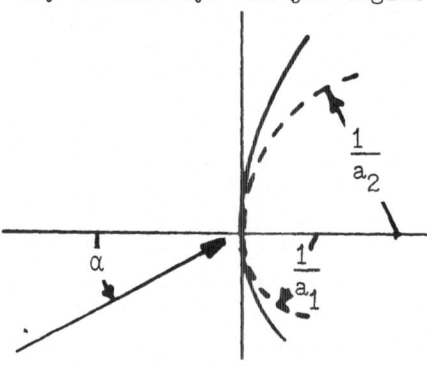

Fig. 6. Junction of two
parabolic cylinders

the scattering from ducts seems to be available, and measurements, at least on reasonably accurate scale models, seem to be nearly essential. Much further work remains to be done on scattering from ducts.

Scattering from corner reflectors can be approximated quite well by assuming optical reflections from the flat surfaces and by examining carefully the changes in polarization produced by the multiple reflections. Ross [49] has had some success in applying the GTD to a dihedral corner reflector as well as on cylinder-flare bodies which produce reflex scattering for some aspect angles.

5.5 Pointed and rounded tips

Backscattering from the tip of a narrow-angle circular cone at angles off the cone axis θ which are less than the cone half-angle can be approximated, for the incident electric vector parallel to a plane normal to the cone axis by [20(p. 119)]

$$G(k) = (-i/k)(\alpha/2)^2 \left[\frac{3 + \cos^2\theta}{4\cos^3\theta}\right] \tag{38}$$

when equation 2.3-53 is put into our notation. More detailed formulations are also available [20(pp. 378-385)]. For θ not very large, equation 38 can be expected to hold for the orthogonal polarization as well. Interestingly, this result is in agreement with physical optics for θ and α both small. For axial incidence backscattering, the RCS found from equation 38 becomes

$$RCS = \lambda^2/(16\pi)\alpha^4 \tag{39}$$

which is essentially the same as the physical-optics result, which has $\tan^4\alpha$ in place of α^4, provided α is small. Backscattering from a sharp tip, especially when short wavelengths are used, is a very small contribution to body RCS or transient response.

Of more practical interest is a cone with a smoothly rounded nose, because shapes of this type are more common and the amount of scattered radiation is significantly greater than that from sharp tips. Formulations for blunt-tip cones are given by Ruck et. al. [20(pp. 385-387)]. In brief, it appears valid to use physical optics for those angles for which the entire tip region of the body is illuminated, even for very small nose radii (see also [50] for further details). In our earlier notation, we get, using elementary physical optics,

$$G(k) = [-a/2 -i/(4k)] + \frac{i}{4k\cos^2\alpha} \exp(-i2ka(1-\sin\alpha)) \tag{40}$$

for the on-axis result. This result leads to an RCS value that
agrees with the RCS given by Weiner and Borison [50(eq. 31)] and
quoted in Ruck, et. al. [20(eq. 6.2-31)]. In equation 40 note that
the quantity in square brackets agrees exactly with the first two
orders of the expansion of the exact result for specular scattering
from a sphere, equation 13. For the off-axis case the result becomes much more complex [20, 50] and we have not evaluated the
physical-optics integral to find G(k); the references give only
the RCS which is not sufficient to get G(k) which requires phase
as well as amplitude. Note, however, that the sphere-cone junction return is generally much smaller than the sphere return, at
least for $a > \lambda/(4\pi)$, and that, in addition, this result is obtained
for axial incidence for which the entire second-order discontinuity
contributes in phase. At off-axis angles of incidence, the join
return should decrease significantly because only a limited region
on the join will contribute. Therefore, for either on-axis or
off-axis incidence we can use just the specular return from the
sphere, even when the radius is quite small.

5.6 Creeping-wave contributions

In addition to being diffracted from surface discontinuities
on conducting bodies, rays incident on smoothly curved surfaces
can propagate along such surfaces. For example, if a circular
cylinder is illuminated normally to the axis, a creeping wave propagates along the curved surface around the rear of the cylinder.
Such a creeping wave radiates continuously as it propagates along
a curved surface, continually shedding rays at a rate depending
upon the local radius of curvature. Because of this continued
radiation, the amplitude of the ray established by a creeping wave
decays exponentially with distance as it propagates (and radiates)
along a geodesic of the surface.

In general, creeping-wave effects are unimportant for backscattering from bodies large relative to the wavelength because of
this exponential decay of the surface-diffracted rays. An exception is the sphere for which an axial caustic exists in the backscattering direction so that appreciable effects still exist for
ka (circumference in wavelengths) values as high as 50 where the
RCS departs from the specular value by 0.03 dB (the variation near
ka=10 is 0.6 dB) [51]. Numerous references discuss the problems
associated with creeping-wave effects [16, 17, 20, 26, 36]. Details of creeping-wave analysis are not given here because of the
complexities of the problem and because a target-imaging radar
will generally use a wavelength much smaller than the target dimensions so that creeping-wave effects will be of relatively little
significance.

6.   EXAMPLES OF SHORT-PULSE TARGET RESPONSES

In earlier sections, we have presented techniques for computing transient responses of radar targets from their CW scattering
characteristics and have also discussed at some length techniques
for approximating the CW responses of various types of targets and
their scattering centers. Now let us briefly consider some of the
characteristics of short-pulse responses of some simple types of
targets. Although an extensive catalog of short-pulse scattering
characteristics cannot be given here, we will mention some of the
most important considerations.

If the scatterer were a single point having uniform scattering
amplitude and phase, the received pulse would be a delayed replica
of the transmitted pulse, the only modification to its shape being
that produced by the bandpass characteristic of the receiver. If
two such scatterers are displaced along the radar line of sight,
the envelope of the received waveform can have several forms depending on scatterer spacing. If the scatterers are far enough
apart so that the response from the first has died out before that
from the second has started, we have completely resolved scatterers
and the response to the pair is simply the sum of the envelopes of
the two responses. Of more interest is the case in which the scatterers are not resolved completely. Here the responses add, but
the envelopes alone cannot be added because the relative RF phases
of the returns must be accounted for.

Figure 7 shows an example of the effects of very small shifts
in scatterer location for scatterers that are not quite resolved.
We have two scatterers with scattering amplitudes 1.0 and 0.9

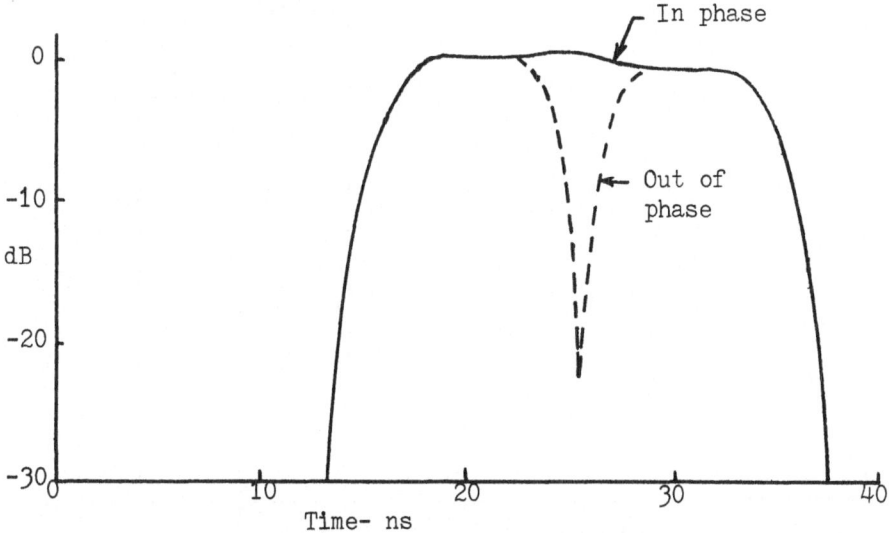

Fig. 7. High-resolution response of two unresolved scatterers

separated by either 1.4390 m (solid line) or 1.4465 m (dashed
line) and observed by a 10-GHz radar with a rectangular 10-ns
pulse and 200-MHz (3-dB) bandwidth, a cosine-squared passband
shape being assumed. The in-phase case produces a single pulse,
whereas the out-of-phase case produces a pair of pulses with a
deep null where the two returns overlap. Thus we have in this
case a very significant change in waveform resulting from a very
small change in scatterer spacing.

Figure 8 shows the response that would be obtained from the
scatterers in Figure 7 if the radar had infinite bandwidth so that
the rectangular input pulse could be recovered. The overlap region is very small, but within it large changes in amplitude can
occur with changes in relative phase. In Figure 7 the overlap
region appeared wider because the pulses were broadened by the
restricted passband.

When the scatterers are much closer together, the interference
effects can lead to waveforms with a single peak for the in-phase
case and to waveforms with pairs of peaks indicating wider scatterer spacing than actually exists for the out-of-phase case.
Such effects are very detrimental to radar imaging applications
and must be considered by the potential user of imaging radars.

Figures 9 and 10 show the results of a two-degree rotation in
the azimuth plane of a circular cylinder. Both analytical and
measured results are shown. The cylinder was 5 inches in diameter
and 15 inches long. Measurements were made on the Calspan FM/CW

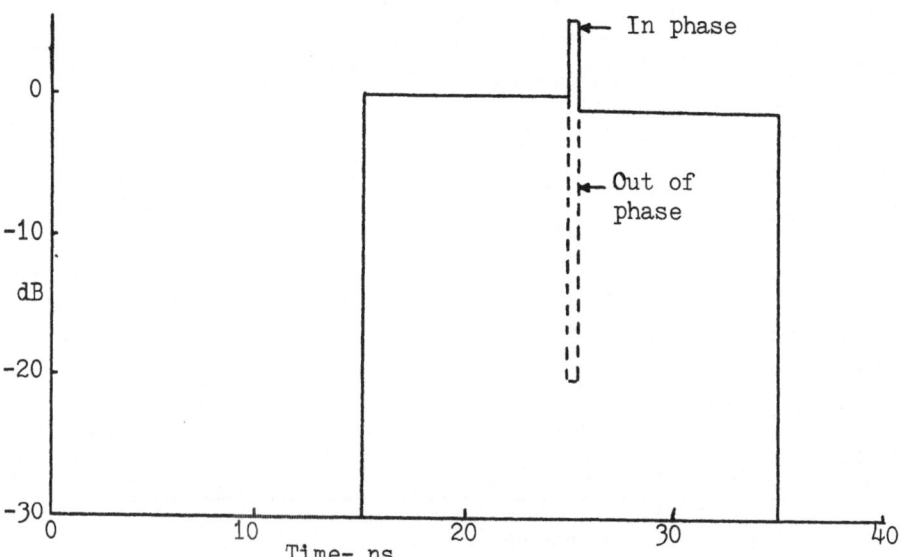

Fig. 8. Response of two unresolved scatterers, infinite bandwidth

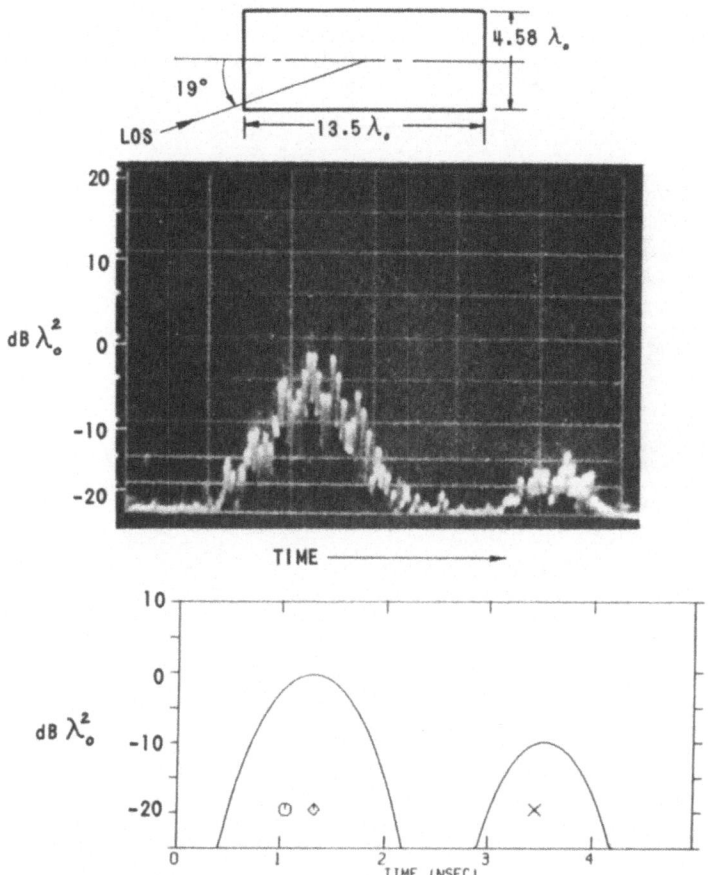

Fig. 9.   Measurement vs. computation for circular cylinder;
          horizontal polarization, aspect angle = 19°

system [35]. The radar had a total swept bandwidth of 3 GHz with
center frequency 10.7 GHz, corresponding in the equivalent short-
pulse case to a 1/3-ns pulse. As shown in Section 8.3, an adjustment in system calibration must be made to allow comparison of CW
and short-pulse radar cross sections. In the present case, the
RCS levels as measured had to be reduced by 8 dB. Figures 9 and
10 are from Reference 38; other examples from that report have
also been published [52]. Two facts stand out: the predicted results show close correspondence with the measurements, and a small
change in target viewing angle can change constructive interference
(Figure 9) into destructive interference (Figure 10). Note that
the rear scattering center is well resolved in each case and that
only the two forward scattering centers show interference effects.
The jagged nature of the measured response is not caused by noise
but results from the fact that the system operates at a fixed PRF

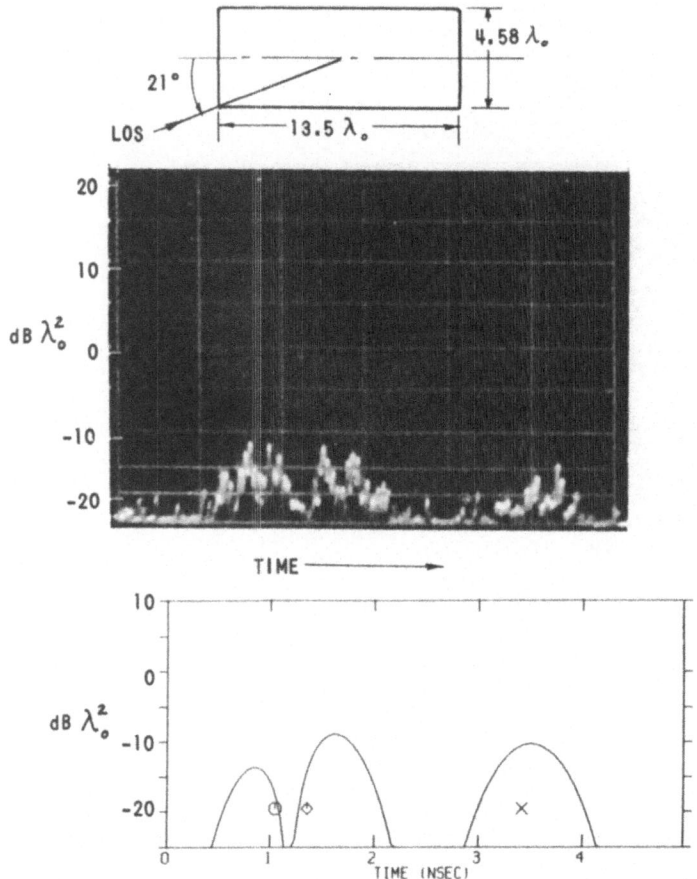

Fig. 10. Measurement vs. computation for circular cylinder;
          horizontal polarization, aspect angle = 21°

and therefore has a line-spectrum output, each line of which produces a separate peak in the spectrum-analyzer output.

It is the relative spacing of unresolved scattering centers
that produces most of the changes in received waveform during
small amounts of target rotation. For large amounts of target rotation, the shadowing of some scattering centers and exposure of
formerly shadowed ones can produce quite significant changes in
observed waveforms. Of particular importance in this regard is
the appearance of specular scatterers which may produce very large
returns over restricted ranges of aspect angles.

Variations in scattering amplitude with frequency over the
passband of the radar can produce some distortion of the incident
waveform even if only a single scattering center is observed.

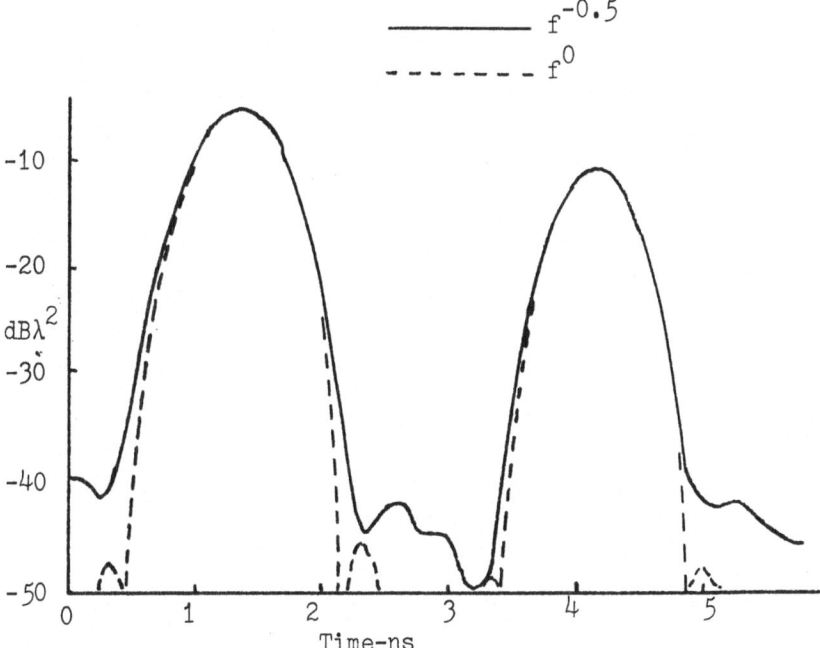

Fig. 11.   Effects of omission of frequency dependence of
           scattering centers for 30-percent-bandwidth system

Such effects tend to be quite small. For specular scattering from
a doubly curved surface, the scattering amplitude is essentially
independent of frequency, so no distortion is produced so long as
the principal radii of curvature exceed the longest wavelength in
the signal. For specular reflection from a cylinder, the transfer
function is proportional to $\sqrt{f}$. Using paired-echo analysis [41],
we find that even for a radar with a 40-percent bandwidth the
largest distortion-producing echo is down 36 dB from the main response [38(pp. 30-31)]. For a planar surface, specular scattering
varies as f and the distortion-producing paired echo is about
twice what it is for a cylindrical surface.

     Edge-type scattering, as at the ends of a circular or rect-
angular cylinder, varies as $1/\sqrt{f}$. Figure 11 shows the transient
response of a circular cylinder to a 30-percent-bandwidth radar
if the $1/\sqrt{f}$ dependence is omitted (dashed line) and if it is included (solid line). The difference is obviously quite small except at very low signal levels. Even for a 60-percent-bandwidth
radar the difference is significant only at low levels (Figure 12).
For more-narrowband systems the effects would generally be negligible.

     Because it is scattering-center location rather than scat-
tering-center frequency characteristics that to a large extent

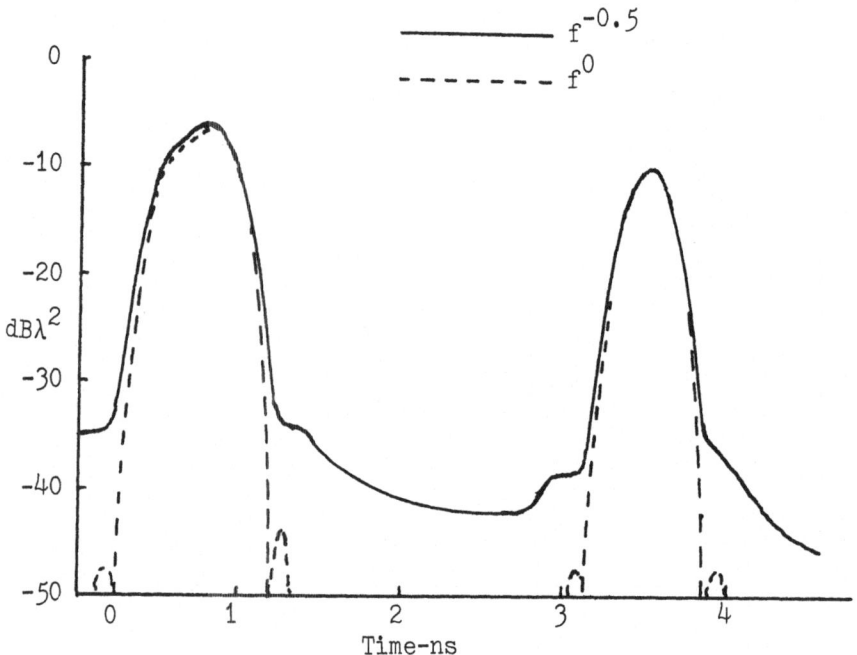

Fig. 12. Effects of omission of frequency dependence of
scattering centers for 60-percent-bandwidth system

govern the short-pulse response of a radar target, identification
of simple targets on the basis of radar returns is very difficult.
We saw above that the response from a cylinder changed drastically
with a very small change in orientation. Figures 13 and 14 show
the responses of a circular cylinder and a cube at 40° aspect
angle to vertically and horizontally polarized radars, the cylinder
having 36-inch diameter and length and the cube having a 36-inch
side. The radar has a 3-GHz carrier, a 2-ns pulse length, a total
bandwidth of 500 MHz, and uses a Taylor-weighted bandpass. It
would obviously be a difficult task to identify the cube or the
cylinder on the basis of such waveforms, particularly when it is
recognized that as the target is rotated the waveforms for both
shapes will undergo considerable variation. Also note that the
relatively large response from the cube results from the fact that
its vertical edges are normal to the line of sight; at more general
aspect angles the response would be of much lower amplitude, but
computations are not available for this case.

The above argument should not be construed as a claim that
complex-target identification is impractical or impossible in all
cases. A large target with numerous, well-separated scatterers,
such as an aircraft, observed by a high-resolution radar capable
of resolving many of the scattering centers might well produce a
radar return yielding sufficient information for target identifi-

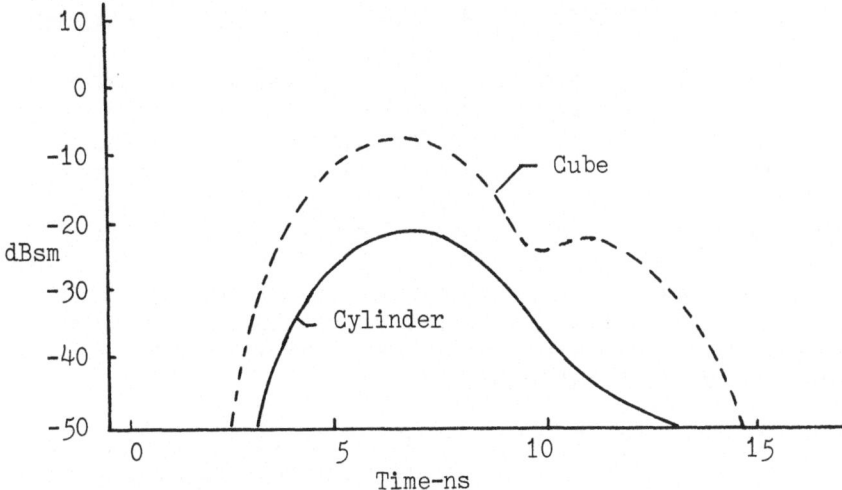

Fig. 13. Responses of circular cylinder and cube at
          40° aspect angle, vertical polarization

cation. Identification of small, relatively simple targets is
much more difficult, although a sufficiently high-resolution radar
may be capable of providing a reasonable estimate of target size.

This section has attempted to indicate some of the determi-
nants of the observed high-resolution radar responses of simple
targets. In summary, the most important such factors are scattering-center amplitude and location and the ability of the radar
to resolve independent scattering centers.

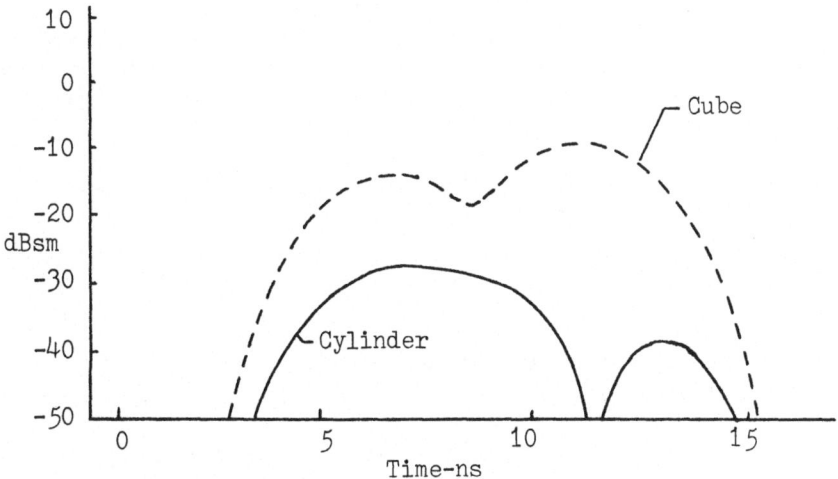

Fig. 14. Responses of circular cylinder and cube at
          40° aspect angle, horizontal polarization

7.   RELATIONSHIPS BETWEEN FM/CW, CHIRP, AND SHORT-PULSE RADARS

Because of the complex spectrum associated with an FM/CW or chirp-radar signal, straightforward application of Fourier analysis is much more difficult than it is for more conventional short-pulse radar such as that assumed in Section 4. It is desirable, and in most cases is quite acceptable, to perform transient analyses for the short-pulse radar case and to use these results as equivalent to those that would be obtained from the FM/CW or chirp radars. One example of such an equivalence has already been demonstrated in Figures 9 and 10 in which computations were made on an equivalent short-pulse system and shown to agree with the results of measurements on an FM/CW radar.

For the purposes of this section, it is convenient to assume a transfer function $G(f)$ which accounts for both the target scattering characteristics and the passband function of the receiver (and, it might be added, a transfer function corresponding to propagation effects). It is also reasonable, in any practical system, to assume some maximum frequency B above which the signal spectrum has essentially no power. We can expand $G(f)$ in a Fourier series as

$$G(f) = \sum_{n=0}^{N} \{A_n \cos(n\pi f/B) + iB_n \sin(n\pi f/B)\} \qquad (41)$$

Using basic paired-echo analysis, we find that for any input $e_i(t)$ the output is

$$e_o(t) = \sum_{n=-N}^{N} \epsilon_n (A_n + B_n) e_i(t+0.5n/B) \qquad (42)$$

where $\epsilon_0 = 1$ and $\epsilon_n = 0.5$, $n \neq 0$. Computation using this formulation would be very difficult because N would typically have to be very large, but here we are only interested in the functional forms involved.

Suppose we have a short-pulse radar whose modulating waveform is

$$f_1(t) = 2BW\text{sinc}(BWt) \qquad (43)$$

where
    BW = total system bandwidth, and
    $\text{sinc}(z) = \sin(\pi z)/(\pi z)$.

Note that BW is the bandwidth around the carrier frequency f_0 and that in equation 40 we require $B > f_0 + BW/2$. Insertion of equation 42 into equation 4 leads to

$$F_1(f) = \text{rect}(f/BW) \qquad (44)$$

where
$$rect(z) = 1, \quad |z| < 0.5$$
$$= 0, \quad elsewhere.$$

Starting from

$$e_i(t) = f_1(t)\cos(2\pi f_0 t) \tag{45}$$

and applying equation 42, we find

$$e_0(t) = 2BW \sum_{n=-N}^{N} \epsilon_n (A_n + B_n)\operatorname{sinc}[BW(t+0.5n/B)]$$
$$x \cos[2\pi f_0(t+0.5n/B)] \tag{46}$$

In practice, we would evaluate $e_0(t)$ using equations 7 and 8 with $F_1(f)$ given by equation 44; note that in equation 7 we would also have $F_1(f+2f_0)=0$ because no spectral folding occurs here and we also have $\phi=\pi/2$. The representation in equation 46 is used here only because it is a more convenient form for comparison of responses of different types of radars.

Two basic types of chirp radar, passive-generation and active-generation chirp, are possible [53]. Passive-generation chirp leads to results identical to those obtained from short-pulse radar and need not be considered further. In active-generation chirp radar, a transmitter is (usually) linearly swept across a band of width BW centered on f_0. For large dispersion factor D (=BWT), the spectrum is nearly rectangular. The transmitted signal is

$$e_i(t) = rect(t/T)\cos[2\pi(f_0 t+0.5BW t^2/T)] \tag{47}$$

and the received signal from a point scatterer is, after compression [53(eq. 20)]

$$e(t) = \sqrt{D} \operatorname{sinc}(BWt)\cos[2\pi(f_0 t-0.5BW t^2/T)+\pi/4] \tag{48}$$

from which we get the waveform received from a complex target as

$$e(t) = \sqrt{D} \sum_{n=-N}^{N} \epsilon_n (A_n + B_n)\operatorname{sinc}[BW(t+0.5n/B)]$$
$$x \cos[2\pi(f_0(t+0.5n/B)-0.5BW/T(t+0.5n/B)^2)+\pi/4] \tag{49}$$

Now compare equation 49 with equation 46. Aside from a scale factor, the only difference is in the argument of the cosine function. The $\pi/4$ phase shift is unimportant as it has no effect on relative phases and can, in any case, be incorporated into equation 8 if desired with no effect on the complex-envelope computation. The quadratic term, which arises from the frequency modulation of the compressed signal, can be written as

$$(0.5BW/T)(t+0.5n/B)^2 = [BW(t+0.5n/B)]\frac{t+0.5n/B}{2T} \qquad (50)$$

If the quantity in square brackets is large, sinc$[BW(t+0.5n/B)]$ becomes small and the amplitude of the term is small, making phase errors unimportant. The factor outside the square brackets is typically small, because after compression times are much shorter than T if the system has a large dispersion factor D. For small dispersion factors, the quadratic phase term in equation 49 will be significant and there is no direct equivalence to equation 46.

Finally, consider an FM/CW radar. The received signal has the form

$$e_r(t) = \sum_{n=-N}^{N} \epsilon_n(A_n+B_n)\text{rect}[\frac{t-\tau+0.5n/B}{T}]\cos[2\pi(f_0(t-\tau+0.5n/B)$$
$$+ (0.5BW/T)(t-\tau+0.5n/B)^2)] \qquad (51)$$

where τ is a delay produced by distance to the target less the length of the delay line in the reference channel. An explicit delay is needed to prevent spectral folding which would make the received signal very difficult to interpret. This signal is mul- tiplied by the transmitted signal, suitably delayed, and the com- ponents around $2f_0$ are filtered out, yielding

$$e_0(t) = 0.5 \sum_{n=-N}^{N} \epsilon_n(A_n+B_n)[U(t+0.5T-\tau+0.5n/B)-U(t-0.5T)]$$
$$x \cos[2\pi(BW/T(\tau-0.5n/B)t+f_0\tau-0.5n/B)$$
$$-BW/(2T)(\tau-0.5n/B)^2)] \qquad (52)$$

where $U(z)$ is a unit step at $z=0$.

In an FM/CW radar, we spectrum analyze this signal, the am- plitude of the spectrum at a given frequency corresponding to the scattering amplitude at a given range. Before attempting to get the spectrum of $e_0(t)$, we note that $\tau-0.5n/B \ll T$ in any practical system so that we can replace $U(t+0.5T-\tau+0.5n/B)$ with $U(t+0.5T)$ with very little error. The spectrum of $e_0(t)$ is thus

$$\tilde{e}_0(f) = (T/4)\sum_{n=-N}^{N} \epsilon_n(A_n+B_n)\text{sinc}[fT-BW\tau+0.5nBW/B]$$
$$x \exp\{i2\pi[f_0\tau-0.5f_0n/B-0.5BW/T(\tau-0.5n/B)^2]\} \qquad (53)$$

The quadratic term in the exponential is negligible in most prac- tical systems, just as it was in the chirp-radar case. We assume the folded-spectrum terms are negligible in the spectrum of the mixed signal. In effect, we thus assume the relative delay to be great enough so the spectrum being observed is displaced along the frequency axis by more than the width of the observed spectrum, a parameter depending on the range constant of the radar and the extent of the target. Such a displacement is needed to permit interpretation of the signal in any case.

We are interested in the amplitude of the spectral line at frequency f; this amplitude is

$$A_{FM/CW} = \frac{T}{4}\{[\sum_{n=-N}^{N} \epsilon_n (A_n + B_n) \text{sinc}[fT - BW\tau + 0.5nBW/B]\cos[n\pi f_0/B]]^2$$
$$+ [\sum_{n=-N}^{N} \epsilon_n (A_n + B_n) \text{sinc}[fT - BW\tau + 0.5nBW/B]\sin[n\pi f_0/B]]^2\}^{0.5}$$

For the true short-pulse radar of equation 46, the amplitude at a given time is

$$A_{SP} = 2BW\{[\sum_{n=-N}^{N} \epsilon_n (A_n + B_n) \text{sinc}[BWt + 0.5nBW/B]\cos[n\pi f_0/B]]^2$$
$$+ [\sum_{n=-N}^{N} \epsilon_n (A_n + B_n) \text{sinc}[BWt + 0.5nBW/B]\sin[n\pi f_0/B]]^2\}^{0.5} \quad (55)$$

which is identical to equation 54, aside from a scale factor, if we transform frequency into time using

$$fT - BW\tau = BWt \qquad\qquad\qquad\qquad\qquad\qquad (56)$$

In summary, we can use a computational approach intended for conventional short-pulse radar for chirp or FM/CW radars under certain conditions that are usually satisfied. It is required that the short-pulse computation be for a flat spectrum rather than for a constant-amplitude pulse, but if the approach given in Section 4.1 is used we need merely specify $F_1(f)$ as in equation 44. Acceptable results often can be obtained using a rectangular-envelope pulse computation with pulse length $1/BW$ if a computer program is only available for this case [36]. The target transfer function naturally is the same for all radar types, but we must use the passband characteristics of the chirp or the FM/CW radar in our short-pulse computation. For either chirp or FM/CW radar, we also require the quadratic phase terms indicated above to be negligible; if this condition is not met, the equivalence of radar types cannot be established. Finally, in the case of FM/CW radar, the system must operate at an offset frequency that prevents spectral folding effects. Note that it has been assumed here that the system and target transfer functions remain fixed during the time the chirp or FM/CW transmitter is swept. If target motion or propagation effects (as, for example, through a turbulent

atmosphere) cause $G(f)$ to vary during the sweep time, the equiva-
lences shown above will not hold.

8.   SPECIAL CONSIDERATIONS FOR SHORT-PULSE IMAGING

8.1 Need for care in use of results verified only for CW scattering

As indicated in Section 4, the transient response of a body
can be computed from the CW scattering characteristics of the body.
This form of computation, based on Fourier analysis techniques,
is exact, but the correctness of the result depends on the validity
of the representation used for the CW transfer function. The CW
response of a target can easily be dominated by one or two scattering centers, and quite large errors in the scattering-amplitude
estimates for other scattering centers which produce small returns
will have little effect on the magnitude of the CW response as
given by the whole-body RCS. If such an erroneous approximation
to the CW response, being justified by the demonstrated correctness
of the whole-body RCS, is used for a transient analysis, large
errors will occur for the poorly described scattering center(s).

An outstanding example of this type of error is the problem
of transient response of a cylinder [11, 52] which was mentioned in
Section 5.2. Although the GTD had been shown to give excellent
whole-body RCS values [4, 7, 21], high-resolution measurements
showed that under some conditions (electric vector normal to the
plane of rotation and angles off axis of less than 40°) large errors occurred in the GTD estimate of the scattering amplitude of
the rear of the cylinder. The RCS of the cylinder is dominated,
for these aspect angles, by the scattering centers at the front
of the cylinder, and the error in the estimate of the contribution
from the rear scattering center is insignificant. The same error
occurs in computations of scattering from a cone, but there the
incorrect scattering amplitude does lead to incorrect whole-body
RCS values [6] because there is no other large scattering contribution to mask the erroneous value.

It is therefore essential that one have better justification
for the scattering-characteristic function used to represent a
target than just the agreement of the computed whole-body RCS
with the measured whole-body RCS.

8.2 Frequency characteristics of scattering-center contributions

In the preceding discussions we have indicated that the frequency variation of scattering-center diffraction is generally
not very important, scattering-center spacing being of more

importance in establishing target scattering characteristics. In some special cases involving resonant scatterers such as antennas or ducts, scattering-center characteristics may vary significantly over the radar bandwidth. The formulation of Section 4.1 is still valid, but the appropriate transfer function, taking into account resonance effects, must be used. It should also be noted that scattering from an antenna depends on its termination as well as on its external structure so that estimation of antenna scattering is a very difficult task.

8.3 Relationship between short-pulse and CW radar responses

Radar cross section is conventionally defined in terms of sinusoidal waves. When a radar transmits and receives pulses, significant amounts of the signal can be removed by the bandpass-filter characteristic of the receiver. Consequently, the level of the received signal after it has been filtered is affected by the CW RCS and the filtering process. Failure to account for the loss in signal energy produced by the filter can lead to apparent disagreement between calculated radar cross section and measured signal level.

In RCS measurements, it is common practice to use a target of known RCS, such as a disk or a sphere, to establish a reference level against which target measurements can be compared. In this case, the effects of the receiver bandpass characteristic are effectively accounted for, since they affect similarly the signal levels from the reference and from the target. Suppose, on the other hand, that the received waveform is computed using the methods discussed above. Here, the effect of the receiver bandwidth is included in the computation and an equivalent signal voltage level is obtained. This signal level will not correspond to that obtained from the measurement, if that measurement has been calibrated as indicated above.

Suppose, for example, that the RCS of a target is C dB. The filter loss associated with the receiver is L dB. The CW RCS of the reference target is R dB. The measurement will produce a reference level of R-L dB and a target level of C-L dB; comparison will then give a target RCS of $(C-L)-(R-L) = C-R$ dB relative to the reference. It is clear that L has no effect on our ability to measure C relative to R and that we can accurately measure the CW value of C, provided we know the CW value of the reference. Computation of received signal level includes the effect of L; therefore, the computations developed above lead to a result C-L and an equivalent signal level, relative to the reference, of C-L-R. It is thus evident that L must be added to the computed result or subtracted from the measured result if agreement is to be obtained. Note that elimination of the bandpass

filter characteristic from the waveform computations would generally not be satisfactory, because the effect of this filter on the
output pulse shape is of interest.

Frequency dependence of the specular reflections (from a large
sphere or flat plate) has little effect on L, so a simple form of
computation suffices to provide a satisfactory estimate of L. If
a small sphere, such as is often used for CW measurements of RCS,
were used, its frequency effects might lead to errors in the estimate of L.

For calibration of RCS measurements, one is interested in the
peak of the pulse received from the calibration target. If we assume this peak to be at the center of the pulse, $t=0.5\tau$, we can
easily find the spectral loss, L, relative to that of an infinite-
bandwidth system, to be

$$L = 20 \log_{10}\left[\frac{2}{\pi} Si(\pi f_c \tau)\right] \tag{57}$$

where

$$Si(z) = \int_0^z \frac{\sin(t)}{t} dt \quad \text{is the sine-integral function.}$$

The infinite-bandwidth system response is found by letting $f_c \tau$
approach infinity and using the relationship $Si(z) \to \pi/2$ as $z \to \infty$.
This result assumes an ideal bandpass filter of total bandwidth $2f_c$.
If the response of the receiver is represented by a Fourier series
over a total bandwidth of $2f_c$, response being assumed zero outside
this band, the response can be found using paired-echo analysis.
For a bandpass characteristic

$$G(f) = A_0 + \sum_{n=1}^{N} A_n \cos(n\pi f/f_c) \tag{58}$$

the loss factor L is given by

$$L = 20 \log_{10} \left\{ \frac{2}{\pi}[A_0 Si(\pi f_c \tau) + \sum_{n=1}^{N} 0.5A_n(Si[\pi(f_c \tau - n)]\right.$$

$$\left. + Si[\pi(f_c \tau + n)])] \right\} \tag{59}$$

which is plotted in Figure 15 for three passband shapes.

8.4   Radar resolution effects on glint

As indicated in Section 6, unresolved scattering centers produce waveforms that depend on the relative phases of the returns
from those scattering centers as well as on the respective amplitudes of the returns. Depending on relative phases, two scattering centers can produce a single peak somewhere between the times
one would expect returns from the two centers or a pair of pulses

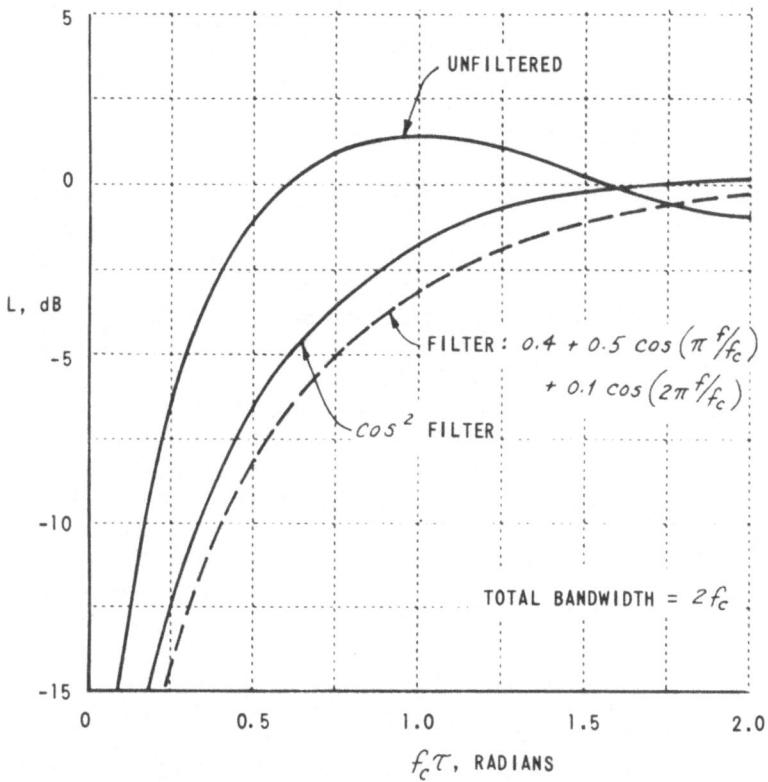

Fig. 15. Loss factor introduced by passband characteristic

separated more widely than would be expected. Unless the scattering centers can be resolved, one faces a great problem in attempting to form an image of the target.

Glint effects which degrade tracking accuracy also depend upon scattering-center location, but here they depend on these locations in both range and cross-range [54, 55]. The techniques discussed in this paper are thus applicable to glint computation. Note that if tracking angles are separately measured in each of a large number of range bins it is possible, in principle, to locate scattering centers in a cross-range as well as in a down-range direction, thus greatly improving target imaging possibilities. Such determination of scattering-center locations is possible only if the scattering centers are isolated in the range bins. If two scattering centers lie in the same range bin, large errors in glint angle can occur, these errors appearing whenever the two contributing signals are out of phase. Such glint errors can be more serious in short than in long range bins, for if there are many scattering centers within the observed range cell the glint effects will tend to average out. Application of cross-range

measurements in conjunction with high-resolution capability, although potentially a valuable tool for target imaging, must be
approached with caution because of these effects.

9. SUMMARY AND CONCLUSIONS

It has been the purpose of this presentation to indicate some
of the primary techniques for estimating the radar scattering
characteristics of common scattering centers and to indicate some
methods for computing the short-pulse responses of targets. Various types of scattering centers have been described, general
methods of RCS analysis have been presented, and explicit formulations have been given for diffraction from several basic types
of scattering centers. Methods for obtaining short-pulse responses from CW responses have also been discussed. Some examples
of short-pulse responses of simple targets have been given. Essential equivalence between FM/CW, chirp, and short-pulse radars
has been discussed and it has been shown that for many cases one
can use a short-pulse computation to represent either the output
of a chirp or an FM/CW radar. Finally, we discussed some special
considerations related to short-pulse radar and its application
to target imaging.

In conclusion, we have shown that techniques currently exist
for the approximation of short-pulse radar responses from many
types of targets, although diffraction-analysis techniques are
not yet capable of accurately predicting the contributions from
as many types of scattering centers as might be desired.

10. ACKNOWLEDGMENTS

I would like to acknowledge the contributions of numerous
present and past associates at Calspan Corporation. Those who
have contributed primarily to analytical diffraction analysis
include R. A. Ross, J. B. Billingsley, R. E. Kell, D. I. Fairbanks,
E. L. Price, H. R. Witt, J. R. Graham, Jr., and the late C. F.
Evans. Those who have contributed primarily to RCS measurement
include A. V. Alongi, D. J. Newton, R. J. Wohlers, C. G. Bachman,
R. V. Gallagher, and W. P. Melling.

11. REFERENCES

1. D. I. Fairbanks, "A Study and Recommended Program to Determine
   the Monostatic Electromagnetic Scattering from a Dielectric
   Rectangular Parallelepiped," Calspan Corporation, Jan. 1970.
2. M. E. Bechtel and D. I. Fairbanks, "Some Analyses of the
   Effects of a Plasma Sheath on Radar Returns from the Rear of

a Reentry Body," Calspan Report No. UB-1376-S-184, Bell
Telephone Lab. Report No. RMAR70-2, April 1970.

3. J. R. Graham, Jr. and M. E. Bechtel, "Radar Scattering Analysis
   of a Dielectric Aircraft Wing," Calspan Report No. AA 2179-
   E-3, Dec. 1968.

4. M. E. Bechtel and R. A. Ross, "Radar Scattering Analysis,"
   Calspan Report No. ER/RIS-10, Aug. 1966.

5. R. E. Kell, "On the Derivation of Bistatic RCS from Monostatic
   Measurements," Proc. IEEE, vol. 53, pp. 983-988 (1965).

6. M. E. Bechtel, "Application of Geometric Diffraction Theory
   to Scattering from Cones and Disks," Proc. IEEE, vol. 53,
   pp. 877-882 (1965).

7. R. A. Ross, "Scattering by a Finite Cylinder," Proc. IEE
   (London), vol. 114, pp. 864-868 (1967).

8. R. A. Ross, H. R. Witt, and J. R. Graham, Jr., "Radar Cross
   Section of a Finite, Rectangular Cylinder," Electronics
   Letters, vol. 3, pp. 336-337 (1967).

9. R. A. Ross, "Radar Cross Sections of Rectangular Flat Plates,"
   Trans. IEEE, vol. AP-14, pp. 329-335 (1966).

10. R. A. Ross, "Small-Angle Scattering by a Finite Cone,"
    Trans. IEEE, vol. AP-17, pp. 241-242 (1969).

11. M. E. Bechtel, "Vertically Polarized Radar Backscattering from
    the Rear of a Cone or Cylinder," Trans. IEEE, vol. AP-17,
    pp. 244-246 (1969).

12. R. A. Ross, "Investigations in Electromagnetic Scattering
    Center Theory," PhD Dissertation, University of Manitoba,
    Sept. 1971; also Calspan Report No. 182, Sept. 1971.

13. R. A. Ross and M. A. K. Hamid, "Scattering by a Wedge with
    Rounded Edge," Trans. IEEE, vol. AP-19, pp. 507-517 (1971).

14. H. R. Witt and E. L. Price, "Scattering from Hollow Conducting
    Cylinders," Proc. IEE (London), vol. 115, pp. 94-99 (1968).

15. J. W. Crispin, Jr. and K. M. Siegel (eds.), Methods of Radar
    Cross Section Analysis, Acad. Press, New York, 1968.

16. B. R. Levy and J. B. Keller, "Diffraction by a Smooth Object,"
    Comm. Pure and Appl. Math., vol. 12, pp. 159-209 (1959).

17. N. D. Kazarinoff and T. B. A. Senior, "A Failure of Creeping
    Wave Theory," Trans. IEEE, vol. AP-10, pp. 634-639 (1962).

18. T. B. A. Senior and R. F. Goodrich, "Scattering by a Sphere,"
    Proc. IEE (London), vol. 111, pp. 907-916 (1964).

19. J. J. Bowman, T. B. A. Senior, and P. L. E. Uslenghi,
    Electromagnetic and Acoustic Scattering by Simple Shapes,
    North-Holland, Amsterdam, 1969. Also Air Force Cambridge
    Research Lab. Report No. AFCRL-70-0047, 15 Jan. 1970 (AD699859).

20. G. T. Ruck, D. E. Barrick, W. D. Stuart, and C. K. Krichbaum,
    Radar Cross Section Handbook, Plenum Press, New York, 1970.

21. R. E. Kell and R. A. Ross, "Radar Cross Section of Targets,"
    Chap. 27 of M. I. Skolnik (ed.), Radar Handbook, McGraw-Hill,
    New York, 1970.

22. J. B. Keller, "Geometrical Theory of Diffraction," Jour. Opt.
    Soc. Amer., vol. 52, pp. 116-130 (1962).

23. R. G. Kouyoumjian and P. H. Pathak, "A Uniform Geometrical Theory of Diffraction for an Edge in a Perfectly Conducting Surface," Proc. IEEE, vol. 62, pp. 1448-1461 (1974).

24. E. F. Knott and T. B. A. Senior, "Comparison of Three High-Frequency Diffraction Techniques," Proc. IEEE, vol. 62, pp. 1468-1474 (1974).

25. R. M. Lewis and J. B. Keller, "Asymptotic Methods for Partial Differential Equations: The Reduced Wave Equation and Maxwell's Equations," New York Univ. Courant Inst. of Math. and Sci. Res. Report No. EM-194, Jan. 1964.

26. V. A. Borovikov and B. Ye. Kinber, "Some Problems in the Asymptotic Theory of Diffraction," Proc. IEEE, vol. 62, pp. 1416-1437 (1974).

27. L. B. Felsen and N. Marcuvitz, Radiation and Scattering of Waves, Prentice-Hall, Englewood Cliffs, N.J., 1973.

28. L. Kaminetzky and J. B. Keller, "Diffraction by Edges and Vertices of Interfaces," SIAM J. Appl. Math., vol. 28, pp. 839-856 (1975).

29. J. Boersma, "Ray-Optical Analysis of Reflection in an Open-Ended Parallel-Plane Waveguide. I: TM Case," SIAM J. Appl. Math., vol. 29, pp. 164-195 (1975).

30. W. V. T. Rusch, "A Comparison of Geometrical and Integral Fields from High-Frequency Reflectors," Proc. IEEE, vol. 62, pp. 1603-1604 (1974).

31. R. F. Harrington, Field Computation by Moment Methods, Macmillan, New York, 1968.

32. C. Klein and R. Mittra, "Stability and Convergence of the Method of Moments Applied to Electromagnetic Problems," Univ. of Illinois at Urbana-Champaign Report No. UILU-ENG-75-2538, Jan. 1975.

33. W. D. Burnside, C. L. Yu, and R. J. Marhefka, "A Technique to Combine the Geometrical Theory of Diffraction and the Moment Method," Trans. IEEE, vol. AP-23, pp. 551-558 (1975).

34. C. L. Bennett, K. S. Menger, and R. Hieronymus, "Space Time Integral Equation Approach for Targets with Edges," Rome Air Dev. Center Report No. RADC-TR-74-132 (AD-785 120), 1974.

35. A. V. Alongi, R. E. Kell, and D. J. Newton, "A High-Resolution X-Band FM/CW Radar for RCS Measurements," Proc. IEEE, vol. 53, pp. 1072-1076 (1965).

36. M. E. Bechtel, J. B. Billingsley, and J. O. Clark, "Investigation of Transient-Response Computation for Simple Radar Targets," Rome Air Dev. Center Report No. RADC-TR-70-99, April 1970.

37. E. M. Kennaugh and D. L. Moffatt, "Transient and Impulse Response Approximations," Proc. IEEE, vol. 53, pp. 893-901 (1965).

38. M. E. Bechtel, "Computation of Transient Responses of Radar Targets," Calspan Report No. UG-2457-E-2, Sept. 1968.

39. R. A. Ross (Lincoln Laboratories of M. I. T.), unpublished.

40. F. G. Friedlander, Sound Pulses, Cambridge Univ. Press, 1958.

41. M. E. Bechtel, "Generalized Paired-Echo Analysis for Band-pass Systems," _Proc. IEEE_, vol. 57, pp. 204-205 (1969).
42. R. E. Hiatt, T. B. A. Senior, and V. H. Weston, "Studies in Radar Cross Section XL-Surface Roughness and Impedance Boundary Conditions," Univ. of Mich. Rad. Lab. Report No. 2500-2-T, July 1960.
43. N. A. Logan, "Scattering Properties of Large Spheres," _Proc. IEEE_, vol. 48, p. 1792 (1960).
44. M. E. Bechtel and R. A. Ross, "Handbook of Scattering Coefficients," Calspan Report No. UF-2457-E-1, Nov. 1967.
45. G. P. Bein, "Fresnel Diffraction of RF Wavefronts," FAA Report No. FAA-RD-73-74, Calspan Report No. AG-5082-E-1, Feb. 1973.
46. H. C. van de Hulst, _Light Scattering by Small Particles_, Wiley, New York, 1957.
47. T. B. A. Senior, "Diffraction Coefficients for a Discontinuity in Curvature," _Electronics Letters_, vol. 7, pp. 249-250 (1971).
48. L. Kaminetzky and J. B. Keller, "Diffraction Coefficients for Higher Order Edges and Vertices," _SIAM J. Appl. Math._, vol. 22, pp. 109-134 (1972).
49. R. A. Ross, "Application of Geometrical Diffraction Theory to Reflex Scattering Centers," IEEE International Antennas and Propagation Symposium Digest, IEEE Cat. No. 68C29AP, pp. 94-99 (1968).
50. S. D. Weiner and S. L. Borison, "Radar Scattering from Blunted Cone Tips," _Trans. IEEE_, vol. AP-14, pp. 774-781 (1966).
51. M. E. Bechtel, "Scattering Coefficients for the Backscattering of Electromagnetic Waves from Perfectly Conducting Spheres," Calspan Report No. AP/RIS-1, Dec. 1962.
52. M. E. Bechtel, "High Resolution Radar Responses of Cylinders and Cones," IEEE International Antennas and Propagation Symposium Digest, IEEE Cat. No. 68C29AP, pp. 100-105 (1968).
53. J. R. Klauder, A. C. Price, S. Darlington, and W. J. Albersheim, "The Theory and Design of Chirp Radars," _Bell System Tech. Jour._, vol. 39, pp. 745-808 (1960).
54. L. Peters, Jr., and F. C. Weimer, "Tracking Radars for Complex Targets," _Proc. IEEE_, vol. 51, pp. 2149-2162 (1963).
55. R. A. Ross and M. E. Bechtel, "Scattering-Center Theory and Radar Glint Analysis," _Trans. IEEE_, vol. AES-4, pp. 756-762 (1968).

SIGNAL PROPERTIES AND TARGET MODELS IN THE VIEW OF TARGET CLASSIFICATION AT A SURVEILLANCE RADAR

K. von Schlachta

Forschungsinstitut für Funk und Mathematik
D 5307 Wachtberg-Werthhoven, Federal Republic of Germany

ABSTRACT.   Typical differences in signal fluctuation between several targets were found by measurements at different surveillance radars. A statistical decision test can be performed for classification purposes. The statistical behaviour of the coherent signal of airplanes is derived from measured data and an example of a classification test between two types of airplanes can be given.

1. INTRODUCTION

It is a well-known fact that discrimination between signals can only be made with the knowledge of at least one typical difference in characteristics. Sufficient information about typical properties of targets and unwanted targets (clutter) is required. Theoretical models derived from physical properties are not the optimum for every technical realization of automatic extraction or discrimination circuits. In examination of a real process only measurements give evidence for example on the actual ratio of fluctuating to steady power of target. The fact that real signals are non-stationary calls for experimental tests.

In the last years measurements of radar signals were performed at different European radar sites [1]. Using L- and S-band surveillance radars of air-traffic control services the digitized coherent signal of different targets was stored on magnetic tape for further evaluation on universal computers. Besides traffic airplanes on airlines all kinds of clutter (ground, weather, sea,

Jeske (ed.), Atmospheric Effects on Radar Target Identification and Imaging, 55–64.

birds) were measured and their typical properties for example on different doppler behaviour was evaluated.

2. USE OF A DECISION TEST

One of the results that may be presented here more extensively is, that discrimination between different types of signals is possible, when a specific test is based on statistical properties. Target and clutter can be distributed into two classes depending on the extent of fluctuation of amplitude respectively of phase shift.

Low fluctuation: - jet planes
 - point shaped echoes from ground, buildings.

Strong fluctuation: - propeller driven airplanes (broadside excepted)
 - bird echoes, if several in one radar resolution cell
 - precipitation clutter.

The fluctuation of targets can be expressed by the probability density function of a typical test criterion, the sample function $g(x)$. A binary test can be performed, we use the Neyman-Pearson criterion for discrimination between one of the two hypotheses H_1 or H_2. R_1 and R_2 are appropriate non-overlapping hypothesis classes [2]. The probability of correct classification for H_1 is:

$$P_K = \int_{R_1} p\left[g(x) \mid H_1\right] dg$$

while the probability of false decision (for the second hypothesis H_2) is

$$P_N = \int_{R_1} p\left[g(x) \mid H_2\right] dg$$

The Neyman-Pearson test is a procedure that minimizes P_N for a fixed P_K. The test rule is a likelihood-ratio test with a threshold η

$$\lambda(x) = \frac{p\left[g(x) \mid H_1\right]}{p\left[g(x) \mid H_2\right]} \gtrless \eta \quad \begin{array}{l} \text{for } H_1 \\ \text{for } H_2 \end{array}$$

The threshold divides the observation space of signals into two subspaces each corresponding to one hypothesis.

For further statements the probability density functions $p(g|H_1)$ and $p(g|H_2)$ have to be known. The following described results of measurements will show the statistical properties of airplane signals. Furtheron the application of a classification test can be demonstrated by use of real measured radar data.

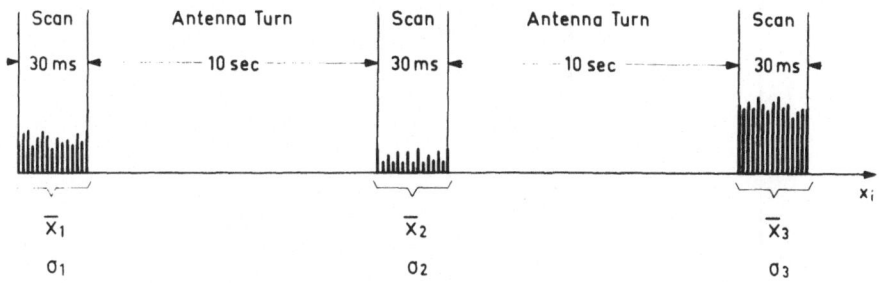

Fig. 1 : Scanning of the target and estimation of fluctuation.

3. THE TEST CRITERION

Fig. 1 is given for description of signal properties and application of test criterion. The target is scanned by a surveillance antenna with a revolution time of 10 sec. Only N=15 samples, the values within the 3 dB beamwidth of the directional antenna, are used for further evaluation.

The first two moments are estimated from the samples x_i (i=1, 2 ... N) by summing up the x_i and x_i^2 as usual in statistics. A criterion for testing the fluctuation of magnitude a for the complex signal $x_i = a_i \cdot \exp j \phi_i$ is:

$$g = \frac{\sigma}{\bar{a}}$$

with

$$\bar{a} = \frac{1}{N} \sum_{i=1}^{N} a_i$$

$$\sigma = \frac{1}{N-1} \sum_{i=1}^{N} (a_i - \bar{a})^2$$

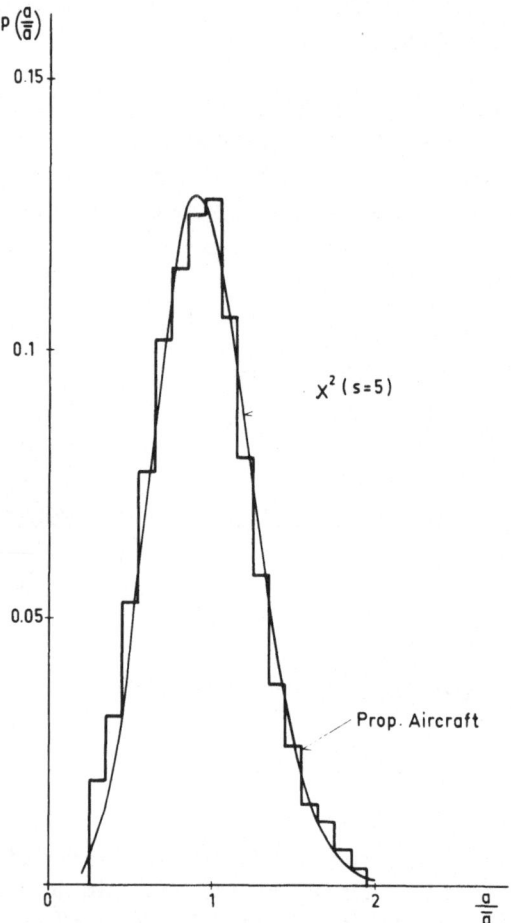

Fig. 2 : Probability density function of amplitude for scans with 15 samples
         (compared with chi-square function)

By normalizing σ on $\bar{a}$ all influences depending on absolute magnitude
differences between hypotheses H_1 and H_2 are suppressed, and only the
ratio fluctuating/steady power of the signal is tested.

For characterisation of phase-shift fluctuation a similar criterion as the
standard deviation σ is used. Because of ambiguity of phase-shift values,
these are summed using the components of a complex phasor $e^{i\Delta\phi}$
with length one.

$$r^2 = \frac{1}{N^2} \left[\left(\sum_{i=1}^{N} \cos\Delta\phi_i \right)^2 + \left(\sum_{i=1}^{N} \sin\Delta\phi_i \right)^2 \right]$$

$$\delta = \arctan\left[\left(\frac{N}{N-1} \cdot \frac{1-r^2}{r^2}\right)^{0.5}\right]$$

where $\Delta\phi_i = \phi_{i+1} - \phi_i$ is the phase shift between successive radar sweeps.

4. MEASURED PROBABILITY DENSITY FUNCTIONS

The probability density functions $p(g)$ and $p(\delta)$ can be calculated, if the statistical properties of the signal are known. Some results of signal analysis [3] may be given in the following characterizing the behaviour of amplitude a and the phase shift $\Delta\phi$ of two different types of airplanes.

Fig. 2: Amplitude density within the scan. The estimated density function of magnitude a_i averaged over several scans along the flight path can be approximated by chi-square functions with higher degrees of freedom or by Rice density functions.

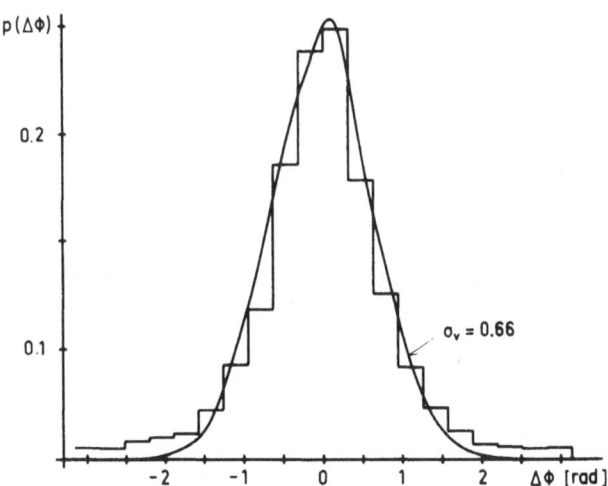

Fig. 3 : Probability density function of phase shift within the scan.
Prop. airplanes on airways (DC 7).
($\Delta\phi_i$ samples from 50 scans are averaged after subtraction of the scan average $\overline{\Delta\phi}$)

Fig. 3: The density of phase shift within the scan can be approximated by a Gaussian density function $N(\overline{\Delta\phi}, \delta)$ where $\overline{\Delta\phi}$ is the mean phase shift induced by radial movement of the target. The scattering angle δ depends on the type of target and the aspect angle. δ is fluctuating during flight of airplanes as can be seen from the next plots.

Fig. 4 shows the track of an observed propeller airplane on different airways.

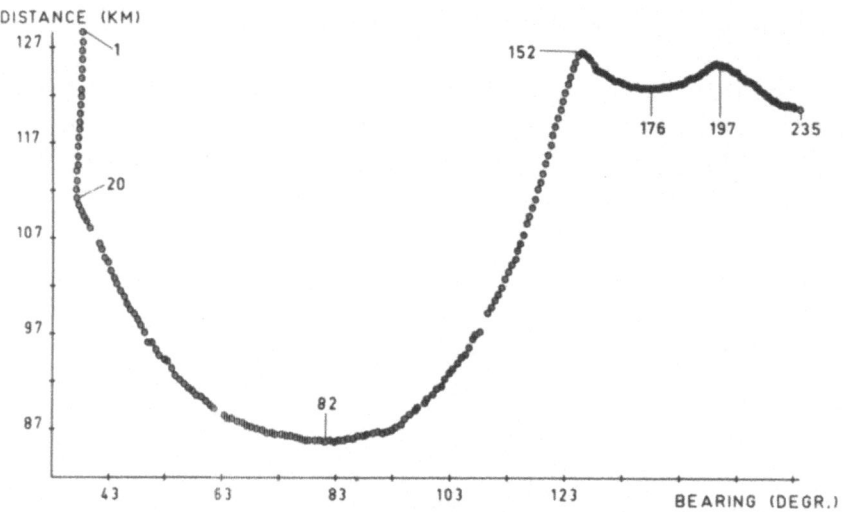

Fig. 4 : Track of an observed propeller airplane (number of antenna turn is referred to Fig. 5)

Fig. 5 indicates the phase shift behaviour during the shown track. The mean phase shift value $\overline{\Delta\phi}$ follows the radial speed of target. The scattering angle δ shows a strong fluctuation from scan to scan. A typical probability density function $p(\delta)$ is given in Fig. 6, which is derived for propeller airplanes on different airways in the range of a surveillance radar (measured at high ratios signal/noise and signal/clutter). The fluctuation properties of g are similar to that of δ.

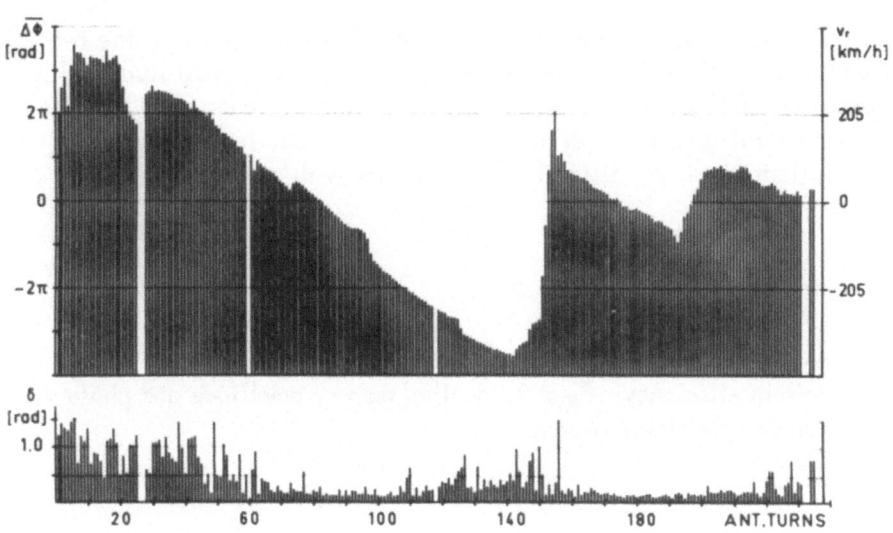

Fig. 5 : Scan average of phase shift and scattering angle δ during flight
on different airways
Prop. airplane, L-band, ant.rot. time 10s, v_r = radial speed.

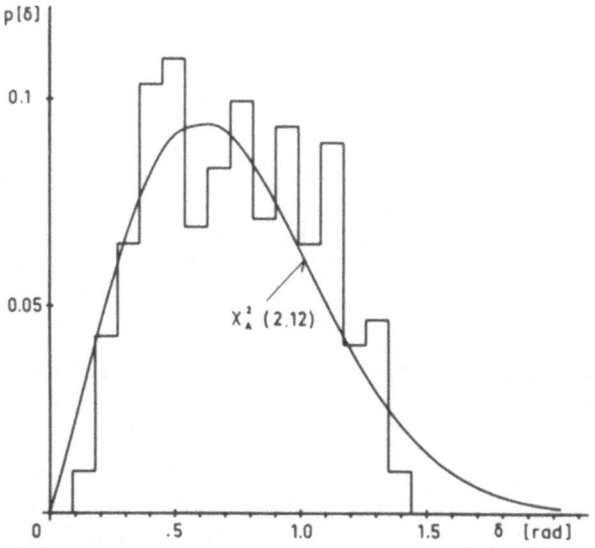

Fig. 6 : Probability density function of scattering angle δ for large prop.
airplanes on different airways (compared with chi-square function).

5. EXAMPLE OF A CLASSIFICATION TEST

For comparison purposes the measured density functions p(g) for the two
hypotheses are plotted into one diagram. Fig. 7 shows typical results. The
probability of classification can be estimated from the data. Fig. 8 gives
the test characteristic for jet airplanes (H_1) and propeller airplanes (H_2).
A classification of H_1 is possible for example with

$$P_K \simeq 0.65$$
$$P_N = 0.1$$

A coherent criterion, using both a and $\Delta\phi$ for classification brings a
small gain in efficiency: $P_K \simeq 0.74$. That means, amplitude and phase shift
deviation are correlated in some way.

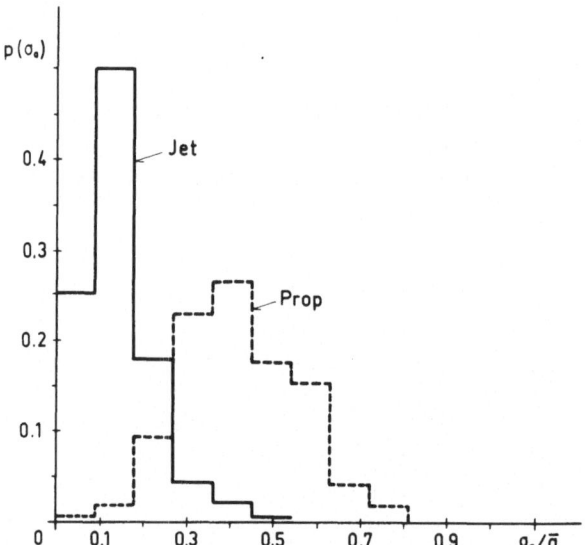

Fig. 7 : Probability density functions p(g) for two types of targets
$$(g = \sigma_a'/\bar{a})$$

This example was given because an adequate quantity of measured data was
available. Other results of our signal analysis give indications that a classification between further kinds of signals is possible. In the same way as
shown a fluctuation test can be applied for discrimination between jet air-

planes and birds or for indication of strong atmospheric turbulence in a calm
environment (an incoherent test is also described in [4]).

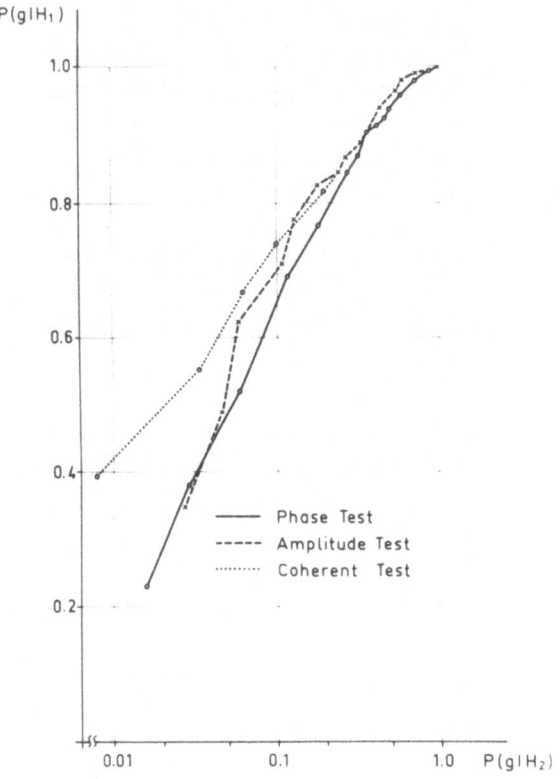

Fig. 8 : Characteristic for the target classification test
      H_1 : Jet airplanes H_2 : Propeller airplanes

The typical of a fluctuation test is to characterize such targets, that are
already detected from the interfered background. In this sense the figures
of false decision probability are sufficient low. A further combination of
the fluctuation test with discrimination procedures based on other criteria
could certainly improve the classification.

REFERENCES

[1] K. von Schlachta   Remarks on Target Models for the Design of
                               Radar Systems
                               IEEE International Radar Conference 1975,
                               Washington D.C., USA, pp. 440-445
                               IEEE Publication 75 CHO 938-1 AES

[2] D. Middleton   Statistical Communication Theory
                               McGraw-Hill Book Comp., New York 1960

[3] K. von Schlachta   Analysis of Radar Signals from Aircraft by
                               Using RADICORD Data
                               Nachrichtentechnische Zeitschrift NTZ-H2,
                               1972, pp. 76-78

[4] D. Atlas   A Method for Radar Turbulence Detection
     R.C. Srivastava   IEEE Trans. on Aerospace and Electronic
                               Systems AES-7, No. 1, January 1971, pp.
                               179-187

ON THE POLARIZATION TRANSFORMING POWER OF RADAR TARGETS

Karl Steinbach

U.S. Army Mobility Equipment Research and
Development Center, Fort Belvoir, Virginia

Abstract. This paper discusses the nature of polarization transforming scattering properties and predicts
target signatures on the basis of simple symmetry con-
siderazations. It further shows that effective target
discrimination can only be obtained by multiple frequency
observations. Experimental data illustrate and confirm
these findings.

The polarization of electromagnetic waves is a well
established concept. One forgets all too easily, however, that this concept appeared relatively late in the
history of science. It was in 1809 that the French
Engineer Malus first observed polarized light as reflected from a glass window.

This discovery was a deciding step toward the understanding of many phenomena. Double refraction, for instance, which had been observed much earlier on crystals,
could now be satisfactorily explained. Soon, the application of polarized light became an indispensable tool
of the natural sciences. Outstanding examples are the
methods of crystal optics, the saccharimeter and the art
of photo-elasticity. The latter is currently widely
used to predict stress distributions for proposed de-
signs of mechanical load bearing structures. Polaroid
sunglasses represent a fine example in our daily life.

Jeske (ed.), Atmospheric Effects on Radar Target Identification and Imaging, 65–82.
All Rights Reserved. Copyright © 1976 by D. Reidel Publishing Company, Dordrecht-Holland.

In the radio frequency range of the electromagnetic
spectrum, experimenters have dealt with polarized waves
right from the beginning. Heinrich Hertz had already
been able to demonstrate anisotropic behavior of a metal
grid, as part of his famous experiments, by the end of
the last century. But more than 60 years went by before
modern radar and communication technology provided the
stimulus for further exploration of such phenomena. Today, numerous excellent publications are available which
deal - from a theoretical viewpoint - the polarization
signatures of radar targets and provide the mathematical
tools for the incorporation of polarization parameters
into radar system design. A British patent application
of 1952 describes a technique for the suppression of
rain drop clutter on the basis of the polarization transforming power of radar targets. Generally, howev r, little
progress has been made toward practical applications
of radar target identification by polarization properties.

This lack of progress can - in part - be explained by
the fact that radar system designers typically do not
have a clear concept of the nature of polarization
transforming properties and are, consequently, not in
a position to readily correlate target shape with specific polarization properties. Unfounded expectations
at an early stage of the design work and subsequent
system failures have tended to discourage further
attempts to fully exploit polarization signatures.

It is a great pleasure for me to report here on some
aspects of our work since it sheds some new light on
the subject matter. I promise not to strain your attention with mathematical equations and solutions to our
specific problem. I will rather focus my discussion on
an elementary analysis of the polarization transforming scattering mechanism, explore briefly its frequency
dependence and illustrate a practical, if somewhat nonconventional application.

The specific application which shall serve here only as
an example, deals with the detection of man-made objects embedded in in-homogeneous media. The target may
be a metallic or non-metallic drainage pipe, an ancient
wall remnant or any other regularly shaped object or
structure. The surrounding medium is assumed to be the
soil characterized by numerous natural inclusions such
as rocks and tree roots, and by surface vegetation.
Needless to say that airborne or space vehicles behind
a weather front or surrounded by chaff represent a
similar situation.

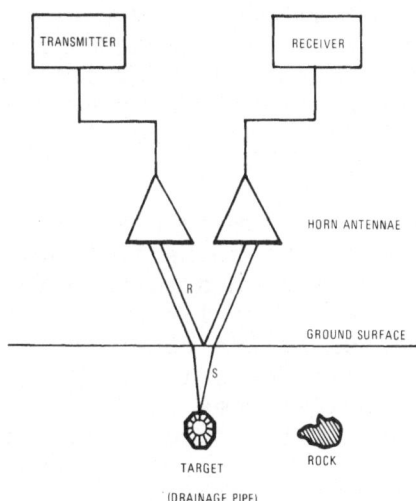

FIG 1 PRINCIPAL ELEMENTS OF THE DETECTION SYSTEM

Figure 1 depicts the key elements of the detection
system and its physical location relative to the target environment. We will concern ourselves with the
following questions:

     a. Can the target echo be separated from soil
surface reflections on the basis of polarization
phenomena?

     b. Are polarization characteristics suitable for
the discrimination between target signatures and echoes
resulting from natural soil inclusions? Or in more
general terms: Do geometrical configurations, shapes
and structures exhibit specific polarization properties that render them distinguishable from randomly
shaped objects and the disordered pattern of several
reflectors?

Let us first examine the sources of unwanted echoes.
To simplify matters, we will assume the antennas to be
linearly polarized and the pointing vector of the incidence wave to be perpendicular to the soil surface.
It is then obvious that reflections from a smooth surface are independent of the polarizational angle. We
want to call this polarization behavior isotropic.
Since the same polarization property is exhibited by

spherical reflectors, we may also expect isotropic behavior from small rocks and from a rough soil surface,
as long as the undulations are small in comparison to
the wavelength. Furthermore, you may agree that a group
of numerous randomly distributed inhomogeneities is
likely to be isotropic. A polarization dependence of
its radar cross section would necessarily require a
somewhat uneven or organized distribution of the individual scatterers.

At this point we may simply postulate that the clutter
producing objects tend to be isotropic. In contrast,
man-made targets are characterized by distinct shapes
and structures which are likely to result in a polarization dependence of the radar cross section. This
anisotropy of the targets is of special interest. I
will show in a few minutes that it is associated with
the power to modify the polarizational state of the incident wave. It is logical, therefore, to use a detection system the receiver polarization of which is orthogonal to the polarization of the transmitter. In the
case of our linearly polarized antennas we simply turn
one antenna by 90° about its axis. The resulting
"cross polarized" system represents then a first measure toward discriminating target signals from clutter
in that only echoes from those objects are received which
have the power to transform at least part of the incident polarization into an orthogonal component.

I would like to direct your attention now to a closer
examination of polarization transforming scattering
properties. Of specific interest with respect to our
detection problem is the question: What causes the rotation of the polarizational plane?

The simplest anisotropic reflector is a straight piece
of wire or dipole. An excitation occurs only if the incident electric field has a component parallel to the
wire. The amplitude of the reflected signal is, therefore, largely dependent on the polarization of the incident wave. The top section of figure 2 illustrates
the correlation between dipole orientation and polarization of the incident wave E. The pointing vector is
assumed to be perpendicular to the plane of the picture. Since the dipole excitation is proportional to
the projection of the incident vector upon the wire,
the wire current I is proportional to cos θ.

The re-radiation caused by this current (center of
figure 2) may be considered to consist of 2 components:

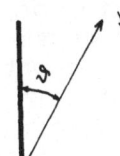

DIPOLE

$\mathscr{E}_i$, INCIDENT VECTOR.

CURRENT IN DIPOLE : $I \sim |\mathscr{E}_i| \cos\vartheta$.

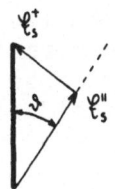

RE-RADIATION : $\mathscr{E}_s^{\shortparallel}$ AND $\mathscr{E}_s^{+}$.

$$\mathscr{E}_s^{+} \sim I \sin\vartheta$$
$$\sim |\mathscr{E}_i| \cos\vartheta \sin\vartheta = \frac{|\mathscr{E}_i|}{2} \sin 2\vartheta .$$

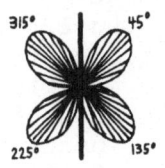

CROSS POLARIZED ECHO AS FUNCTION OF ϑ

FIG 2 POLARIZATION PROPERTY OF DIPOLE

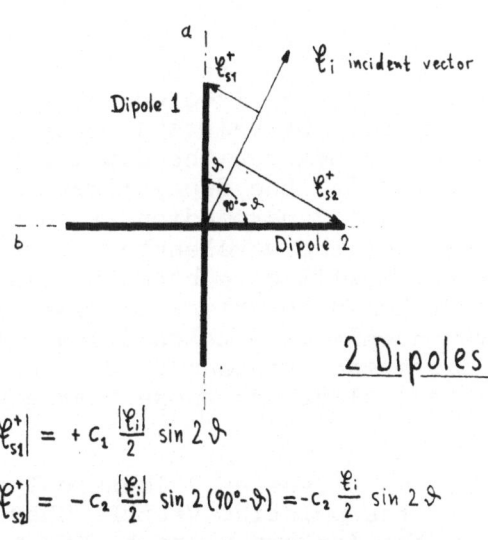

$$\left|\mathscr{E}_{s1}^{+}\right| = + c_1 \frac{|\mathscr{E}_i|}{2} \sin 2\vartheta$$

$$\left|\mathscr{E}_{s2}^{+}\right| = - c_2 \frac{|\mathscr{E}_i|}{2} \sin 2(90°-\vartheta) = -c_2 \frac{\mathscr{E}_i}{2} \sin 2\vartheta$$

FIG 3 TWO PERPENDICULAR DIPOLES

E_s'' parallel to the incident polarization, and the orthogonal component E_s^+ . This latter component is obviously proportional to the product of wire current and
the sine of the angle θ. Through substitution of I by
the value given above, one obtains for the cross polarized re-radiation a magnitude which is proportional
to sin 2θ. In polar coordinates, this provides a perfectly symmetrical four leaf clover leaf pattern as
shown in the bottom part of figure 2. The cross polarized
signal reaches its max. values for the angles θ = 45,
135, 225, and 315°. It vanishes whenever the incident
polarization is parallel or perpendicular to the wire.
This is quite obvious for these conditions permit
either no excitation of the wire at all or cause the
total re-rediation to be parallel to the incident
polarization.

Figure 3 illustrates a reflector which consists of 2
perpendicular dipoles. One can readily see that the
cross polarized re-radiation components E_{s1}^+ and E_{s2}^+
are 180 out of phase. The different orientations of the
dipoles relative to the incident vector have no effect
for sin 2θ is equal to sin 2 (90-θ). The 2 components
may, however, differ in magnitude by their respective
reflection coefficients and C_1 and C_2. If the 2 dipoles
are identical, the 2 echoes E_{s1}^+ and E_{s2}^+ cancel each
other and the total cross polarized echo is zero for
all angles.

It is of some interest to ask what happens if the 2
dipoles are not located in the same plane but displaced in the direction of wave propagation. Obviously,
in this case the 2 excitations are not in phase and
consequently, the 2 cross polarized echoes are not
exactly 180 out of phase. The relative phase is now a
function of the distance between the dipoles measured
in wavelength. As a result, the depolarization power
of a fixed arrangement of 2 perpendicular, displaced
dipoles is a function of the wavelength or frequency
of the incident wave. I will come back to this frequency dependence of the depolarization power later on.
At this point I would only like to call your attention
to a significant difference between 2 identical dipoles
positioned in the same plane and those that are displaced:

The planar dipole cross possesses 2 identical planes of
symmetry parallel to the pointing vector. They are
marked in fig. 3 by the letters a and b. The non-planar
arrangement of 2 identical dipoles also possesses 2

such planes of symmetry. These symmetries are, however, not identical. The difference lies in the sequence of the 2 dipoles relative to the wave propagation. You will recognize the significance of these different symmetry properties in a few minutes.

A wire rolled into a spiral as shown in Figure 4 is a somewhat unique example of an anisotropic reflector. In this case the scattering cross section for linearly polarized waves is not a function of the incident polarization, but the reflected wave is circularly polarized and contains, therefore, always a cross polarized component.

All three examples, the single dipole, the crossed dipoles and the spiral show clearly that the polarization transformation is due to the fact that the induced currents are conditioned by the physical configuration of the reflector. A change of the polarizational angle of the incident wave affects the amplitude of these currents, but not their direction. Consequently, whenever the incident vector is not in line with the current supporting element (s) of the reflector, a cross polarized component is generated - provided, of course, that an excitation is possible at all.

The depolarization power of any arbitrarily shaped three-dimensional object can be understood by a similar, though somewhat more complex mechanism. In place of the current along the wire one finds specific current distributions and wave modes within the scattering object. They are a function of its shape and structure as well as of the incident polarization. Similar to the well-known modes in wave guides, a variety of modes can usually be excited in a given reflector and the prediction of its polarization properties is therefore not as straight forward as in the case of simple wire configurations. It is, however, possible to determine the polarization properties on the basis of simple symmetry considerations - without knowledge of specific wave modes.

Figure 5 shows the geometrical cross sections of various three-dimensional objects. The dotted lines identify planes of symmetry parallel to the poynting vector of the incident wave. We can readily distinguish 3 classes of reflectors: the examples shown in the top row are characterized by 2 or more identical planes of symmetry. The center row represents objects which posses 2 different planes of symmetry or - as in the case of the iso-

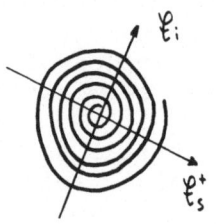

FIG 4 SPIRAL REFLECTOR

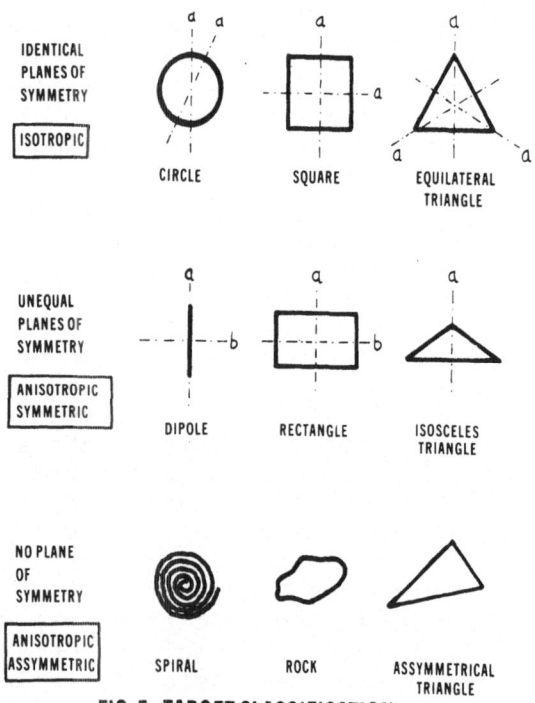

FIG 5 TARGET CLASSIFICATION

sceles triangle - only one plane of symmetry. The last class of reflectors, at the bottom, is characterized by the lack of any symmetry.

The polarization behavior of the round cylinder (the first example at the upper left) can readily be predicted on the basis of its rotational symmetry: the scattering cross section is obviously independent of the po-

larizational angle of the incident wave. A cross polarized component cannot be generated since the incident polarization coincides always with a plane of
symmetry. Any imaginary cross polarized component would
have to be phased to the right of this plane as well as
to the left and therefore cancel itself.

I have already discussed the first examples of the 2
other classes, namely the straight wire or dipole and
the spiral. They are both anisotropic with the difference, that the dipole possesses symmetry properties,
the spiral does not.

It is not difficult to prove that the three examples of
each class exhibit identical scattering properties, or
more specifically that the polarization transforming
power is directly and unambiguously related to symmetry
properties of the reflector. The time does not permit
me to discuss this in detail, but I wish to illustrate
briefly the thought process. Let us first take a look
at those reflectors which possess at least 2 identical
planes of symmetry: The incident field vector can in
this case always be described by 2 components which fall
into these planes of symmetry. The relative phases and
amplitudes of the 2 components -- and with them the polarization of the reflected wave -- remain then unchanged. This is independent of the polarizational
angle of the incident wave, just as in the case of the
round cylinder.

The situation is quite different if the 2 planes of
symmetry are not identical or if only one such plane
exists. The incident vector can then not be described
by 2 components which remain invariant with respect to
each other. Consequently, the reflector possesses in
this case a polarization transforming power. The cross
polarized echo is only zero when the incident polarization happens to be parallel or perpendicular to a
symmetry plane. This behavior results in the cloverleaf-
pattern of the cross polarized signal which we observed
on the straight wire.

The same sort of reasoning leads to the conclusion that
reflectors lacking any symmetry possess generally a
depolarization power. Their cross polarized echo has
no points of zero amplitude but may depend on the polarizational angle of the incident polarization.

The results of these considerations are not at all obvious. Two examples are shown in Figure 6. Who would

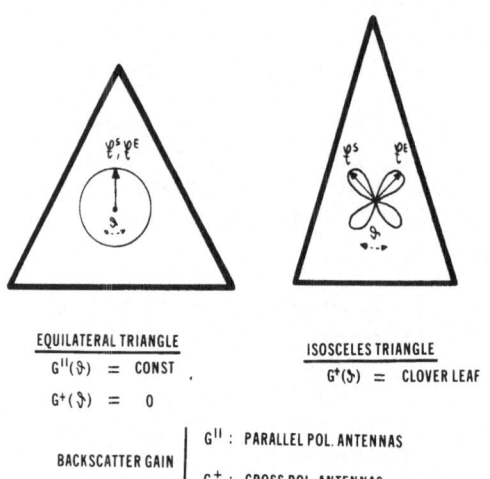

EQUILATERAL TRIANGLE
$$G^{||}(\vartheta) = CONST$$
$$G^{+}(\vartheta) = 0$$

ISOSCELES TRIANGLE
$$G^{+}(\vartheta) = CLOVER LEAF$$

BACKSCATTER GAIN | $G^{||}$: PARALLEL POL. ANTENNAS
| G^{+} : CROSS POL. ANTENNAS

FIG 6   POLORIZATION PROPERTIES GOVERNED BY SYMMETRY

expect that the back-scatter gain of an equilateral
triangle is independent of the polarizational angle?
and that this triangle does not have any polarization
transforming power?

Equally surprising is the clover leaf pattern of the
cross polarized echo from an isosceles triangle. One
might perhaps be tempted to seek a common denominator
for the 3 corners of the reflector and the 90° symmetry
of the cross polarized antenna system. On that basis
one might expect a 12-fold periodicity of the return
signal. It was especially this reflector shape that
caused us great conceptual difficulties. Our delight
was great when we discovered that its polarization properties are governed by such a simple principle, namely
that of symmetry.

In order to verify these findings I have derived an
exact equation which describes the cross polarized signal
as a function of polarizational angle θ, see Figure 7.
The magnitude V^{+} is the voltage measured at the ter-
minals of a cross-polarized receive antenna. It is a
function of the 2 constants A and B which are related
to the shape of the reflector, and of the polarizational
angle θ. The latter is measured with respect to a reference plane on the reflector.

$$|V^+(\vartheta)| = \sqrt{(1+AB)^2 \sin^2(2\vartheta - \vartheta_0) + (A+B)^2 \cos^2(2\vartheta - \vartheta_0)}$$

ϑ = ANGLE OF INCIDENT POLARIZATION

A, B = CONSTANTS RELATED TO SHAPE

ISOTROPIC REFLECTOR	A = -B = 1	$|V^+(\vartheta)| \equiv 0$
ANISOTROPIC SYMMETRICAL REFLECTOR	A = -B ≠ 1	
ANISOTROPIC ASSYMMETRICAL REFLECTOR	A > 0 B > 0	

FIG 7 RIGOROUS DESCRIPTION OF CROSS POLARIZED SIGNAL

For isotropic reflectors is A = -B = unity; and consequently V+ = zero for all angles θ. For anisotropic
symmetrical reflectors is also A = -B but not equal to
unity. In this case, only the second term of the radi-
cal becomes zero and the signal exhibits the perfect
cloverleaf pattern. Assymmetrical reflectors are
mathematically identified by positive values of both
form-factors and the resulting signal is as shown in
the bottom part of Figure 7.

The equation describing V+(θ) does not take frequency
changes into account and we have to return, for a minute,
to intuitive reasoning in order to gain some inside into
the frequency dependence of these signal patterns.

At the upper left corner of Figure 8, you see an assymmetrical arrangement of two color-coded sticks (red
appears grey, blue appears black in the reproduction).
Suitable optical filters will cause one or the other to
become invisible as shown on the right. As a result, the
assymmetrical arrangement appears in both of these cases

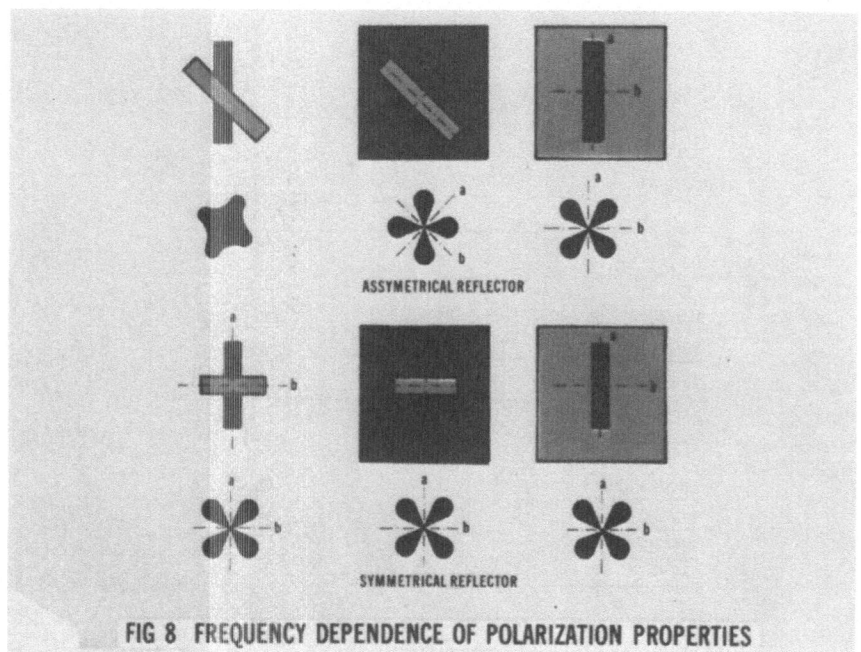

FIG 8 FREQUENCY DEPENDENCE OF POLARIZATION PROPERTIES

to be symmetrical. Furthermore, the orientation of the symmetry planes "a" and "b" seems to be dependent on whether a blue or a red filter is used.

Let us now assume that the 2 colored sticks represent 2 dipoles of different lengths. When illuminated with a radio frequency outside the resonance of either, the cross polarized signal pattern would be that of an assymmetrical reflector. A suitable frequency change, however, may cause the response of one of the dipoles to be enhanced while the response of the second dipole may become negligibly small. The signal pattern will then exhibit the zero points, characteristic of symmetrical reflectors. The orientation of the cloverleaf will be determined by the orientation of the resonating dipole, that is to say it will change with frequency.

A symmetrical arrangement of 2 sticks or dipoles as shown in the lower half of Figure 8, behaves quite differently. Here is the orientation of the symmetry planes, and accordingly the orientation of the cloverleaf pattern independent of frequency. Frequency changes can only affect the signal amplitude.

These observations disclose an important property of polarization signatures. They show that assymmetrical reflectors may very well exhibit the characteristics of symmetrical targets; but they indicate also that such apparent or false symmetries can be recognized by multiple frequency operation.

The time does not permit me to explore this topic further with you. I prefer to use the remaining few minutes to show some experimental data which illustrate the potential of radar target discrimination by polarization properties.

Figure 9 shows the various targets used in our experiments. I will present data on items A, E and F.

To conduct measurements on targets actually embedded in the soil, we used the experimental set up shown in Fig. lo. The cross polarized transmit and receive antennas are shown stationary on the left, rotating on the right. The entire system is suspended from overhead tracks and can be moved to any point over the soil.

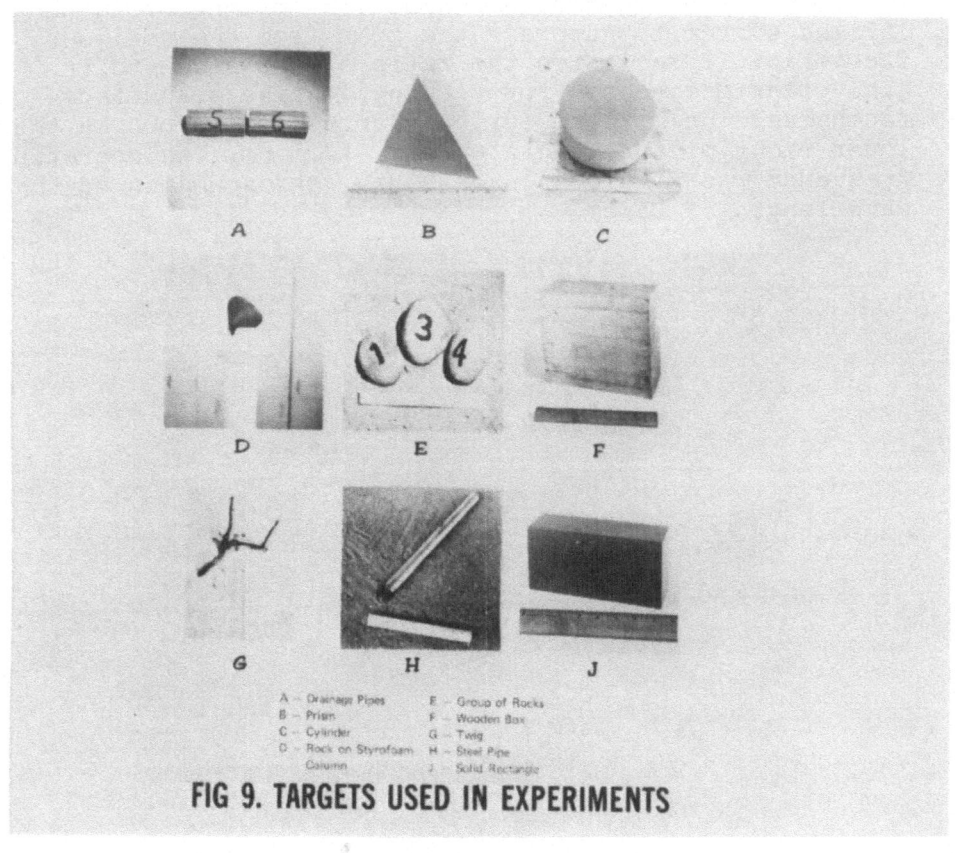

A — Drainage Pipes   E — Group of Rocks
B — Prism   F — Wooden Box
C — Cylinder   G — Twig
D — Rock on Styrofoam   H — Steel Pipe
    Column   J — Solid Rectangle

FIG 9. TARGETS USED IN EXPERIMENTS

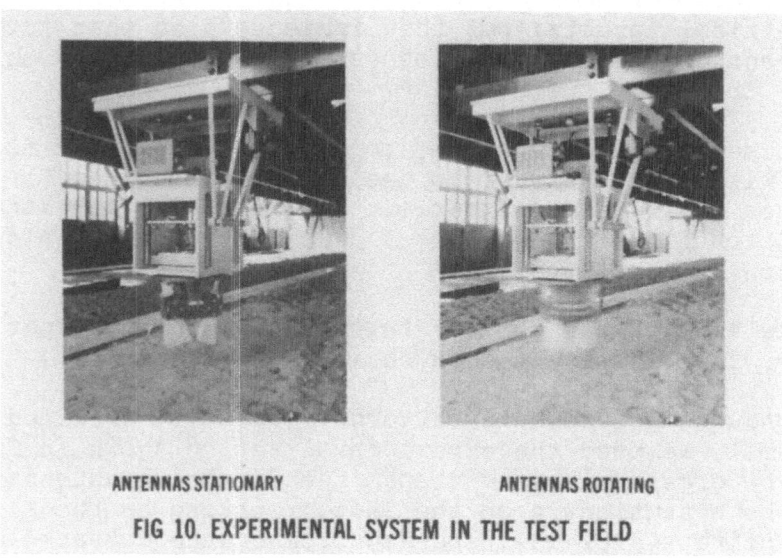

ANTENNAS STATIONARY   **ANTENNAS ROTATING**

FIG 10. EXPERIMENTAL SYSTEM IN THE TEST FIELD

First, we investigated the capability of the cross-polarized system to suppress the echo from the soil surface. Fig. 11 indicates the ratio of crossed - to in
line polarized echoes for varying degrees of surface
roughness. The maximum surface depressions shown in the
lower right picture, were approx. 12.5 cm. The operating
frequency was in all 4 cases 2GHz, corresponding to the
wave length of 15 cm.

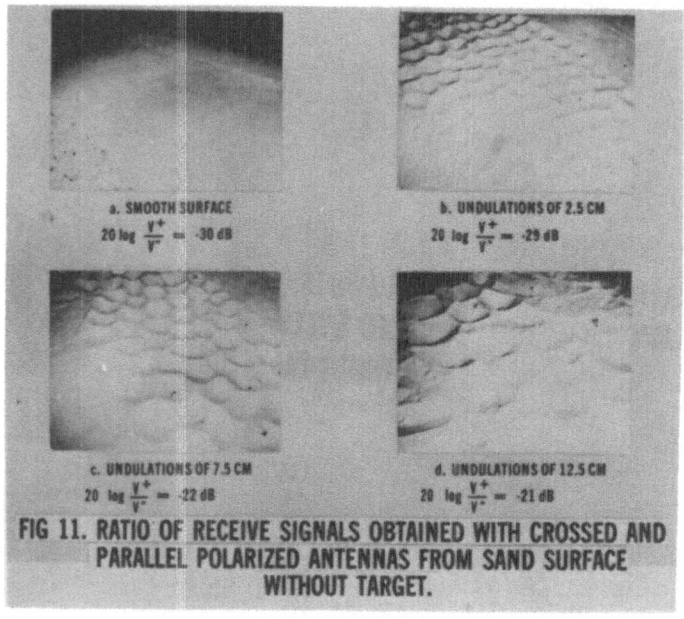

a. SMOOTH SURFACE
$20 \log \frac{v^+}{v^-} = -30 \, dB$

b. UNDULATIONS OF 2.5 CM
$20 \log \frac{v^+}{v^-} = -29 \, dB$

c. UNDULATIONS OF 7.5 CM
$20 \log \frac{v^+}{v^-} = -22 \, dB$

d. UNDULATIONS OF 12.5 CM
$20 \log \frac{v^+}{v^-} = -21 \, dB$

**FIG 11. RATIO OF RECEIVE SIGNALS OBTAINED WITH CROSSED AND
PARALLEL POLARIZED ANTENNAS FROM SAND SURFACE
WITHOUT TARGET.**

Figure 12 shows typical oscilloscope displays of the cross polarized signal as a function of the rotation of the antenna system. The horizontal axis represents one full rotation of the antennas. The 2 top pictures correspond to antenna positions where no targets were buried. The symmetric signal patterns displayed in the center are the return from buried clay pipes. Buried rocks could not be located at all during these tests. In order to obtain the 2 bottom signals, it was necessary to place three large rocks on top of the soil surface. Furthermore, their relative positions had to be carefully selected to obtain sufficient signal amplitude. The specific pattern shown are, in part, due to antenna near field effects which I will not further discuss. Of primary interest is the fact that the rock signatures are quite irregular and thus distinguishable from the clay pipe signals.

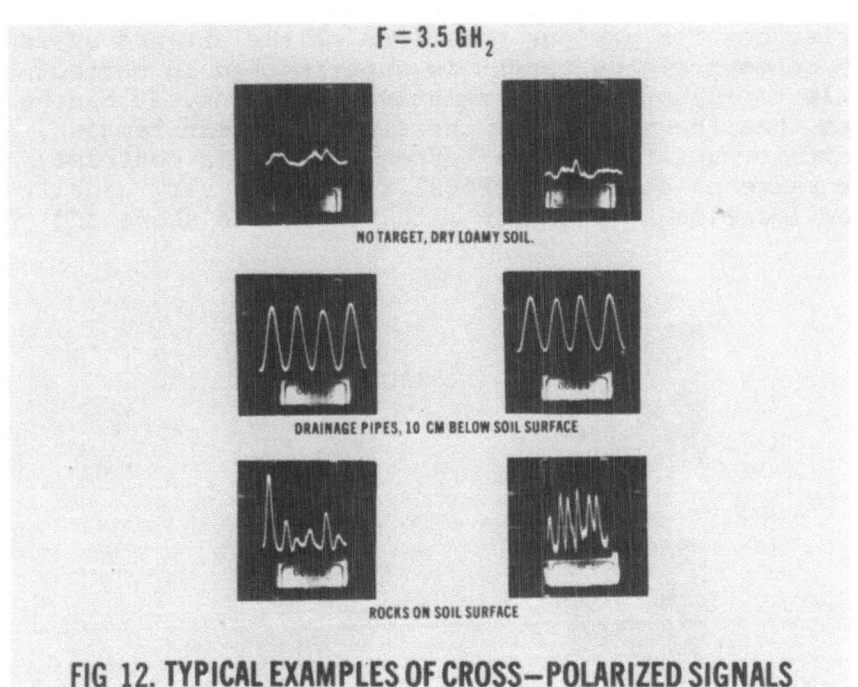

$$F = 3.5 \ GH_2$$

NO TARGET, DRY LOAMY SOIL.

DRAINAGE PIPES, 10 CM BELOW SOIL SURFACE

ROCKS ON SOIL SURFACE

FIG 12. TYPICAL EXAMPLES OF CROSS—POLARIZED SIGNALS

Figure 13 illustrates the additional signal discrimination capability obtainable through frequency modulation.
The two oscillograms on the left represent echoes of
buried clay pipes, while the ones on the right are from
a group of rocks. For the frequency of 2.96 GHz (top
pictures) the rocks happened to respond almost like a
symmetrical object and, as you see, are hard to distinguish from the clay pipe.

The two bottom displays are the result of multiple frequency operation. The transmitter was, in fact, rapidly
swept from 2.6 to 3.6 GHz resulting in the superposition
of the signal pattern obtained at various frequencies.
The signal pattern of the clay pipes is, as anticipated,
independent of frequency and maintains, therefore, its
characteristic zero points. In the case of the rock
group, the zero positions of the signal shift with frequency. This results in the smeared-out pattern shown
at the bottom right.

Figure 14 shows the signal of a rectangularly shaped
buried box for various positions of the antenna system.
The contour of the target is superimposed in correct
scale to illustrate the relative positions. It can be
seen that the phasing of the signal pattern remains
constant over the entire target region. In contrast,
the patterns of assymmetrical reflectors vary usually
from location to location with respect to shape and

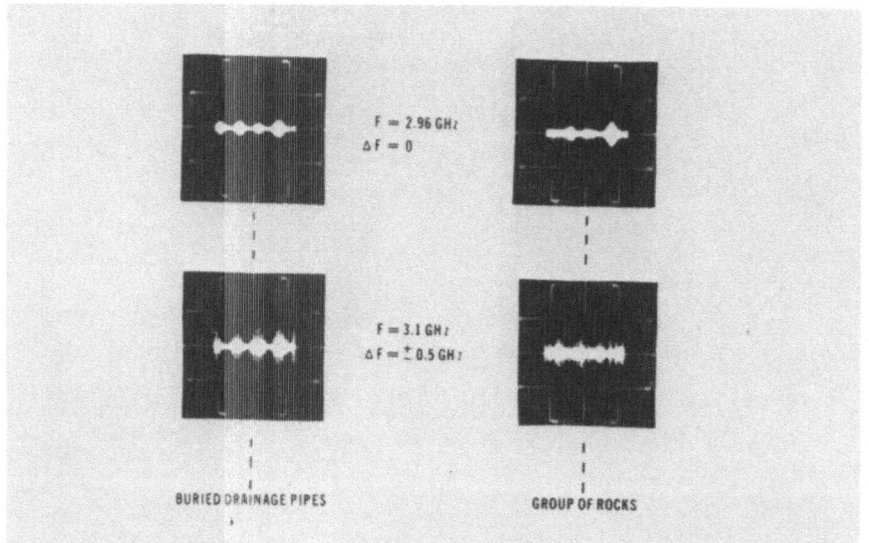

FIG 13. SIGNAL DISCRIMINATION OBTAINABLE THROUGH FREQUENCY MODULATION

phasing along the horizontal axis. This is demonstrated by the example shown in Figure 15 which represents the signatures of three rocks located on top of the soil surface.

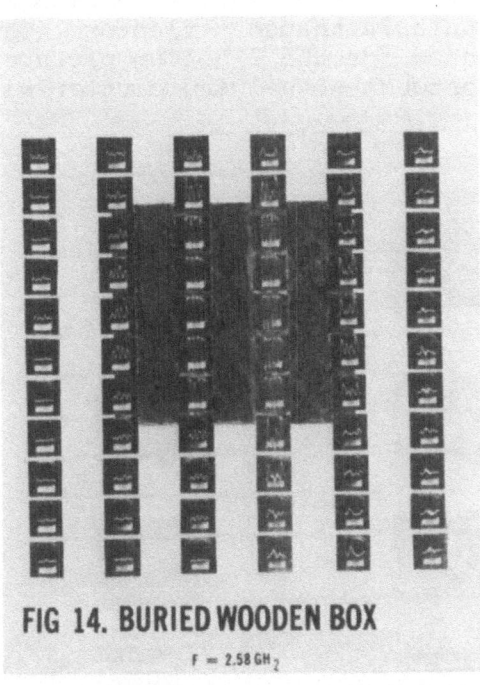

FIG 14. BURIED WOODEN BOX

F = 2.58 GH$_2$

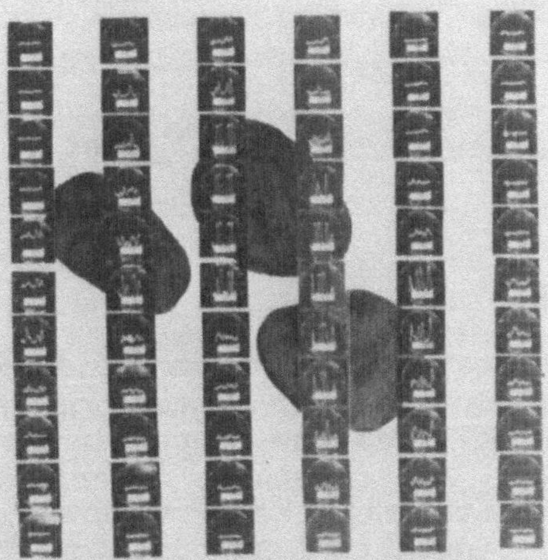

FIG 15. 3 ROCKS ON SAND SURFACE

F = 3.15 GH$_2$

Figure 16 finally, demonstrates various signal display techniques taking rotation of the incident polarization, frequency sweeping and position changes of the antennas into account. The pictures on the left represent a symmetrical target, those on the right a disordered group of several irregularly shaped reflectors. The contourogram technique used for the 2 bottom pictures emphasizes the difference between signal and clutter with particular clarity.

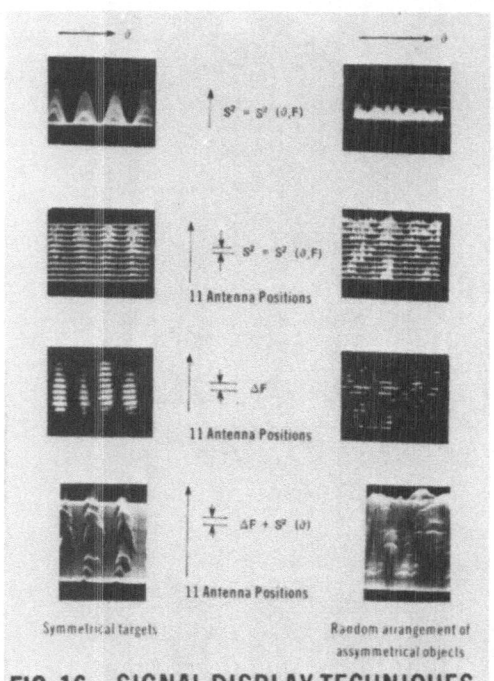

FIG 16. SIGNAL DISPLAY TECHNIQUES

ACKNOWLEDGEMENTS:

This paper represents excerpts from U.S.A. MERDC Report 2065 entitled "Nonconventional Aspects of Radar Target Classification by Polarization Properties", Fort Belvoir, Va., Jun 1973. All reference material is listed in this report. A Secretary of the Army Research and Study Fellowship provided essential support for the investigations. The author wishes to thank Professor emeritus Dr.-Ing. habil. H. Tischner, Technical University Hannover, FRG, for technical guidance and many fruitful discussions.

SHORT-PULSE AND WIDE-BANDWIDTH HIGH-RESOLUTION RADARS

Anthony V. Alongi

Calspan Corporation, Buffalo, New York, U.S.A.

1. INTRODUCTION

1.1 Historical Background of Short-Pulse High-Resolution Radar

High-resolution radar is a more recent specialized development within the broad area of radar technology which includes
many different types of systems designed for many different
applications. In 1886, Henrich Hertz first demonstrated that
electromagnetic waves could be reflected by metallic and dielectric bodies. Early work, [1] such as performed by Hülsmeyer of
Germany in 1903, A. H. Taylor and L. C. Young of NRL in 1922, and
L. A. Huland, also of NRL, in 1930 showed the detection of aircraft with radio waves. U. S. developed radars for shipboard use
were tested on warships in 1939 and installed on major ships of
the fleet in 1941. U. S. Army radars originated in 1938 with
the SCR-268 an anti-aircraft fire control radar. In Britain, because of the impending war, radar development was quite rapid, so
that in 1938 they had several chain home radar stations in operation. Their development of the magnetron as a source of microwave power during the war years was a noteworthy achievement.
Germany simultaneously carried out independent radar developments
also during the war. After World War II further advances were
made in transmitter tubes and in lower noise receivers. The high-
power klystron amplifier traveling wave tubes and wideband backward wave oscillators contributed greatly toward advancing radar
state-of-the-art. At Calspan, the application of the high-power
klystron amplifier in 1961 led to the development of a high-power
radar having a transmitter with fifty megawatts of peak power.
This system has been employed in radar cross section measurements
of earth satellites, space vehicles, aircraft and extraterrestrial
objects. Concurrent with radar development was the growth of

Jeske (ed.), Atmospheric Effects on Radar Target Identification and Imaging, 83–156.

interest in target backscatter. The need for a better understanding of the backscattering mechanisms necessitated increased
resolution of radar systems and a search for the generation and
transmission of very wide bandwidth signals.

Developments in high-resolution have been measured in meters
and ultra-high-resolution radar measured in centimeters and has
led to applications in target identification and diagnostics of
backscattering media. Inferences may now be made even on the
nature and internal structure of dissipative dielectric bodies.
Target identification is of obvious interest in military applications, while diagnostics are of interest both in military and
civilian use.

Much of the early work in high-resolution originated with
the development of radar cross section measurement techniques.
Personnel, such as the author and others in organizations such as
Convair, Micronetics, Sperry Rand and Calspan developed high-
resolution measurement facilities.

Measurements performed in these facilities on simple geometrical bodies and on scaled models of relatively complex radar
targets has permitted the observation of scattering centers and
has demonstrated their sensitivity to the frequency, aspect angle
and polarization of the illuminating electromagnetic energy.

The approaches taken in the development of high-resolution
radar centered about means for direct generation of short pulses
and the generation and transmission of wide bandwidth signals
with some type of frequency modulated waveform. Hively's [2] investigation in short-pulse work started in 1959 and led to a
search for sub-nanosecond pulsed sources which included the direct
pulsing of an oscillator tube, ferrite switching of a CW signal,
gating the grid and/or helix of a traveling wave tube, and diode
switching of a CW signal. It was soon concluded that direct
pulsing of the oscillator and ferrite switching of CW would not
produce sub-nanosecond pulses.

Investigations then included switching with traveling wave
tubes. Beck and Mandeville [3] of Bell Telephone Laboratories
had achieved pulses of several nanoseconds duration by sweeping
the helix voltage of a TWT through its synchronous region while
a CW signal was applied to the input of the TWT. During the short
time interval when the helix voltage was at or near its synchronous value, the tube would amplify, thus forming a short RF
pulse at its output. These results did not produce pulses shorter
than six nanosecond duration.

Other direct short pulse generation techniques reported by
Hively were diode switching of CW sources as suggested by
Robert Garver [4] of the Diamond Ordnance Fuze Laboratory. These
results were not acceptable due to the limited cut-off range of the

switching diodes. The Sylvania Company's work on diode burnout investigations suggested the use of the spike leakage through a TR tube [5] as a source of short pulses. Pulses of 0.7 nano-second duration with peak powers of 100 watt were achieved. Pulfer and Whetford [6] of Canada developed another short-pulse generation technique in 1961. They found that a short RF pulse could be produced at the output of a TWT when input of the TWT was subjected to a video pulse. Application of this technique led to the generation of sub-nanosecond pulses from L-band through K_a-band with peak powers of up to half megawatt at X-band. On the other hand, the author at Calspan, then known as Cornell Aeronautical Laboratory, and his associates worked with wide bandwidth linear FM as the transmitted signal source, and in 1962 achieved a 3-inch range resolution as measured at the received signal half power points with range sidelobes down 35 dB.

In 1966, Prinsen and Tripp [7] developed a nanosecond X-band Coherent Pulse Doppler radar which was employed for Hypersonic Wake Studies. This short-pulse radar has been applied to applications such as chaff cloud measurements, radar cross section measurements and projectile velocity measurements [8].

A more recent development in the short-pulse radar tech-nology has been the development of a downward looking radar that radiates a single cycle of RF having a nanosecond duration. This system was described by Alongi [9] in Edinburgh, Scotland in 1973.

1.2 Defining High-Resolution

In conventional pulse radar systems range resolution is defined

$$\Delta R = \frac{C\tau}{2}$$

where

ΔR = increment of range in meters
C = velocity of light, 3×10^8 meters per second
τ = pulse duration in seconds

Usually the range resolution increment is considerably greater than the maximum length of the target such that the return signal for a distant target appears to have originated from a single source. A high resolution radar system generally develops a transmitted waveform that provides ΔR values approximately the same or smaller than the target length. Resolution now entails the actual separation of the scattering sources of a complex tar-get or the generation of interference from scattering sources separated less than the range resolution capability of the system. To demonstrate these concepts for a simple pulse radar system, I refer to the expression derived by Rothman, Guthard and Morita [10] for the scattering from extended laboratory targets separated by a distance d which is less than the pulse width of the radar

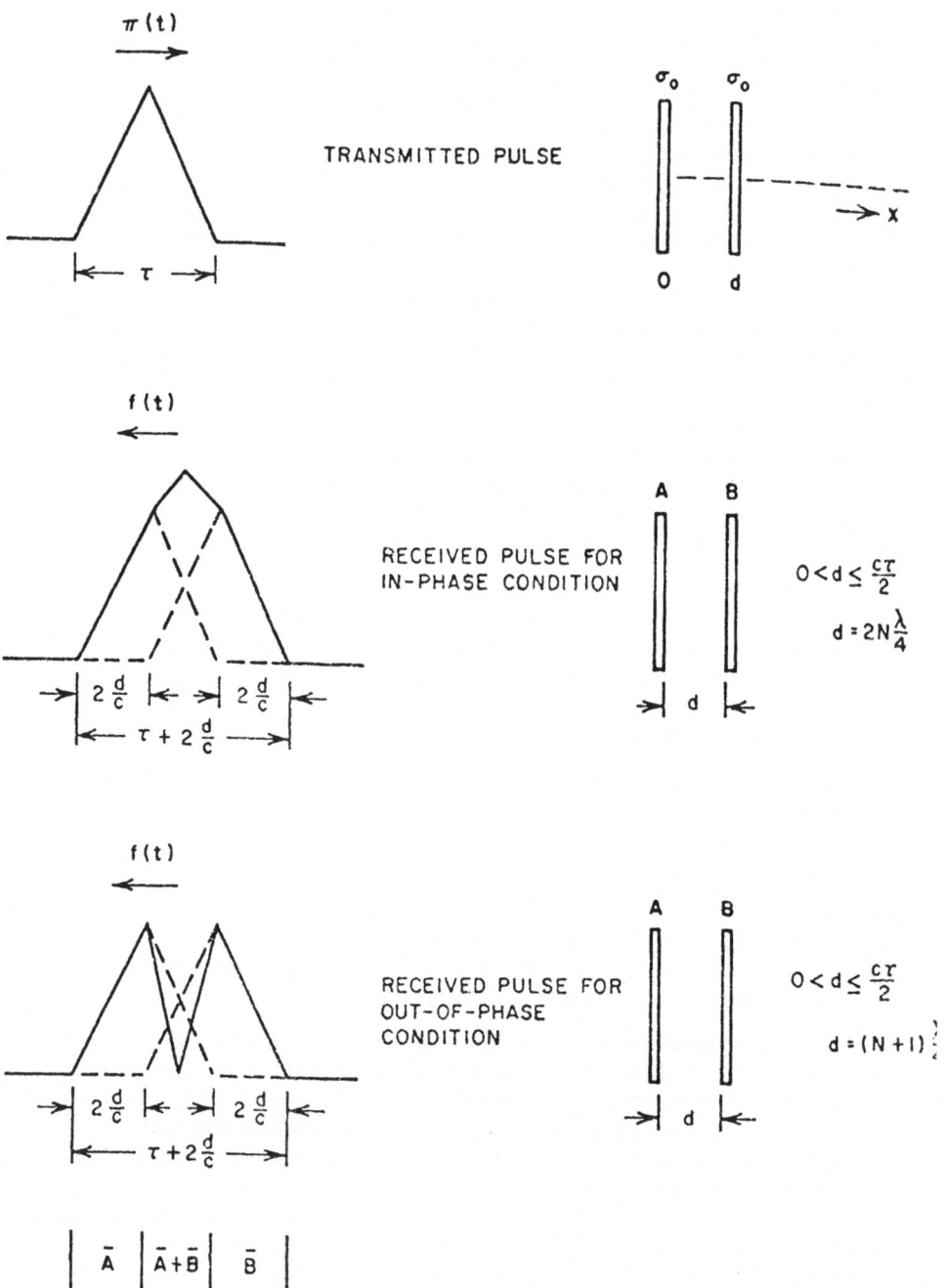

Figure 1 - RECEIVED PULSE AMPLITUDE FROM TWO DISCRETE SCATTERERS

$$f(t) = \sigma_o' \, \pi(t) + \sigma_o \, \pi(t - \frac{2d}{c}) e^{-2j\beta d}$$

where
   σ_o' =   the backscatter from a dipole of a dipole array
   $\pi(t)$ =   the illuminating electromagnetic pulse
   c =   volocity of light
   β =   propagation constant along the target

Figure 1 illustrates the expression for the received pulse amplitude from two interfering targets.

If d $> c\tau/2$, then f(t) is simply two triangular pulses with
a peak-to-peak separation of (2d/c) seconds. When d $< c\tau/2$,
the two pulses interfere and the pulse amplitude as a function of
time depends on β d, the relative phase between the scatterers.
Two of the possible configurations for f(t) are shown in Figure 1
for the condition d $> c\tau/2$ and for (2βd), an odd or even multiple of π radians. The total pulse width is consequently greater
than the transmitted pulse width by (2d/c) seconds. The first
(2d/c) seconds of the return pulse represents scatter from the
second dipole alone. In the overlap region or the center portion, the amplitude represents the vector sum of the scatter
from both dipoles.

2.   PROBLEMS COMMON TO HIGH-RESOLUTION RADAR

There are at least three basic types of high-resolution
radar: FM/CW, chirp, and short-pulse. Each of these types of
radars may take on several different forms. Generally, the choice
between these basic types is made on the basis of the ease of
implementation for a particular application. Direct short pulse
has the advantage of relatively simple design and construction,
although a short-pulse system will inevitably be power limited.
FM/CW is considerably more complex to build, has the advantage
of a very high duty cycle, thereby offering the advantage of
high average power over the short pulse system. The transmitted
waveform must have stringent requirements on the amplitude and
phase behavior with time. Chirp radars generally employ the
same waveform as FM/CW, i.e., linear FM, the principal difference
being that the FM/CW radar does not employ expansion or compression filters.

A classic problem in FM/CW and chirp radars is the range and
velocity resolution ambiguity. In the case of a single pulse,
range resolution can be improved by making the pulse shorter
while range rate resolution or Doppler shift is improved by making
the pulse longer. For the pulse train case in selecting the
pulse repetition frequency, the ambiguities in range will be determined by the pulse repetition frequency. Since this topic has
been covered extensively in the literature [11], it will not be
repeated here.

Transmission of nanosecond pulses or of the equivalent bandwidth in FM/CW or chirp systems requires a careful consideration
of the sources of waveform distortion during the system design.
In a pulse system the transmitter must have low range or time
sidelobes in order to permit the isolation and observation of the
scattering sources of a complex target. (It should be remembered
that a nanosecond or sub-nanosecond waveform will permit resolution of two equal scatterers separated on the order of fifteen
centimeters.) High range sidelobes will tend to mask the radar
echo arising from weaker scatterers located near a strong scatterer.

Chirp or FM/CW radar systems require the same careful attention to the waveform sidelobes in the frequency domain since they
have effects identical to range or time sidelobes as in the pulse
radar system. Range sidelobes arise from the nonlinear phase or
amplitude response of radar system elements, or from improperly
generated transmitter waveforms having phase terms other than
linear and quadratic or non-constant amplitude such as in the
case of the linear FM system. Other sources of range sidelobes
arise from frequency-dependent multiple reflections within the
system. Pulse distortion of short pulses or waveform distortion
in the case of wide bandwidth linear FM also arise from the use
of components with insufficient bandwidth or the use of components that have frequency-dependent delay. The latter effect will
cause pulse stretching. Of particular interest to the short
pulse imaging radar designer is the dispersive effect of the
atmosphere on the transmitted waveform.

2.1 Pulse Distortion in Linear Microwave Components

Most microwave components can introduce dispersion and
attenuation in a wide bandwidth transmission since the complete
complex propagation constant is a function of frequency. Mathematical models employed to evaluate distortion are generally
oversimplification of the actual systems, bends, discontinuities,
rotary joints, periodic structures give rise to frequency dependent reflections which will only further degrade the transmitted
waveform. Since a practical imaging radar employs both active
and passive microwave circuitry, it is instructive to consider
the response of these devices when subjected to a short pulse
having a Gaussian envelope.

Although pulses with many different envelope shapes might be
considered, attention will be confined here to pulses which have
Gaussian envelopes. This choice is based on several considerations. First, this form of pulse is convenient for analysis so
that results are often obtainable in a form which is easily
evaluated. Second, it is a pulse shape which is practical to
approximate quite closely in practice with microwave equipment,
so that experiment and theory could be closely integrated. And
finally, this pulse shape is not only practical to attain but

also desirable since it is economical of bandwidth for a given
pulse length. The majority of the energy is confined to a finite
range of the frequency spectrum centered on the carrier frequency.

    In this section, attention will be directed first to passive
microwave devices and then to active microwave devices. In principle, the methods of analysis of passive devices and also of
linear active devices are well known (e.g., Fourier integral
transforms). The problems which exist are concerned with treating a device with a particular characteristic in a manner such
that the pulse response is easily evaluated. For nonlinear active
devices, however, there are no general methods of analysis which
are applicable. Each device must be considered individually in
an attempt to obtain the response to pulses. Naturally, the
analysis of the pulse response of nonlinear devices is at a much
more elementary level than is that of linear devices. One question which is of paramount importance for nonlinear devices is
whether the associated circuit elements or the inherent non-
linearity is of more importance in causing pulse degradation.
The relative importance of these two factors will, of course,
differ for different devices, but this must be evaluated in
assessing the usefulness of a particular device for nanosecond
pulse or wide bandwidth applications.

2.1.1 Linear Passive Microwave Devices

    Fourier Transform Method

    For a linear microwave device the principle of superposition
applies, and if the CW response over a sufficiently broad band of
frequencies is known, then the pulse response can often conveniently be determined using Fourier integral transforms. In
effect, this method decomposes the input time-varying pulse into
its corresponding frequency spectrum, computes the response of
the device to each frequency separately, and superposes these frequencies at the output to give the desired time-varying output
pulse.

    The Fourier transform pair used here is:

$$f(t) = \frac{1}{2\pi} \int_{-\infty}^{+\infty} F(\omega) e^{j\omega t} d\omega, \tag{1}$$

$$F(\omega) = \int_{-\infty}^{+\infty} f(t) e^{-j\omega t} dt, \tag{2}$$

where t is the time and ω the radian frequency. $F(\omega) d\omega$ represents the amplitude of the CW signal at frequency ω. The CW
response of the device to a signal varying as $\exp(j\omega t)$ is taken
as $H(\omega) \exp(j\omega t)$. The amplitude of the output signal at frequency ω is then

$$G(\omega) d\omega = H(\omega) F(\omega) d\omega. \tag{3}$$

The output pulse as a function of time is

$$g(t) = \frac{1}{2\pi} \int_{-\infty}^{+\infty} G(\omega) e^{j\omega t} d\omega. \tag{4}$$

The central problem of the analysis of linear systems is usually to represent the transfer function, $H(\omega)$, in a form that is both suitable and convenient for performing the inverse transformation of Equation 4.

A number of analyses have been made of the dispersion of various types of pulses in waveguides. One of these analyses is that by Forrer [12] who considered a transmission structure, infinitely long in the z direction with no reflections, with the application of an electric field with a Gaussian pulse at $z = 0$. The response of the structure to a CW signal applied at $z = 0$ is

$$H(\omega) = e^{-\gamma(\omega)z} = e^{-\alpha(\omega)z - j\beta(\omega)z}. \tag{5}$$

The input pulse at $z = 0$ is taken as

$$f(t) = e^{-at^2} e^{j\omega_0 t} \tag{6}$$

$$F(\omega) = \sqrt{\frac{\pi}{a}} \, e^{-\frac{(\omega - \omega_0)^2}{4a}} \tag{7}$$

Forrer develops the propagation constant $\gamma(\omega)$ in a Taylor series about the carrier frequency of the input pulse ω_0, and retains terms to the second order.

$$\alpha(\omega) = \alpha_0 + \alpha_1(\omega - \omega_0) + \frac{\alpha_2}{2}(\omega - \omega_0)^2 \tag{8}$$

$$\beta(\omega) = \beta_0 + \beta_1(\omega - \omega_0) + \frac{\beta_2}{2}(\omega - \omega_0)^2 \tag{9}$$

where

$$\alpha_0 = \alpha(\omega_0), \quad \beta_0 = \beta(\omega_0)$$

$$\alpha_1 = \left(\frac{d\alpha}{d\omega}\right)_{\omega = \omega_0}, \quad \beta_1 = \left(\frac{d\beta}{d\omega}\right)_{\omega = \omega_0}$$

$$\alpha_2 = \left(\frac{d^2\alpha}{d\omega^2}\right)_{\omega = \omega_0}, \quad \beta_2 = \left(\frac{d^2\beta}{d\omega^2}\right)_{\omega = \omega_0} \tag{10}$$

Under these circumstances, the inversion, Equation 4, is readily performed and reference should be made to Forrer's paper for the details of the results. In the event that the structure is lossless and propagating, i.e., $\alpha = 0$, then the response at a position z is given by

$$g(\mathfrak{z},t) = \frac{e^{\dfrac{a(t-\beta_1 \mathfrak{z})^2}{1+4a^2\beta_2^2 \mathfrak{z}^2}}}{\sqrt[4]{1+4a^2\beta_2^2 \mathfrak{z}^2}} \cos\left[\omega_o(t-\beta_1 \mathfrak{z})\right.$$

$$\left. + \frac{2a^2\beta_2 \mathfrak{z}}{1+4a^2\beta_2^2 \mathfrak{z}^2}(t-\beta_1 \mathfrak{z})^2 - \frac{1}{2}\tan^{-1}(2a\beta_2 \mathfrak{z}) + (\omega_o\beta_1 - \beta_o)\mathfrak{z}\right].$$

$$(11)$$

The zero order phase constant term, β_o, leads to a phase shift only. The first order term, β_1, gives a time delay of the pulse (note that is the reciprocal of the group velocity at $\omega - \omega_o$). The second order term, β_2, causes an increase in the pulse length, a decrease in the pulse amplitude (this must occur for a passive lossless system if the pulse lengthens, as energy is conserved), frequency modulation of the signal, and a further phase shift. The frequency modulation contained in the argument of the cosine term of Equation (11) is such that a fixed position, the r-f frequency increases with time for $\beta_2 > 0$, and decreases with time for $\beta_2 < 0$. The latter case occurs for conventional waveguides, for example.

In general, the presence of losses and a consequent non-zero value for α will lead to attenuation of the pulse amplitude, further pulse lengthening, and additional phase shift. Forrer presents the results of the calculation of the response of X-band waveguide to nanosecond pulses and shows curves for the attenuation, pulse lengthening, and rate of frequency modulation. One conclusion is that for pulses shorter than about three nanosecond at X-band, conventional waveguide produces severe degradation. For shorter pulses, transmission systems with smaller curvature of the β versus ω characteristic (smaller β_2) such as ridged waveguide or strip line would be desirable. This conclusion can, of course, be scaled to other frequency bands. The application of Forrer's results to other types of transmission systems, so long as they are infinite and reflectionless, is straightforward.

2.1.2 Distortion in Linear Active Devices

Characteristics of linear active devices can be explored in the same manner as for linear passive devices since superposition applies and the Fourier integral transforms are applicable. For some nanosecondradars the most important amplifiers are the traveling wave tubes because of their inherent wide bandwidth. Other tubes will not be considered because they are narrow band or nonlinear.

The broad bandwidth of TWT's, one or two octaves for low

level tubes and approximately twenty percent for megawatt tubes
makes them useful in short pulse applications. The broad bandwidth and high gain is accomplished by long electrical lengths of
the tube. For some tubes this may be as much as ten thousand degrees. The consequences of this long electrical length are a
comparatively long time delay for pulse transmission and pulse
degradation caused by the cumulative effects of departures from a
linear phase versus frequency characteristic for the traveling
wave tube circuit.

All traveling wave tubes incorporate a periodic circuit with
which the electron beam interacts, Figure 2a. However, most low
level traveling wave tubes, for which linear operation is economi-
cally feasible, use helices as their r-f circuit element. The
ω-β diagram for a helix is shown in Figure 2b. For these tubes,
the operating point is not at a point of inflection so that the
technique of expanding β in a Taylor series about ω_o can be
carried out conveniently.

Essentially this technique was employed by McIsaac and
Itzkan [13] to analyze the response of a typical traveling wave
tube to a Gaussian input pulse. The tube gain and phase shift
data, reported by Bachman [14] and presented in Figure 3, were
obtained by McIsaac by means of an analog computer analysis of
the parameters shown. Both the gain in dB, and the phase shift,
were approximated by the parabolas shown in the figure. The
bandwidth is about 30% at the 3-dB points.

Because of the gain of the tube, the propagation constant
associated with it is complex, with a negative real part to
account for the gain. Approximating the gain and phase curves by
parabolas, as in Figure 3, is equivalent to expanding the propagation constant in a Taylor series and retaining terms to the
second order. Thus, this is the same approach as that discussed
earlier. In conformity with the results obtained there, one
would expect that the curvature of the amplitude and phase
characteristics would cause pulse lengthening, a decrease in
pulse amplitude, and frequency modulation. This is, indeed,
found as shown in Figures 4 through 7.

The response of this traveling wave tube was calculated for
a Gaussian input pulse which was one nanosecond long between points
where the envelope was down by 40 dB (0.275 nanoseconds between
the 3-dB points) with a carrier frequency in the neighborhood of
10 GHz. The ratio of the frequency for maximum gain to carrier
frequency of the input pulse is termed η , and this parameter was
varied to explore the effect of changing the carrier frequency.

Figure 4 shows the envelope of the output pulse versus time
for various values of η . The input pulse, multiplied by ten, is
shown for comparison. It is seen that the pulse delay can be
decreased by decreasing the input carrier frequency (increasing η).

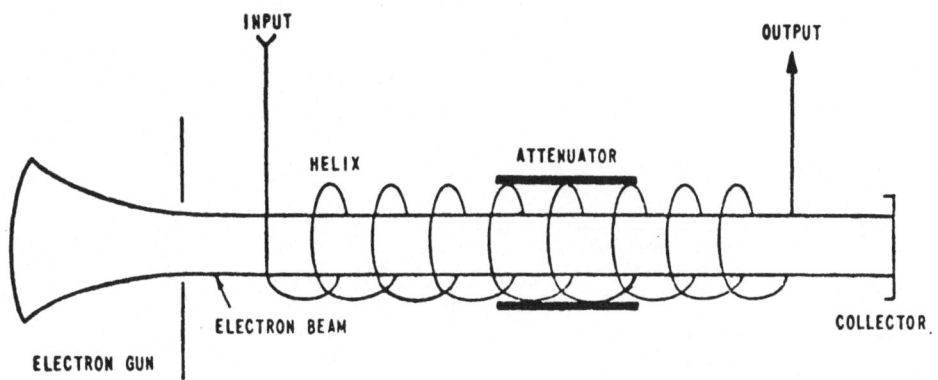

(a) SCHEMATIC DIAGRAM FOR TUBE

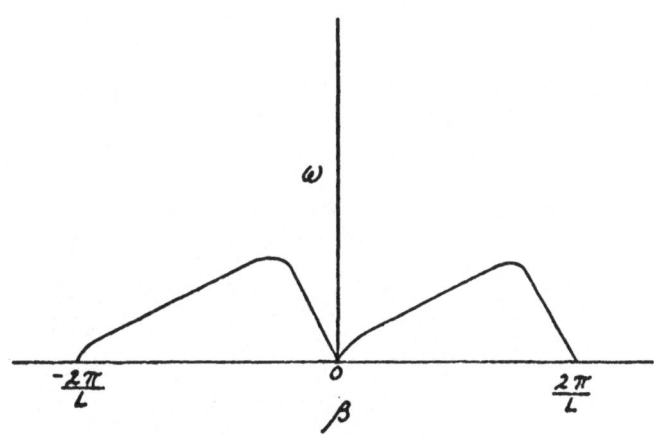

(b) DIAGRAM FOR A HELIX

Figure 2   TRAVELING WAVE TUBE

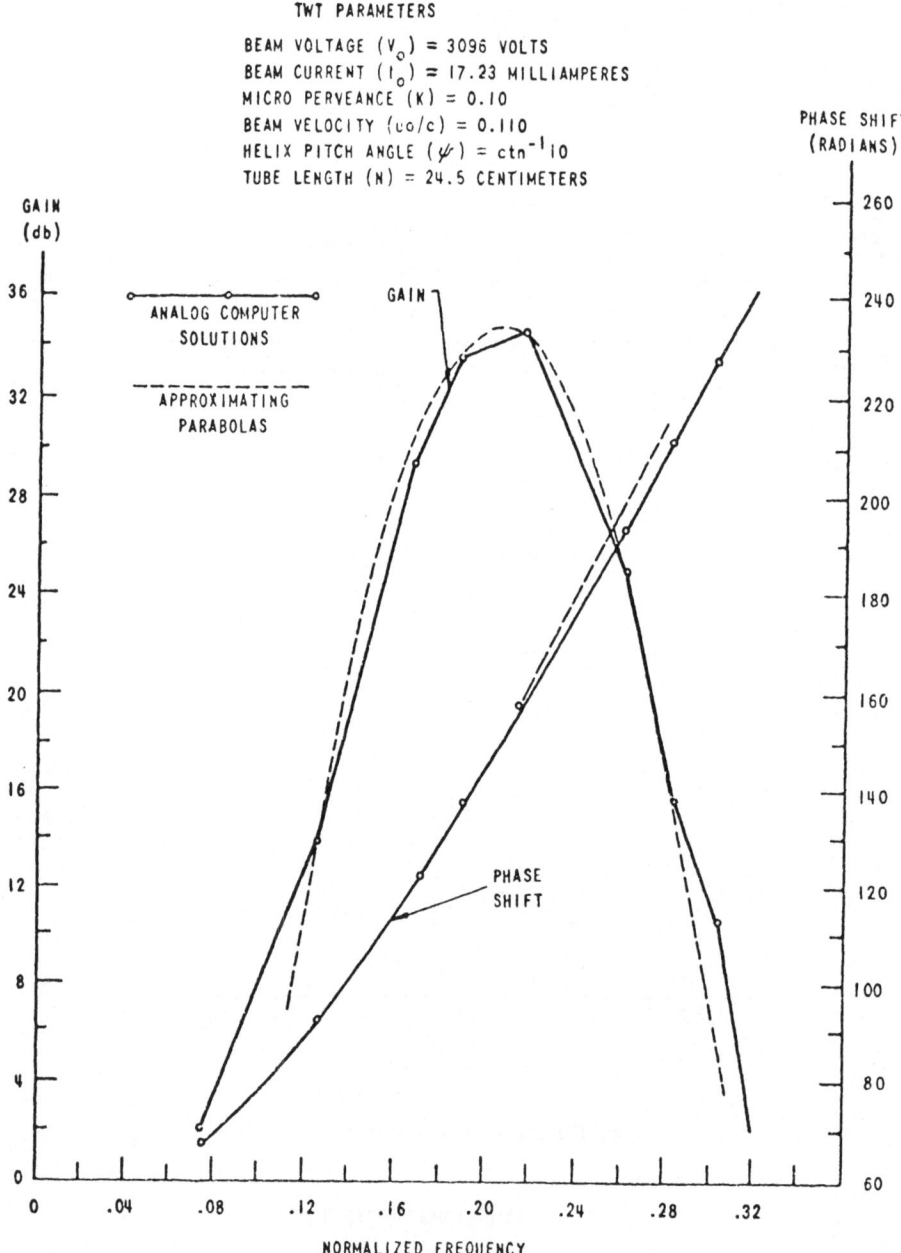

Figure 3

GAIN AND PHASE SHIFT vs NORMALIZED FREQUENCY FOR TWT AMPLIFIER

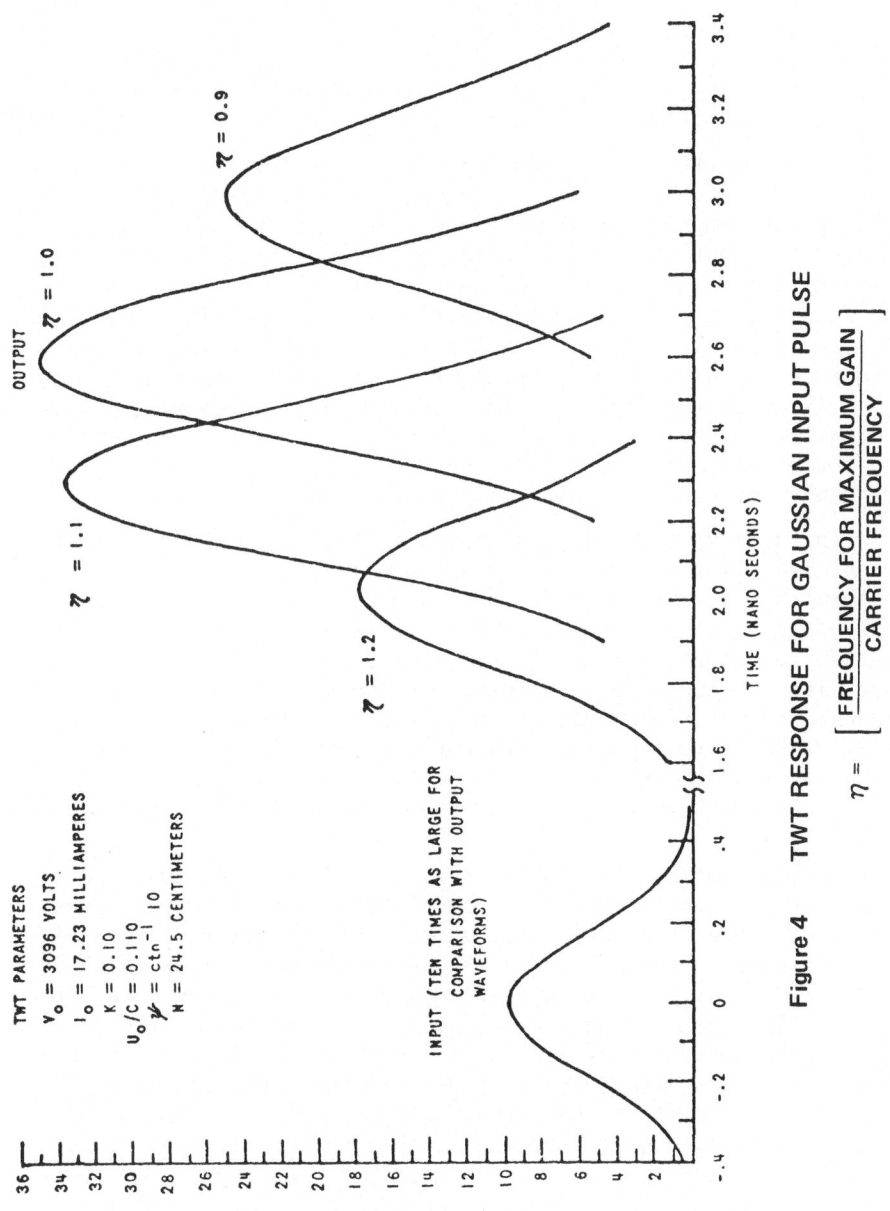

Figure 4   TWT RESPONSE FOR GAUSSIAN INPUT PULSE

$$\eta = \left[\frac{\text{FREQUENCY FOR MAXIMUM GAIN}}{\text{CARRIER FREQUENCY}} \right]$$

If η is increased too far, however, the pulse amplitude falls off
sharply. Figure 5 points out specifically how the pulse delay
depends on η. This difference in pulse delay arises because of
the change in curvature of the phase characteristic at the carrier
frequency. It would appear that one could control the time delay
through a traveling wave tube by controlling the input carrier
frequency, or by varying the beam voltage which determines the
frequency for maximum gain.

The parameter η also controls the pulse spreading, as shown
in Figure 6 which gives the pulse width between 3-dB points.
(This pulse lengthening is rather severe, the pulse having increased by about 80% in length for η changing from 1.0 to 0.8.)
This again is a consequence of the curvature of the phase shift.
And finally, an indication of the frequency modulation introduced
into the pulse is presented in Figure 7. Here the instantaneous
frequency at three points in the pulse is shown as a function of
η. Although the degree of frequency modulation decreases with
increasing η (again because of the change in the curvature of
the phase characteristic) still at $\eta = 1.1$ the total change in
frequency between 3-dB points is nearly 1 GHz.

In summary, the analysis indicates that, at X-band, conventional traveling wave tubes are capable of providing reasonably
faithful amplification of nanosecond pulses. The indications are
that appreciable shortening of the pulses would cause severe
degradation of the pulse response, however. Also, it appears
that with particular attention to the shape of the phase versus
frequency curve that certain characteristics, such as the time
delay or the output pulse width could be optimized. By proper
frequency scaling techniques, the conclusions reached here in regard to X-band traveling wave tubes may be scaled to other frequency bands and appropriate length pulses.

In addition to the non-ideal nature of the amplitude and
phase shift characteristics of traveling wave tubes which cause
the pulse degradation described above, certain other sources of
pulse distortion exist. One of these is the attenuator which is
present internally in all high-gain traveling wave tubes to prevent oscillation. If this attenuator is not matched perfectly to
the circuit, then reflections will result which will be amplified
by the gaining sections of the tube. If the input or output
coupler is not well matched, then multiple echoes of the main
pulse will be produced. Because of the customarily long electrical length for the tube, these multiple echoes would be observed as a pulse train of nanosecond pulses. This problem can
be prevented by careful attention to the matching of the input
and output couplers and the internal attenuator over the whole
frequency band of interest during the design of the tube.

Two other types of distortion may result if the electon beam
itself is pulsed. First, trouble may arise if the electron beam

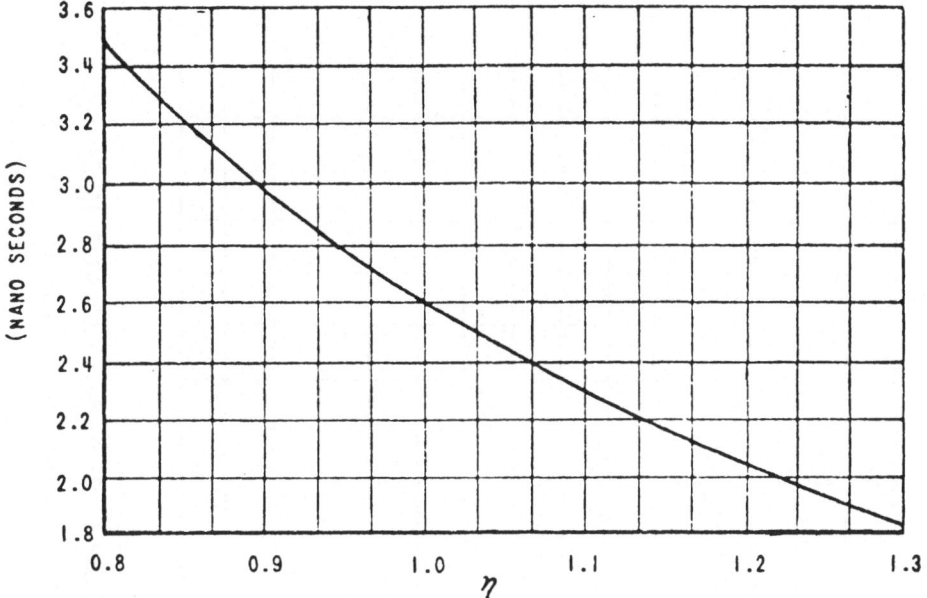

Figure 5   TIME DELAY OF OUTPUT PULSE VS η FOR TWT AMPLIFIER

$$\eta = \left[\frac{\text{FREQUENCY FOR MAXIMUM GAIN}}{\text{CARRIER FREQUENCY}} \right]$$

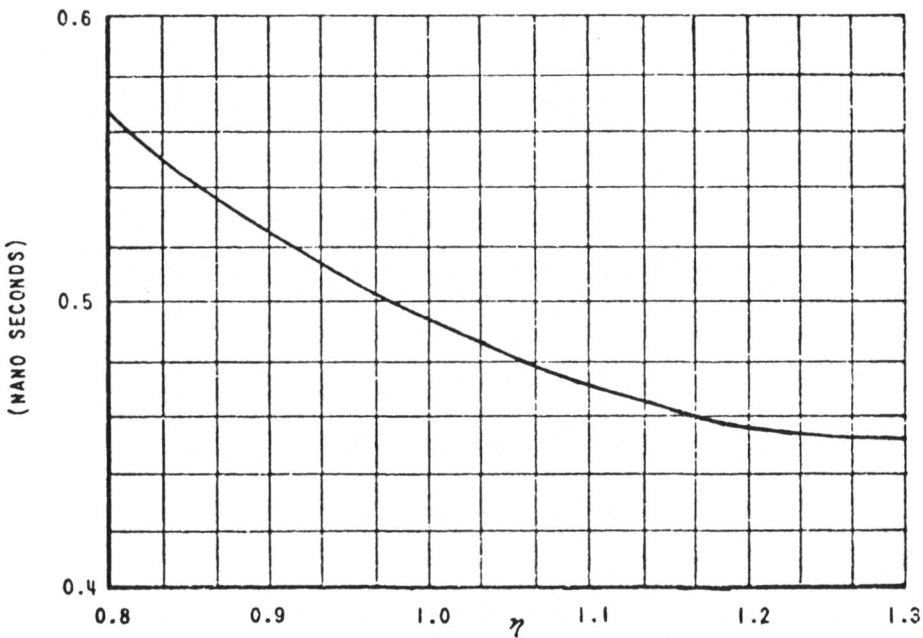

Figure 6   OUTPUT PULSE WIDTH VS η FOR TWT AMPLIFIER

$$\eta = \left[\frac{\text{FREQUENCY FOR MAXIMUM GAIN}}{\text{CARRIER FREQUENCY}} \right]$$

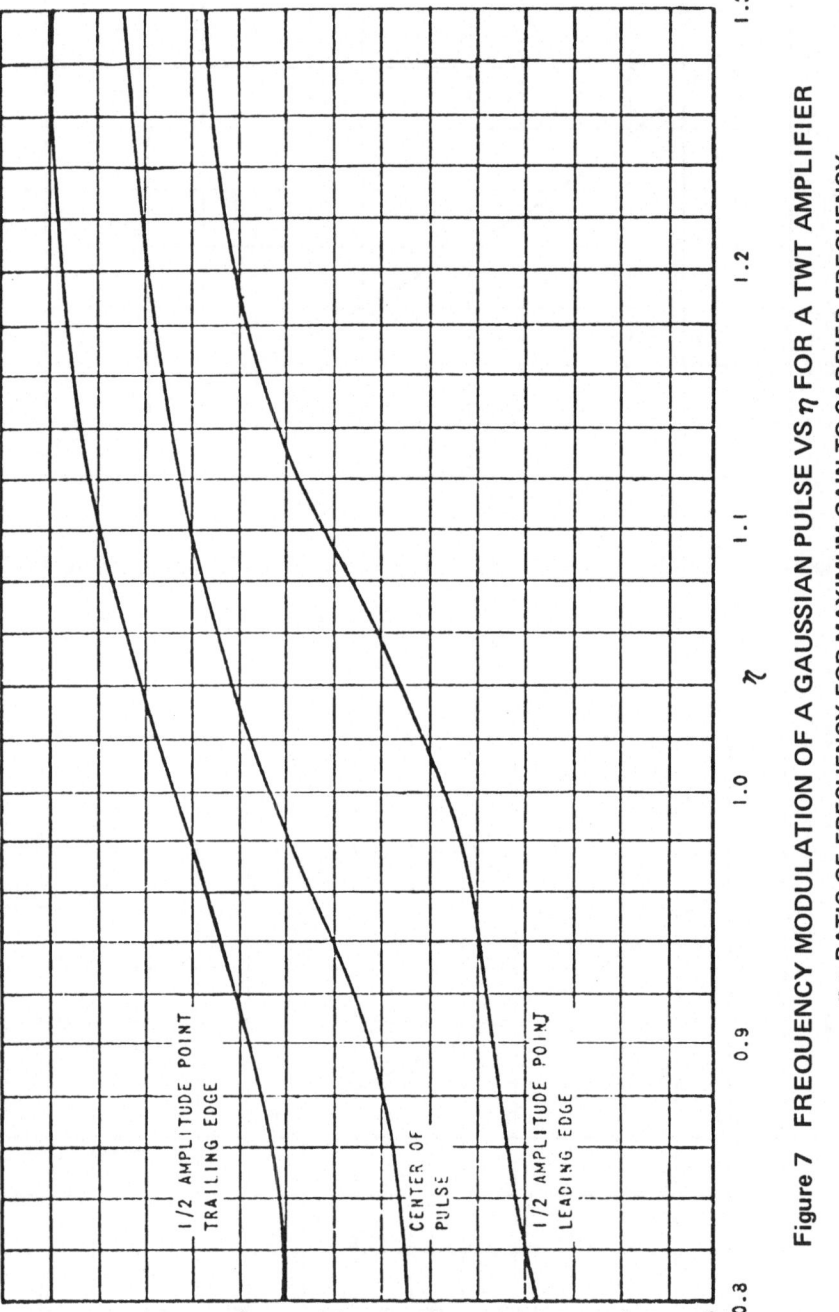

Figure 7   FREQUENCY MODULATION OF A GAUSSIAN PULSE VS η FOR A TWT AMPLIFIER

η = RATIO OF FREQUENCY FOR MAXIMUM GAIN TO CARRIER FREQUENCY

pulses are used to help define the r-f pulse length. Because of
space charge forces, a bunch of charges tends to spread out as
they travel from the electron gun to the collector and thus
lengthen in time as well as space. Thus, if the beam pulse defined the r-f pulse, there would be some lengthening of the r-f
pulse together with a decrease in amplitude, and some frequency
modulation.

Finally, short pulses of the electron beam may cause shock
excitation of the helix. The attenuator should damp out any resulting oscillations in the normal frequency range of the tube.
However, unless special precautions are taken, the effectiveness
of the attenuator decreases markedly at frequencies far below the
normal operating range. Hence, the helix may be shock excited
and oscillate at a low frequency, such as the frequency where the
whole helix is a half wavelength long. This would at the least
cause an undesirable modulation of the output signal. And if the
oscillation amplitude is appreciable it may change the effective
helix voltage (and hence the beam velocity) enough to seriously
reduce the gain in the normal operating range.

2.2 Pulse Distortion in Antennas

An antenna can be considered as a form of microwave coupler or
transition section since energy is transferred from one propagation system, transmission line, to another propagation system,
free space. To obtain effective antenna performance the energy
must transfer efficiently from the bounded system to the un-
bounded system while meeting the high resolution system requirements for low waveform distortion and low range sidelobes. In
addition, the antenna must meet the conventional requirements
such as directivity, polarization, and beamwidth. In order to
preserve the resolution developed by a wide bandwidth waveform
it is necessary that the antenna minimize its internal reflections which give rise to range or time sidelobes. Pulse stretching arising from insufficient bandwidth and nonlinear phase response of the antenna will also degrade system resolution and
must be controlled. Another type of distortion called pulse
fidelity, which is the modification of the pulse shape within the
pulse resolution cell, does not really perturb system performance
and hence, is not considered a governing factor in antenna design.
The inability of an antenna to transmit DC makes it impossible to
transmit and receive a signal without this particular form of
distortion.

In the design of the short pulse downward looking radar mentioned earlier, the transient response was examined of several
types of wideband antennas. The experimental data of Figures 8
through 13 illustrate the three basic forms of distortion that
destroy high-resolution radar performance. Figure 8 shows a
nanosecond pulse fed to an antenna. Figure 9 displays the signal
received when both receiving and transmitting antennas have

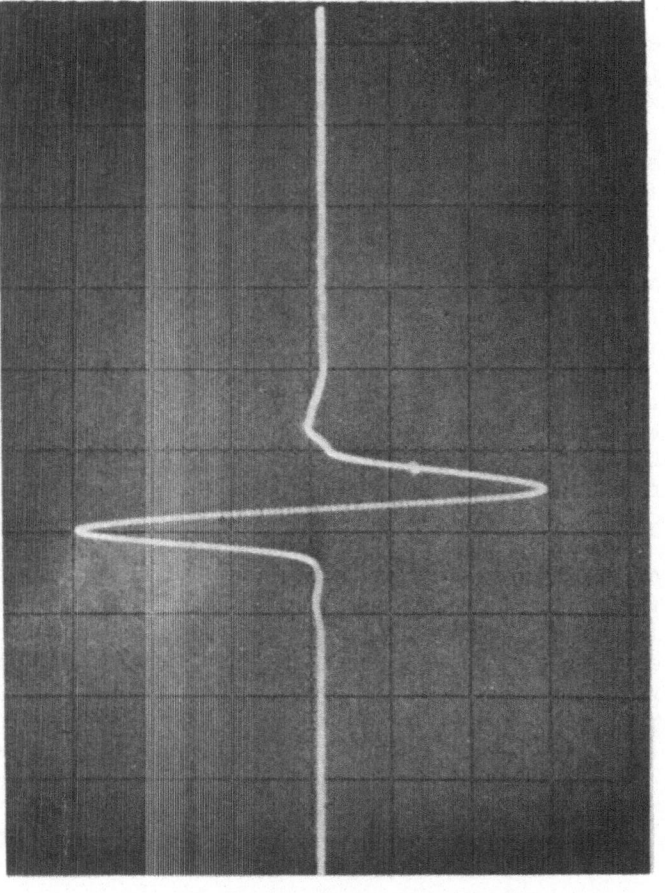

Figure 8 – TRANSMITTER PULSE

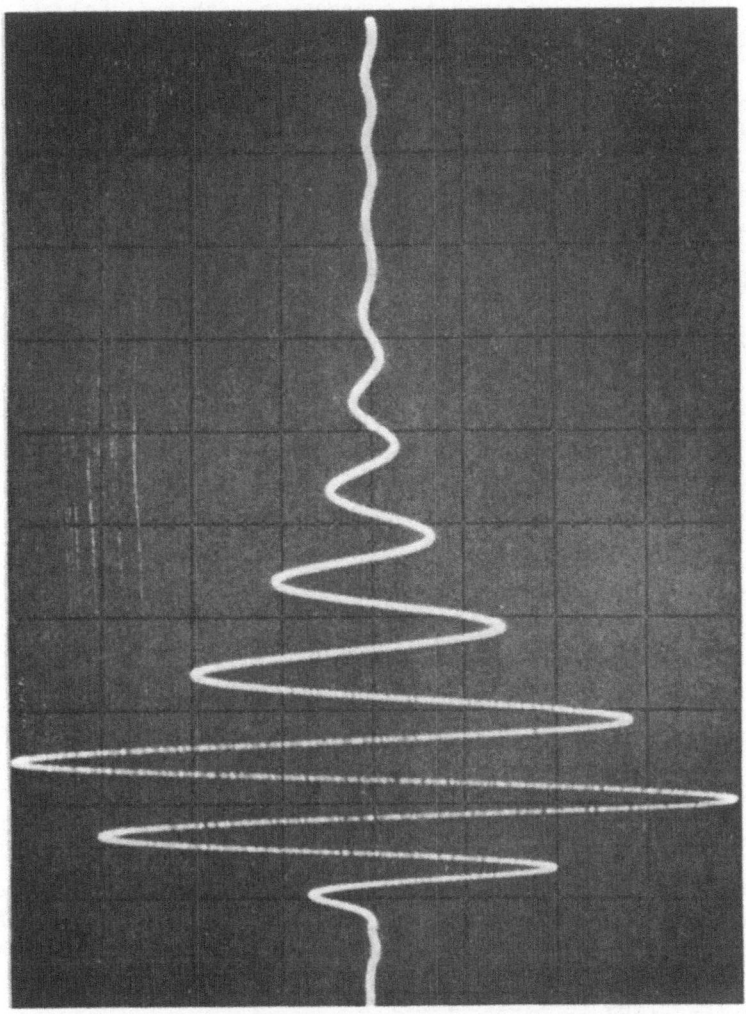

RECEIVED PULSE
1 ns/cm
0.5-V PEAK-TO-PEAK

Figure 9 – EFFECTS OF LIMITED BANDWIDTH ANTENNAS (PYRAMIDAL HORNS)

inadequate bandwidth coupled with nonlinear phase response. The increase in pulse width is quite dramatic. The antennas used in these tests were "broadband" pyramidal antennas.

Dispersion distortion, as well as the effects of internal reflections, are shown in Figure 10. Here, a broadband bifilar helix antenna was used for both reception and transmission. Note that the signal is "spread", even though the amplitude response of the antenna was adequate for the spectrum of the signal as illustrated in Figure 11.

The effects of internal antenna reflections are clearly shown in the data of Figure 12. Here, the antenna was a wideband open-sided TEM horn fabricated in consonance with the analysis of Piefke [15,16]. The high range sidelobes displayed can be traced directly to the internal reflections of the antenna (Figure 13). Some pulse stretching can also be observed and is caused by the curved surfaces of the antenna (Figure 14). There are techniques that may be employed in antenna design to minimize the effects of the illustrated distortions and they will be discussed later.

2.3 Dispersive Effects of the Ionosphere

A dispersive medium is one in which the local refractive index is a function of frequency. The presence of free electrons in the ionosphere modifies the refractive index such that:

$$\eta = \left[1 - \frac{Ne^2}{m\omega^2\epsilon_o}\right]^{1/2} \tag{12}$$

where η = refractive index
N = local electron density (number per cubic meter)
m = mass of electrons 9×10^{-31} kilograms
e = charge of electron 1.59×10^{-19} coulombs
ω = angular radio frequency
ϵ_o = permittivity of free space = 8.854×10^{-12} farads/
meter.

For radio frequencies higher than 100 MHz, the first two terms of the binomial expansion serve to represent this expression:

$$\eta \approx 1 - \frac{1}{2}\frac{Ne^2}{m\omega^2\epsilon_o} \tag{13}$$

A model of the ionosphere, which might be considered "typical" for a mid-latitude station around local noon time, indicates that the integrated electron density measured in a meter-square column vertically upwards to an altitude of 1000 kilometers can be as high as a few times 10^{17} electrons.

If one considers the propagation of a pulse wave of frequency ω at an angle θ_o to the zenith through such a medium, the wave

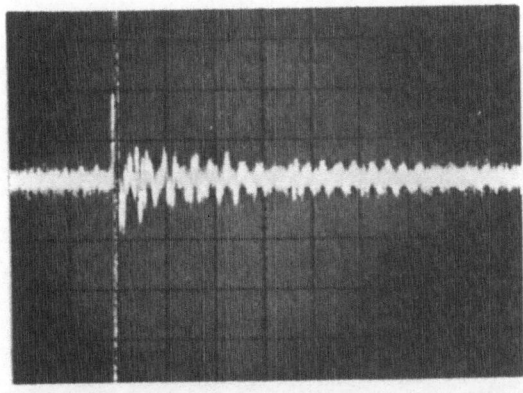

a. RECEIVER WAVEFORM
10 mV/cm
10 ns/cm

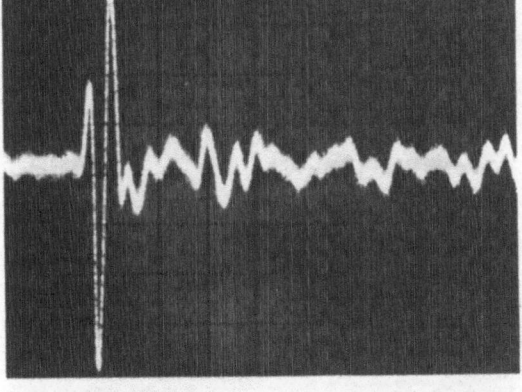

b. RECEIVER WAVEFORM EXPANDED
10 mV/cm
2 ns/cm

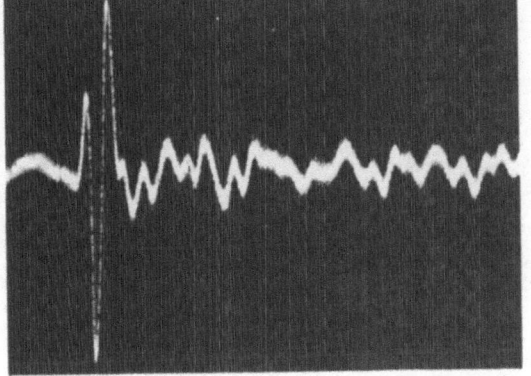

c. RECEIVER WAVEFORM EXPANDED
10 mV/cm
2 ns/cm
TRANSMITTING AND RECEIVING
ANTENNAS INTERCHANGED

Figure 10 - RECEIVED SIGNAL WITH BIFILAR HELIX ANTENNA

ANTHONY V. ALONGI

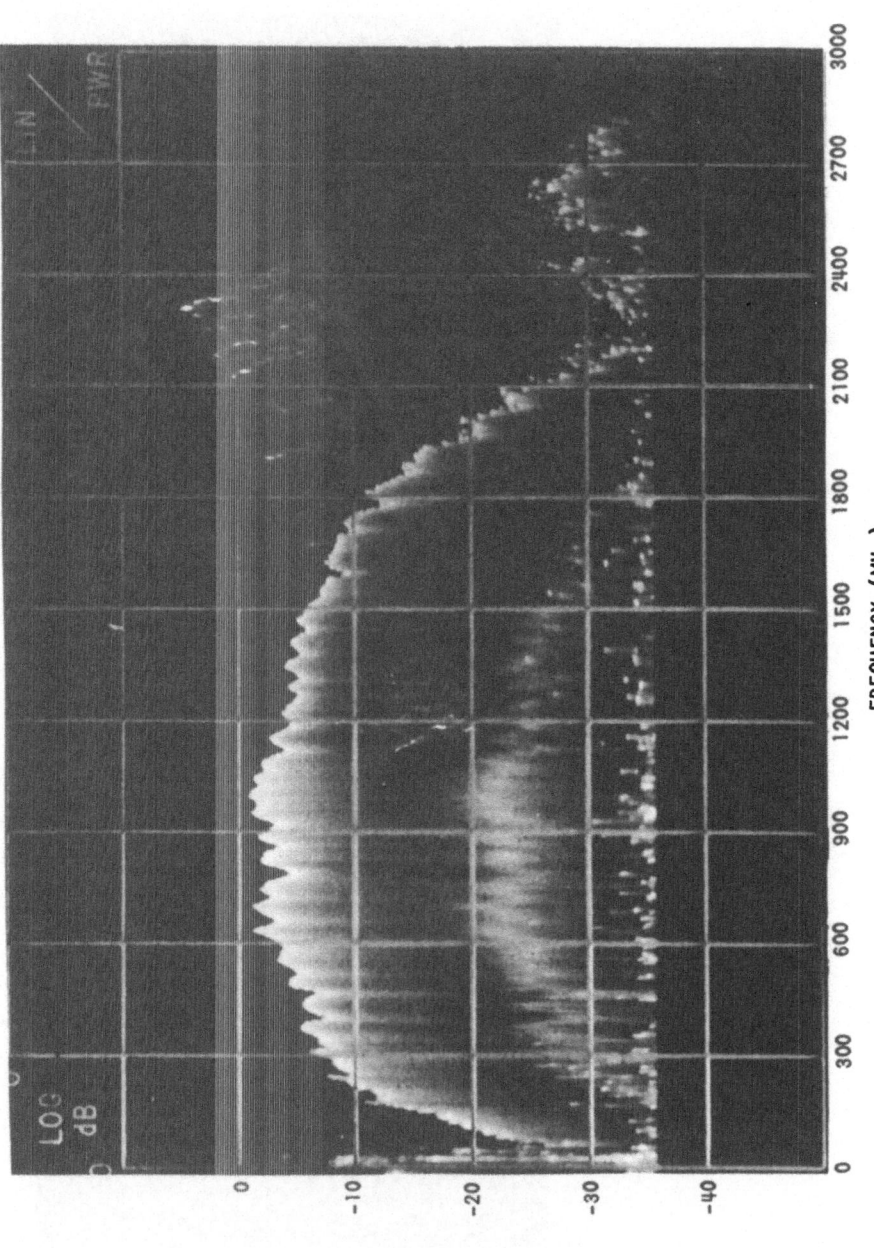

Figure 11 - SPECTRAL RESPONSE OF TRANSMITTER SIGNAL

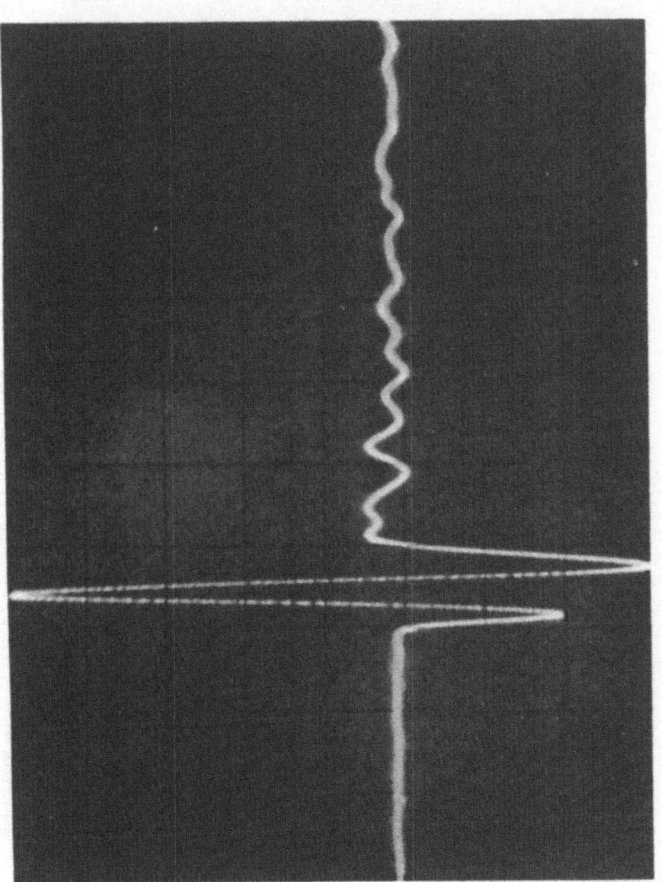

Figure 12 - TRANSMISSION MEASUREMENTS WITH 3/4-SCALE ARCSIN ANTENNA

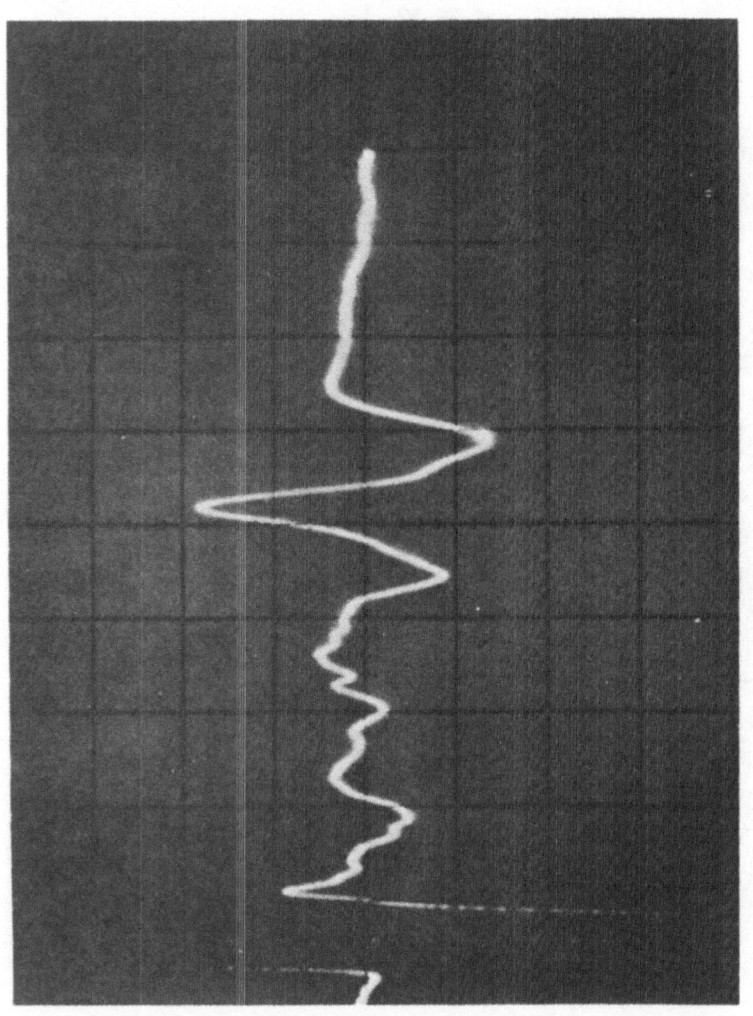

2 ns/cm
(1 cm 1S 0.1 Vmax
or –20 dB)

Figure 13 – REFLECTOMETER MEASUREMENT WITH 3/4 SCALE ARCSIN ANTENNA

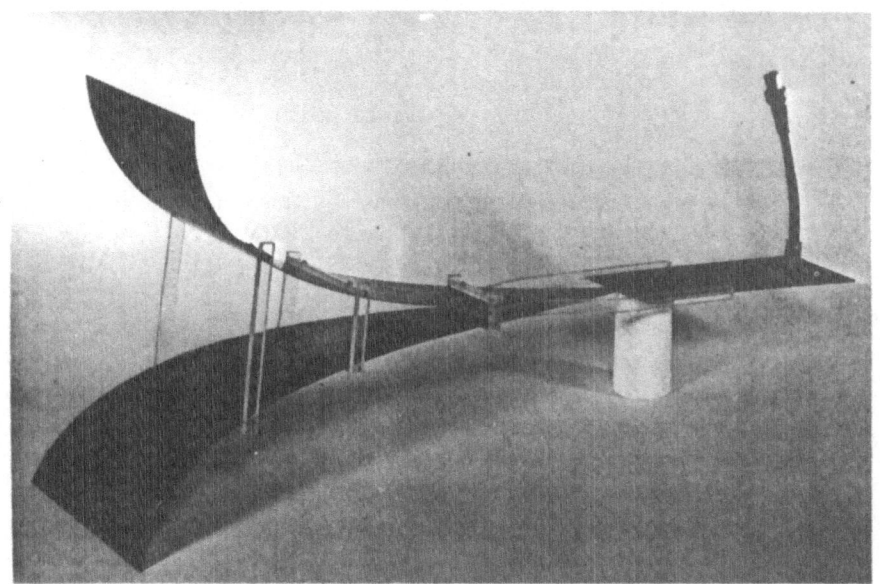

(A) ELONGATED ARCSIN ANTENNA WITH RUMSEY BALUN: HIGH END REFLECTIONS,
HIGH ANTENNA CROSS SECTION (MULTIPLE BOUNCE)

(B) 3/4-SCALE ARCSIN ANTENNA WITH DIRECT FEED: HIGH END REFLECTIONS,
HIGH ANTENNA CROSS SECTION (MULTIPLE BOUNCE)

Figure 14 - ARCS IN ANTENNAS

is of the form:

$$E(t) = \frac{A}{R} \exp\left\{i\left[\omega t - K_0\left(y\sin\theta_0 + \int_0^z q(z,\omega)\,dz\right)\right]\right\} \tag{14}$$

where $E(t)$ = electric field at a point (y, z),
        A = amplitude factor,
        R = distance from origin to point, (y, z),
        ω = angular frequency,
        t = time,
        $K_0 = 2\pi/\lambda$ = propagation constant in free space,
        θ_0 = angle of propagation with respect to zenith,
        $q(z)$ = vertical component of the propagation vector in
               the medium;

and

$$q(z,\omega) = \cos\theta_0\left[1 - \frac{N(z)e^2}{2m(\omega\cos\theta_0)^2\epsilon_0}\right] \tag{15}$$

where $N(z)$ = the electron density at an altitude z .

The fact that the vertical component of the propagation vector is a function of frequency, is indicative of the dispersive nature of the ionosphere.

A pulse of radio energy before traveling through the ionosphere can be decomposed into a spectrum of waves of differing frequencies, each propagating with the form of the above equation. The plane wave components can be summed at any given time to illustrate the effects of ionospheric dispersion upon pulse shape.

Computations of the dispersive effects of the ionosphere on Gaussian shaped pulses have been performed [17]. Assuming a pulse whose amplitude was of the form $\exp\left\{-\frac{t}{2\sigma^2}\right\}$. The spectrum of such a pulse is of the form $\exp\left\{-\frac{4(f-f_0^2)}{B^2}\right\}$, where f_0 is the carrier frequency and $f_0 + \frac{B}{2}$ are the frequencies at which the spectrum is equal to $1/e$ of its peak value. Calculations of the final shapes of transmitted pulses were made for ionospheric paths having total integrated electron densities of 10^{16}, 10^{17}, and 10^{18} electrons per square meter, for carrier frequencies of 1000 to 4000 MHz, and various initial pulse lengths.

The conclusion to be drawn from the data is that for uncorrected, distortion-free transmissions of bandwidths up to 500 MHz, carrier frequencies in excess of 4000 MHz are required for propagation along paths with an integrated electron density of 10^{18} electrons per square meter. It must be pointed out that lower carrier frequencies could utilize a wider transmission bandwidth if simple corrective filters are used in the receiver.

3. EXAMPLES OF HIGH-RESOLUTION RADARS

High-resolution radars for imaging will require the ability

to resolve the predominant scattering sources of complex targets. Nanosecond and sub-nanosecond resolution of 15 centimeters or less have been developed primarily for radar cross section measurements and to perform target backscatter diagnostics. Direct short-pulse and linear FM/CW radars have been built to satisfy the requirements for those applications. A description of such systems is presented [6,9].

3.1 FM/CW High-Resolution Radar

By introducing a wide bandwidth modulation within a long pulse several advantages are incurred along with some disadvantages. As noted previously, long pulses take advantage of a high duty cycle ratio which reduces peak power requirements for the same detection capability. The wide bandwidth improves range resolution but doppler resolution is also increased with use of the long pulse, thereby generating ambiguities in range and velocity. Weighting in time or frequency is employed to reduce the effect of range sidelobes. A small reduction in the signal-to-noise ratio is incurred with the use of weighting. Linear FM is preferred as the source of modulation because of the ease in generation, although a nonlinear FM may be utilized that employs built-in weighting. An example of a high-resolution radar is shown in the Delta radar which was constructed by the author and his associates in 1962. This system was developed as a radar cross section measurement facility in order to permit separation of scatterers as close as three inches apart at the signal half-power point.

A functional diagram of the Delta linear FM radar is shown in Figure 15 and characteristics given in Table 1. A linear FM signal is generated by an X-band BWO and amplified by a wideband TWT. Part of the transmitted power is sent toward the target to be measured through a wideband coupler and radiated from parabolic antenna, while the remaining power is attenuated and passed through a nondispersive coaxial delay line, amplified by a TWT and employed as the local oscillator signal for the balanced mixer. The transit time difference at the receiver of the energy propagated through the delay line and the signal reflected and received from the target determines the beat difference frequency of the signal out of the mixer. This signal is in the audio frequency band and consists of spectral lines separated by the sweep-repetition rate frequency. The amplitude of the spectral lines is proportional to the square root of the radar cross section of the scatterer. A synchrodyning process of phase modulation of the receiver TWT is employed to translate the audio band of signals to an intermediate frequency, in order to avoid intense low-frequency noise arising from detection of the local oscillator signal.

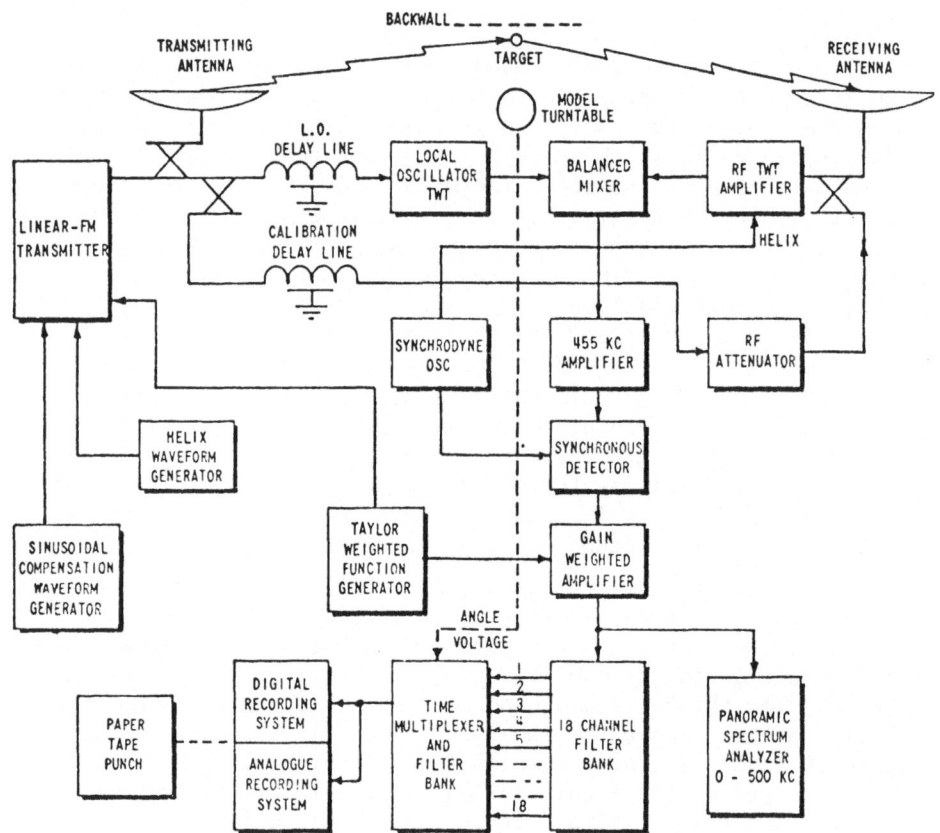

Figure 15   DELTA RADAR - FUNCTIONAL DIAGRAM

Table 1 - SYSTEM PARAMETERS

Center Frequency	9.950 GHz (adjustable)
Fractional Bandwidth	35%
Transmitted Power	100 mW
Range Resolution (adjustable)	see Figure 19
Radar Range Constant at 35% BW	73 Hz/in. or 1.7"/spectral line
Sweep Repetition Range	83.3 Hz, 100 Hz, 123 Hz
Operating Range to Model Target	100 ft.
Fine Range Coverage at 30% BW	27.5 in.
Number of Range Filters	18
Individual Filter Bandwidth	125 Hz
Antenna Beamwidth (each)	3°
Antenna Polarization	horizontal
Minimum Discernible Single Target for 100 mW Transmitted Power	+ 5 dB S/N -- $10^{-4} \; \lambda^2$

The range and amplitude information signal is recovered in the audio region by synchronously mixing the i-f signal with the same signal that is applied to modulate the receiver TWT. The audio signal is then amplitude-modulated by a Taylor weighting function. This modulation serves to shape the spectrum to reduce the signal spectrum sidelobe level, thereby improving the visibility of weak target close to a strong one.

The basic range information delivered by the Delta radar is the energy distribution among successive prf harmonics and is therefore in the form of a power spectrum plot of harmonic power versus frequency. Frequency is then interpreted in terms of discrete range-marker position. Any form of spectrum analyzer capable of indicating the separate levels of the discrete prf harmonics can be used to obtain and display these data.

3.1.1 Design of Linear FM Waveform

An FM/CW ranging system requires that a linearly swept frequency be employed in order that the signal beat frequency be proportional to range. Relations between reference signal and the target return signal are shown in Figure 16, while the corresponding beat difference signal is depicted in Figure 17. The amplitude-weighted beat difference signal is shown in Figure 18. If there are distortions present in the frequency sweep then the beat frequency from an ideal scatterer at some fixed range will not be a single frequency as given in Figure 17, but will contain harmonics related to the distortion components. If the phase distortion of a swept BWO should increase with time, then it would be advantageous to minimize the delay between the reference signal and the target return signal. In the present design, the possibility of nonlinear phase distortion was thought to be great enough to demand special effort at minimizing the relative delay.

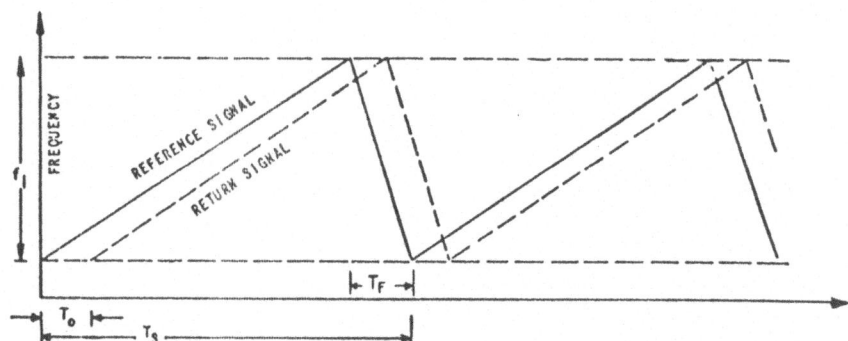

Figure 16   REFERENCE SIGNAL – RETURN SIGNAL RELATIONSHIP

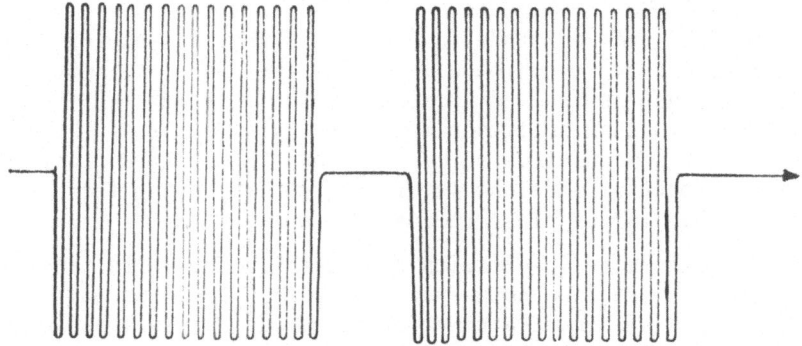

Figure 17   BEAT FREQUENCY (f_b) FOR A DELAY T_o

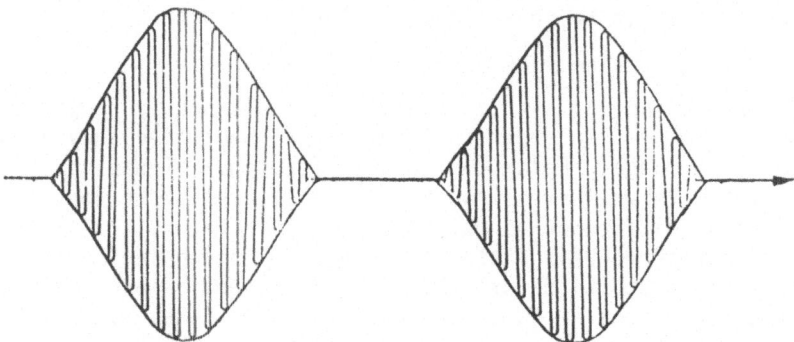

Figure 18   AMPLITUDE WEIGHTING OF BEAT FREQUENCY (f_b)

This is accomplished by sending the mixer reference signal through a coaxial delay line which, in effect, positions the range gate so that the path delay difference between target and reference signal is in the order of 5 nanoseconds. In addition, to obtain linear FM from the BWO, the waveform imposed on the helix must correspond to the voltage frequency characteristic for the specific tube employed. This waveform characteristic is approximately parabolic, but for optimum obtainable linearity, the voltage-to-frequency characteristic must be experimentally determined. Then, with the assumption that the desired frequency-time relation is a linear function, an appropriate voltage time waveform is tailored to the experimentally determined voltage-frequency characteristic and applied to the helix of the BWO.

Additional improvement in linearity may be accomplished by injecting phase and amplitude modulation in the BWO such that periodic distortions are, in effect, partially compensated. These distortions show up in the signal as harmonics of the sweep rate frequency. By injecting modulation terms as these harmonic frequencies with the proper phasing, these effects are reduced. In the Delta system, the sinusoidal compensation waveform generator develops the eleventh harmonic of the sweep repetition frequency. This sinusoidal signal was then separately injected on the cathode of the backward wave oscillator and the helix of the power amplifier TWT.

3.1.2 Range Resolution

The radar range resolution of an FM/CW system having a unity time bandwidth product may be expressed as

$$\Delta_R = c/2K\Delta f \tag{16}$$

where
Δ R   is resolvable range separation,
C   is the speed of light (3×10^8 meters/second),
Δ f   is the swept r-f bandwidth,
K   being a constant accounting for system imperfections and always less than unity.

The factor K includes the spectral broadening or decreased resolution resulting from nonuniform amplitude time history, such as is imposed deliverately by amplitude weighting for sidelobe reduction. The weighting employed in the present radar contributes a factor of 0.83 towards a total, experimentally determined K of 0.68. Figure 19 presents a plot of range resolution versus r-f bandwidth for the ideal, weighted case.

3.1.3 The Effects of Dispersion on Resolution

If transmission lines in the radar system possess appreciable dispersion, so that delay is a nonconstant function of frequency, then linear FM will be converted into nonlinear FM upon passage

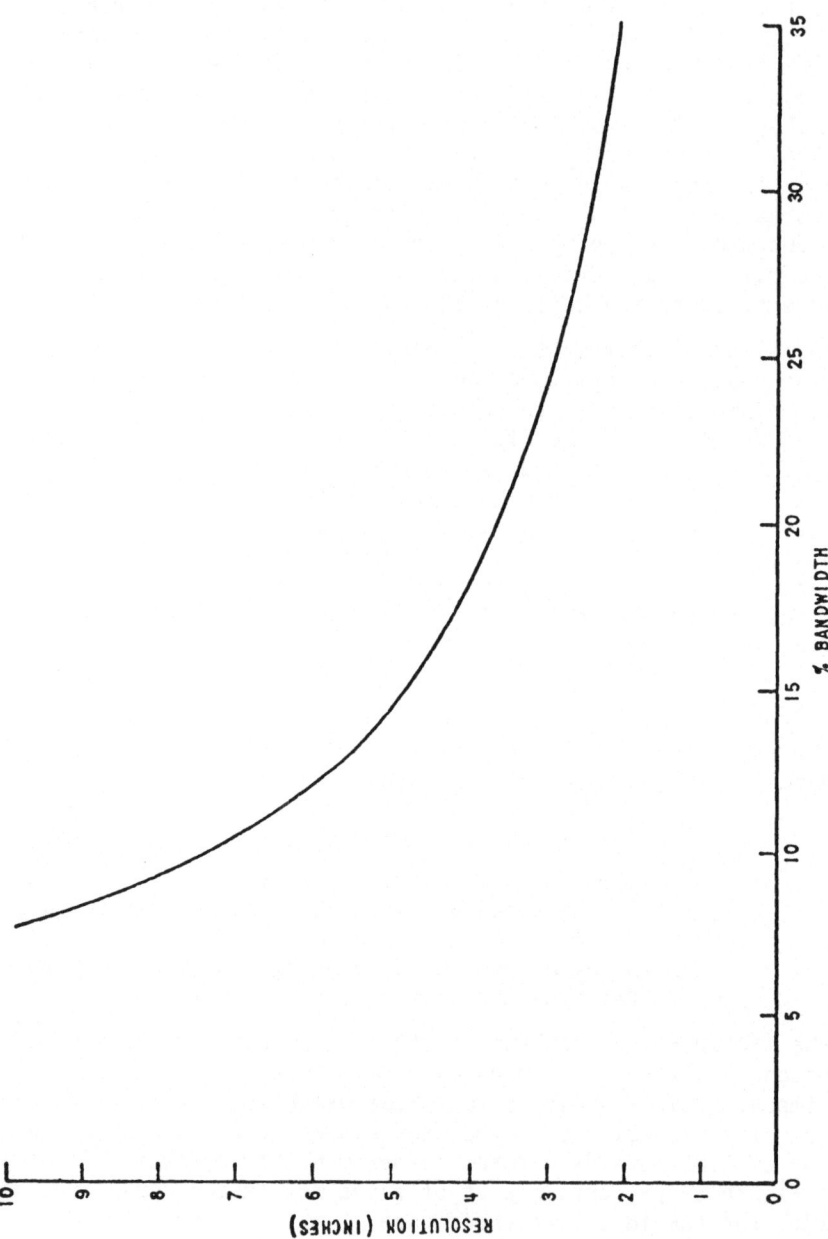

Figure 19   RANGE RESOLUTION vs FRACTIONAL BANDWIDTH

through the line. The resulting nonlinearity causes a decrease in the imperfection factor K. In the present radar design, three possible sources of this type of distortion were considered and investigated. These were the following:

a. Waveguide runs of equal length in local oscillator and receiving legs

b. Waveguide runs of unequal length

c. Coaxial line.

For the case of equal lengths of waveguide, it was assumed that a slope deviation of one-tenth of one per cent from the mean linear sweep rate was the allowable limit. The relation between operating frequency and waveguide length to just reach this limit for 1" x 1/2" X-band waveguide and for an assumed sweep rate of 4×10^5 Hz/sec is given in Table 2.

Table 2   Waveguide Length for 0.1% of Error in Frequency Sweep Rate

f_o (GHz)	Total Length of Waveguide (feet)
8	5,350
9	13,020
10	23,950
11	38,200
12	56,400

The actual value of waveguide length employed in the radar is less than 50 feet. For waveguide runs of unequal length, differences in guide length of approximately 3/4" will cause a cyclic frequency shift on the return signal equivalent to one tenth of its own shift spectral breadth (at 3 dB points). Thus, waveguide length unbalance must be kept to a very few inches.

Coaxial cable might introduce frequency spectral broadening because it is used in the present radar to generate a delay in the reference LO side in an unbalanced manner as shown in Figure 20. Estimates of the change in propagation velocity that is produced by a +20% change in frequency about a 10 GHz center frequency, which have been based on published cable data, show the fractional change in propagation velocity to be less than 1 part in 10^8, therefore no serious dispersion effects were expected from the coaxial cable. Attenuation in coaxial cable is a smooth function of frequency making it acceptable as a delay source, Figure 21.

3.1.4 Factors Affecting Signal-to-Noise

For a given target cross section, antenna gain transmitted power, receiver noise figure and target range, the received signal

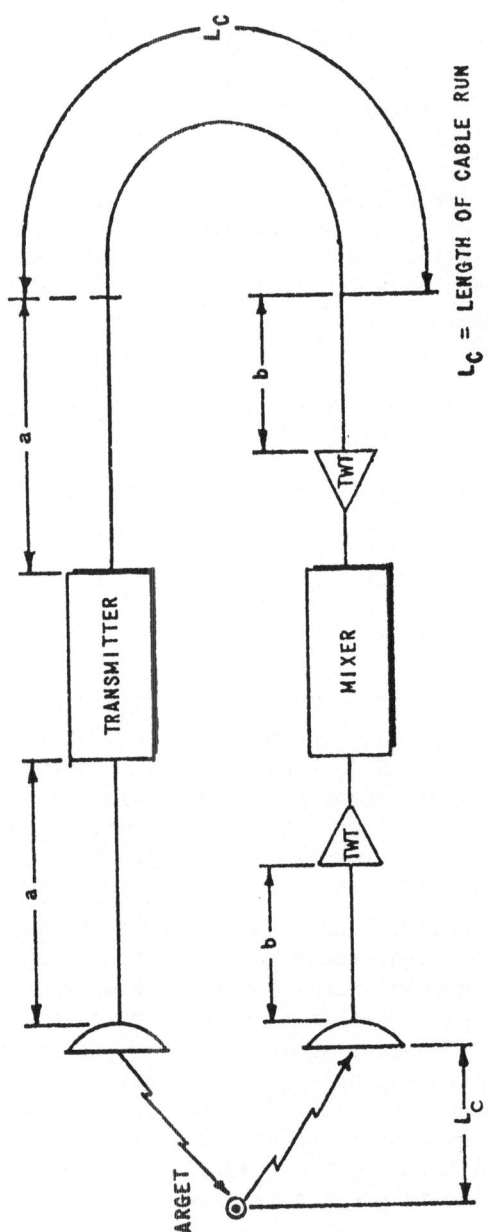

Figure 20   TRANSMISSION PATHS – DISPERSION EFFECTS

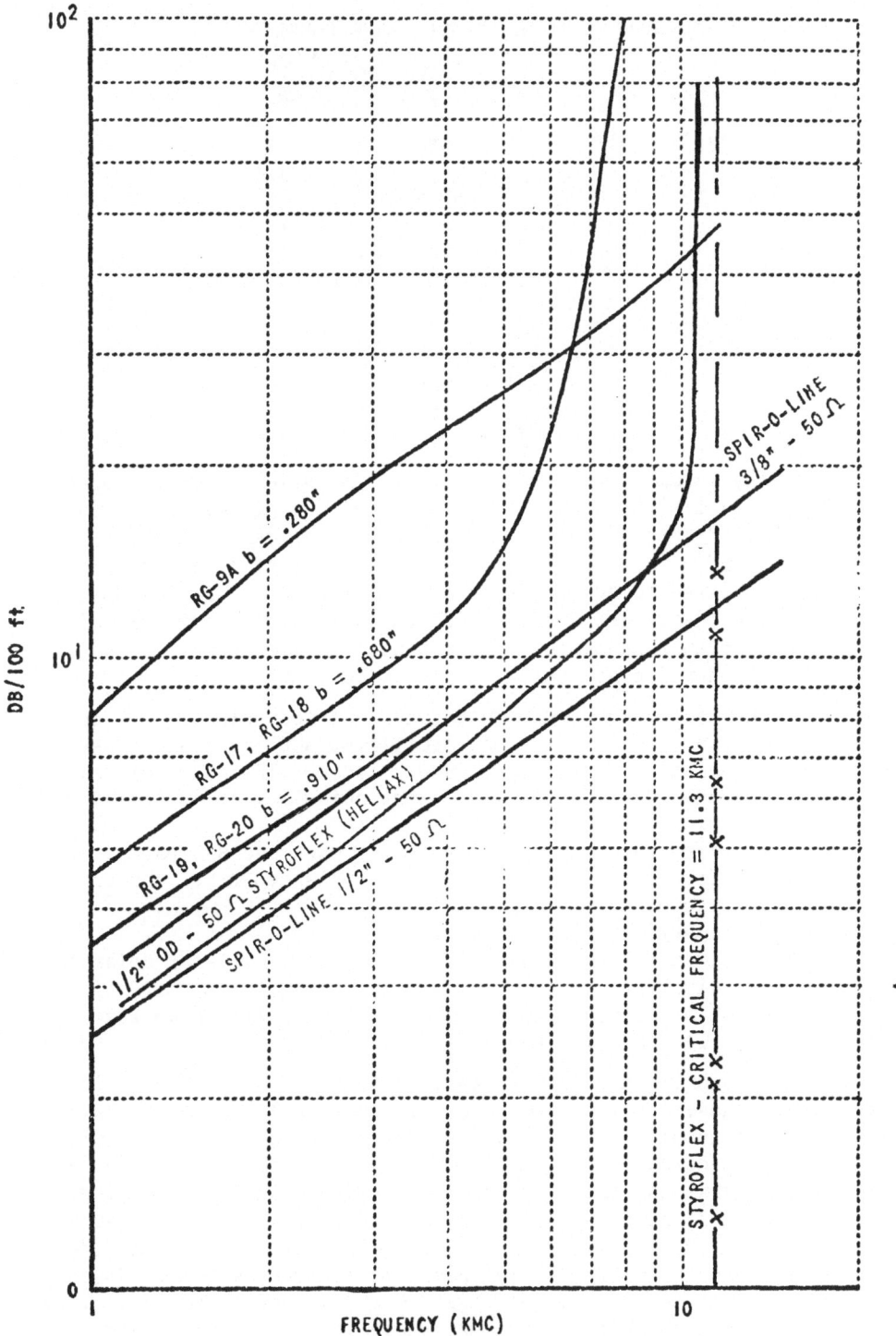

Figure 21 - COAXIAL CABLE ATTENUATION vs. FREQUENCY

can be calculated from the radar range equation

$$\frac{P_r}{P_t} = \frac{G_1^2 \lambda^4 \left(\frac{\sigma}{\lambda^2}\right)}{(4\pi)^3 R^4} \tag{17}$$

The signal-to-noise ratio becomes

$$S/N = \frac{G_1^2 \lambda^4 \left(\frac{\sigma}{\lambda^2}\right) P_t}{(4\pi)^3 R^4 KTNB} \tag{18}$$

where

$$G_1 = 10^3$$
$$\sigma/\lambda^2 = \text{normalized radar cross section}$$
$$\lambda = 0.3 \text{ meters}$$
$$P_t = 1 \text{ watt}$$
$$R = 33 \text{ meters}$$
$$kTNB = 18^{-18.3} \text{ watts}$$
$$N = 13 \text{ dB (for TWT noise alone)}$$
$$B = 125 \text{ Hz.}$$

With the parameters listed, a unity S/N ratio would be produced by a target with a normalized cross section of -77 dB. This calculated capability has not yet been realized with the experimental radar, but the system sensitivity is still extremely high when compared to equivalent short pulse radars. If we consider that the energy per sweep of the FM/CW radar is nearly 10 millijoules, then a nanosecond pulse would require 10 megawatts of power to develop an equivalent millijoule/pulse.

One factor which may introduce unwanted noise is the detection of fluctuations in local oscillator amplitude and phase about their mean values during each sweep cycle. This type of noise has been called "swept noise." Swept noise may arise from any of the following sources:

    a.   frequency dependent traveling wave and BWO parameters,
    b.   frequency sensitivity of the balanced mixer conversion ratio,
    c.   imperfect mixer balance,
    d.   internal reflections in the local oscillator delay line circuitry,
    e.   time modulation of the BWO and traveling wave tube electrode voltages.

It may be shown that quite small variations in local oscillator signal phase and amplitude can produce very large amounts of noise in comparison to the relatively weak target signal; this noise, in the spectral region 2 to 4 kHz, has been measured as 60 dB greater than the usually considered kTNB noise (with a noise figure of 13 dB), and thus produces an effective noise figure of 73 dB. Obviously, this extreme noise level must be avoided. In the present design it was possible to translate the target signal from the 2 to 4 kHz region where the swept noise is intense to the neighborhood of 456 kHz, where the swept noise spectral density is very much less. The translation is accomplished by sinusoidal phase moduation of the received signal prior to mixing with the local oscillator signal. One sideband in the vicinity of 456 kHz which is subsequently recovered by synchronous demodulation contains the desired target intelligence, again in the 2 to 4 kHz spectral region, but now with a much smaller contribution from swept noise. Measurement of the S/N ratio for this synchrodyne configuration has shown an overall noise figure of 35 dB; thus synchrodyning has improved noise figure by 38 dB.

Noise may also be contributed by leakage between the antenna and by multiple reflections from various parts of the room which couple the receiving and transmitting antennas. These effects are minor contributors to the present noise level.

The radar sensitivity now permits observation with 5 dB S/N of targets with normalized cross section as small as -40 dB, and measurements with 15 dB S/N of targets having -30 dB normalized cross sections.

3.1.5 Signal Spectrum Sidelobes and Sidelobe Reduction

Signal spectrum sidelobes, which represent range ambiguities, arise from various causes. The simplest ideal target which is frequency independent and behaves as a nondistributed scatterer would develop sidelobes that are approximately 13.2 dB below the main lobe, if no weighting were employed. The idealized signal shown in Figure 17 (a pulsed sinusoid) would have the spectral power distribution given in equation (19).

$$C^2(\omega) = E_0^2 \left(1 - \frac{T_F + T_0}{T_S}\right)^2 \left\{\frac{\sin\left[(\omega_B - \omega)\frac{T_S}{2}\left(1 - \frac{T_F + T_0}{T_S}\right)\right]}{(\omega_B - \omega)\frac{T_S}{2}\left(1 - \frac{T_F + T_0}{T_S}\right)}\right\}^2 \left\{\frac{\sin M \frac{\omega T_S}{2}}{M \sin\left(\frac{\omega T_S}{2}\right)}\right\}^2 \quad (19)$$

This is the familiar $\frac{\sin x}{x}$ spectrum amplitude function with line spectra consisting of harmonic multiples separated by $1/T_S$ Hz and maximizing in amplitude near the signal beat frequency f_b. The sidelobes of such a function are -13.2 dB. It is imperative for quantitative inspection of complex targets with weak scatterers near much stronger ones to reduce the sidelobe level such that the sidelobe of the stronger target will not mask the weaker one. In the Delta radar this need has been met by providing time amplitude weighting circuitry. This incorporates a photoforming function generator which develops an electrical signal from a single-valued real functional mask inserted into its viewing frame. The weighting then occurs in the time domain as compared to the frequency weighting used in chirp type systems. Table 3 displays the characteristics of various weighting functions which may be used. Triangular and Cosine weighting produce a sidelobe to mainlobe ratio of -20.5 and -19.2 dB respectively which is not a great improvement over constant weighting. Cosine-squared weighting develops a -25.7 sidelobe to main lobe while the Taylor weighting with $\bar{n}$ = 6 develops a -40 dB sidelobe to

TABLE 3   CHARACTERISTICS OF WEIGHTING FUNCTIONS

TYPE OF AMPLITUDE MODULATION	MAIN LOBE	1st SIDE LOBE	SIDE LOBE/MAIN LOBE RATIO
CONSTANT $A(t) = 1$ $-\frac{T}{2} \leqslant t \leqslant \frac{T}{2}$	0 db	-13.2 db	-13.2 db
TRIANGULAR $A(t) = \begin{cases} 1 + \frac{2t}{T}, \frac{t}{2} \leqslant t \leqslant 0 \\ 1 - \frac{2t}{T}, 0 \leqslant t \leqslant \frac{T}{2} \end{cases}$	-6 db	-26.5 db	-20.5 db
COSINE $A(t) = \cos \frac{\pi t}{T}, -\frac{T}{2} \leqslant t \leqslant \frac{T}{2}$	-3.9 db	-23.1 db	-19.2 db
COSINE SQUARED $A(t) = \cos^2 \frac{\pi t}{T}, -\frac{T}{2} \leqslant t \leqslant \frac{T}{2}$ TAYLOR-TCHEBYCHEFF $A(t) = 1 + 2 \sum_{m=1}^{\bar{n}-1} F_m \cos 2\pi m \frac{t}{T}$, $-\frac{T}{2} \leqslant t \leqslant \frac{T}{2}$	-6 db 0 db	-31.7 db * **	-25.7 db * **
HANNING $A(t) = \frac{a_o}{2} + \Sigma a_n \cos \frac{2\pi n t}{T}$			-32 db

The coefficient F_m, supplied by A.C. Price of Bell Telephone Laboratories, are as follows:

*

For A -40 db sidelobe ratio $(\bar{n} = 6)$

$F_1 = 0.39116$
$F_2 = -0.0094524$
$F_3 = +0.0048817$
$F_4 = -0.0016102$
$F_5 = +0.00034703$

**

For A -47.5 db sidelobe ratio $(\bar{n} = 8)$

$F_1 = 0.44615707$
$F_2 = 0.0061897357$
$F_3 = 0.0036959014$
$F_4 = -0.0019122769$
$F_5 = 0.00085612793$
$F_6 = -0.00031987097$
$F_7 = 0.000075788521$

main lobe. A slight disadvantage in employing Taylor weighting is that the weighting flattens out and remains level at -40 dB. Hanning weighting has higher initial sidelobes (-32 dB) than the -40 dB of Taylor weighting, but Hanning weighting continues to drop off to at least -70 dB. This is at the expense of some additional main lobe broadening over Taylor weighting (a factor of 10% at the -20 dB level). Different forms of weighting have distinct advantages depending upon the characteristics of the signal return from the various types of scatterers being investigated. The photoforming method of weighting function generation allows the choice of any desired weighting function to take advantage of these different properties.

Spectrum broadening and high sidelobe levels may also arise from inadequately linear FM and from reflections within the system which make its amplitude and phase transmission vary with frequency. In the case of nonlinear FM the distortion terms arising from the nonlinearity appear as time-varying phase and amplitude modulation of the signal received from the target. Then simple modulation theory can be used to predict the resulting target sidelobe pattern.

Another important cause of sidelobe generation is differing frequency-dependent delay distortion between the local oscillator reference signal and the received signal. The microwave path through all waveguide components is matched in both arms of the system, and as it was previously pointed out, the coaxial cable can be considered to have a constant velocity for frequencies of interest. A likely source of delay perturbation is slight differences between the two traveling wave tubes.

3.2 Short-Pulse High-Resolution Radars

Short-pulse high-resolution radar systems may be constructed in several different ways. One approach employs a low PRF high peak power source. Another approach employs a high PRF low peak power source.

Earlier systems employed the low PRF, high peak power technique with the short pulse originating from the spike leakage from a TR tube or direct amplification of the high frequency components of a video pulse applied directly to a traveling wave tube. For example, Figure 22 shows a functional block diagram of the Micronetics short pulse radar employing the TR tube spike leakage as the pulse source. The basic microwave pulses are generated by a magnetron dirven by a hard tube modulator. The RF pulses from the magnetron may have a duration on the order of 0.1 microseconds.

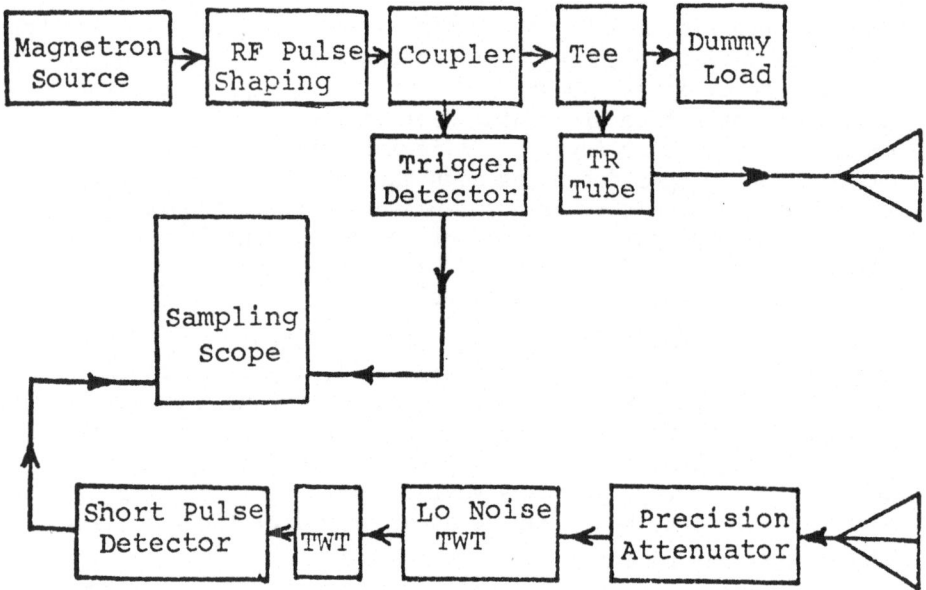

Figure 22 - BLOCK DIAGRAM OF MICRONETICS' SUB-NANOSECOND
SCATTERING RADAR

These pulses are shaped and passed through to a TR tube. The
spike leakage from the TR tube become the source of nanosecond
pulses. The balance of power is absorbed in a dummy load. The
received signal is amplified by a TWT, detected and displayed
on a sampling oscilloscope.

Present work on high power, short pulse generators has employed spark gap techniques. These sources, sometimes called
Hertzian generators, are capable of producing sub-nanosecond
pulses with megawatt peak power [18]. Low power, high PRF nanosecond pulse generators are readily available from solid state
sources. At Calspan, step recovery diodes were employed to generate nanosecond long video pulses, which were then used to modulate an X-band CW RF source with high speed microwave diode
switches. This system, shown in Figure 23, operates as a pulse
doppler radar in the following manner.

The transmitter operates at a frequency of 9.2 GHz supplied
by a klystron oscillator. Frequency of the klystron is stablized
through the use of an FEL Model 113A oscillator synchronizer. One-
nanosecond (nominal) X-band pulses are generated by passing this
9.2 GHz through a waveguide/diode switch which is driven by nanosecond video pulses at a 2-MHz repetition frequency.

The waveguide/diode switch is shown in Figure 24. Four type
1N263 microwave mixer diodes are used as the microwave switching

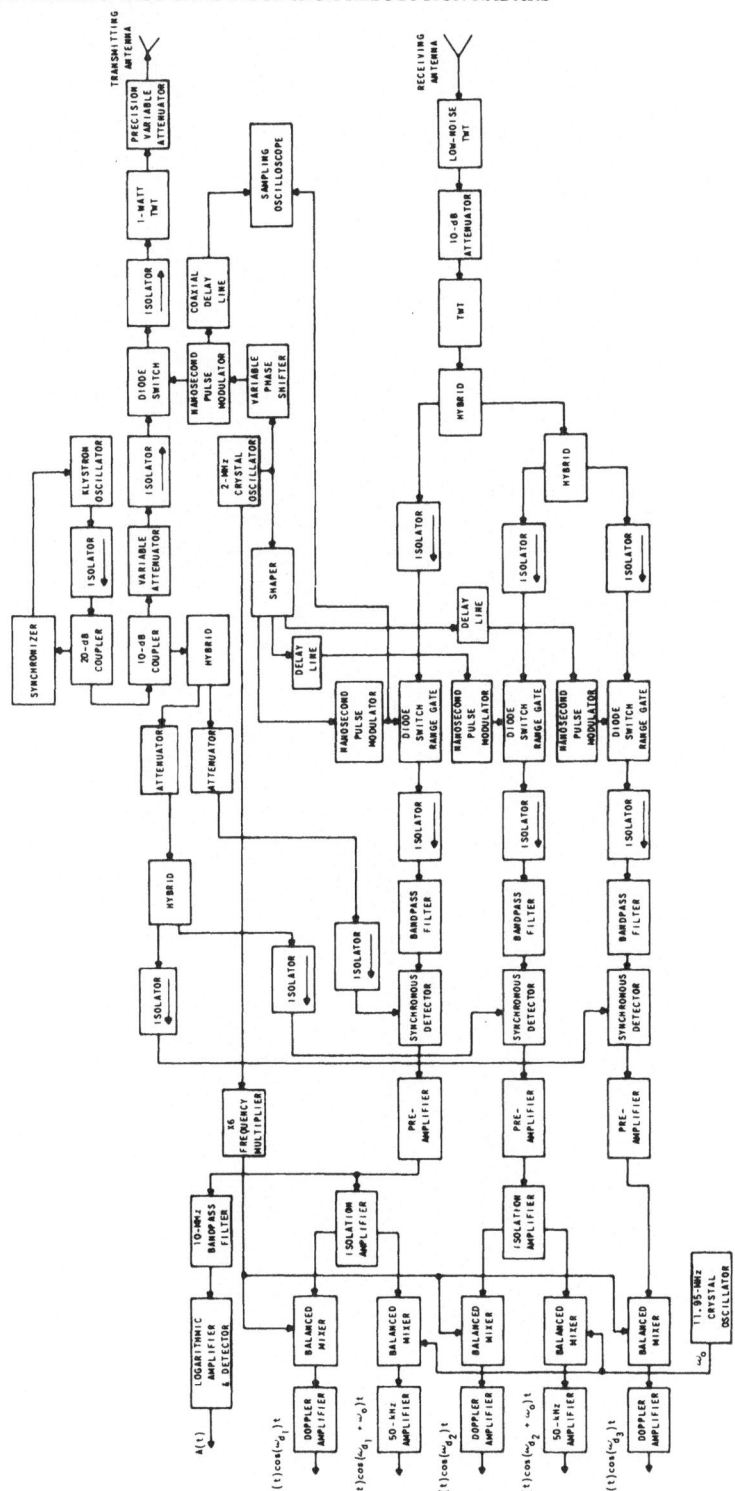

Figure 23 – SHORT PULSE COHERENT DOPPLER RADAR

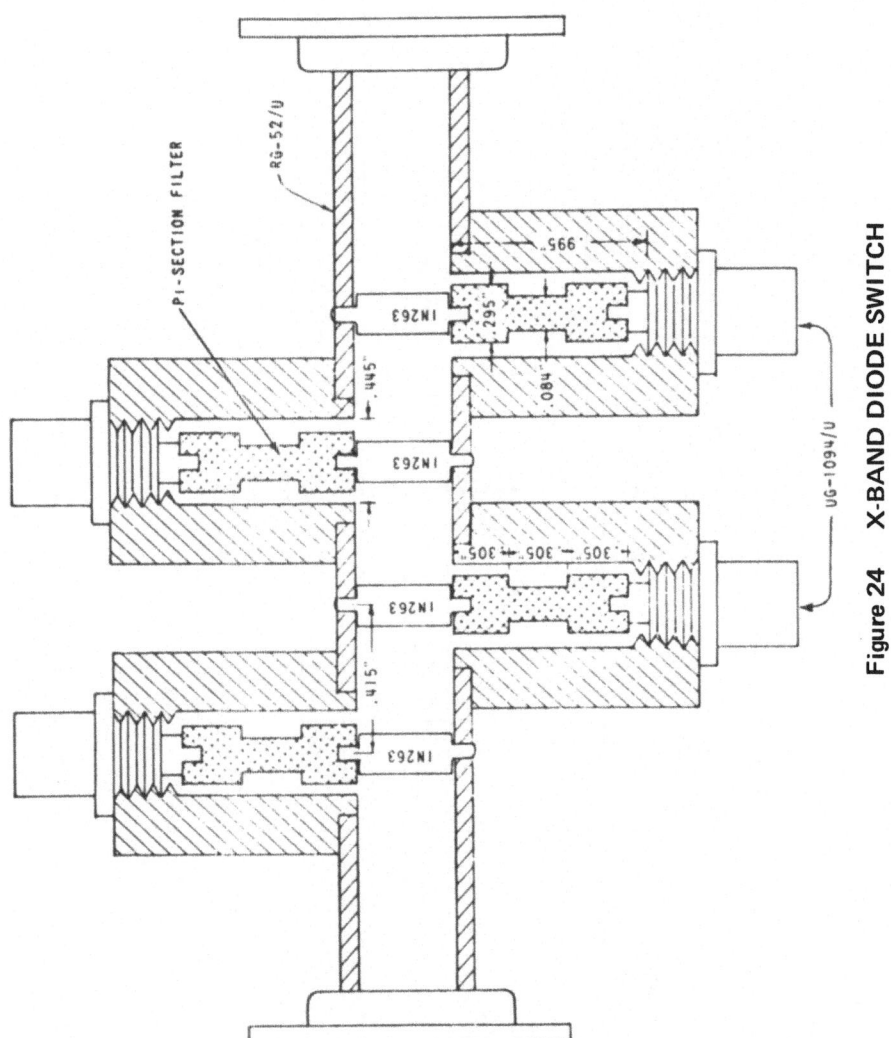

Figure 24   X-BAND DIODE SWITCH

elements. The diodes are centered in the broad wall of standard RG-52/U X-band waveguide and are spaced $\lambda_g/4$ apart (λ_g = guide wavelength). Video pulses from a 2-MHz source are applied to the switching diodes through low-pass pi-section coaxial filters. These filters provide a virtual short at X-band without signi-ficantly affecting the video pulse rise time or shape. Each diode is reverse biased at approximately 1.0 volt in the off-state and draws a current of 30 mA in the on-state of the switch. Insertion loss of the switch is approximately 6 dB in the on-state and it provides a 50-dB isolation in the off-state. The maximum input power is 10 mW.

The nanosecond video driver pulse is generated by means of a step-recovery diode and a short-circuited length of RG-58/U co-axial cable which serves as a pulse-forming network as shown in Figure 25. A positive pulse applied to transistor Q1 causes diode D1 to change from a forward to reverse bias condition. At a critical current threshold, the reverse current through D1 (and the pulse forming line) is suddenly interrupted by the snap action of the diode, producing a voltage pulse across the ter-minals of the pulse forming line. A voltage divider network con-sisting of 1/2-watt carbon resistors feeds the pulse to the four diodes of the waveguide switch. Four lengths of RG-58/U cable are used between the voltage divider network and the waveguide/diode switch, and cable lengths are trimmed to compensate for signal propagation time delay between the diodes of the switch.

The X-band pulses from the waveguide diode switch are ampli-field to 1-watt peak power by a traveling wave amplifier and fed to the transmitting antenna via semi-rigid coaxial cable to avoid the dispersive effects of waveguide, which would cause broadening of the pulses.

The types of antennas used with the radar are dictated by test requirements. Separate horn antennas for transmitting and receiving (as shown in Figure 23) have been used in the shock tunnel and ballistic-range applications thus far.

The received signal is fed from the antenna via coaxial line to a broadband X-band amplifier consisting of two cascaded traveling wave tubes (TWT). The output of the second TWT is divided through a hybrid: one output of the hybrid goes through the isolator to range gate switch 1; the output of the other arm of the hybrid goes to a second hybrid, where the signal is divided again between range gate switches 2 and 3. Isolators are in-stalled between the hybrid outputs and range gate switches. The range gate diode switches and nanosecond drivers are identical to those used in the transmitter.

The transmitter and receiver gates are driven by a common 2-MHz crystal oscillator source. A variable-phase shifter be-tween the 2-MHz source and the transmitter modulator is used to

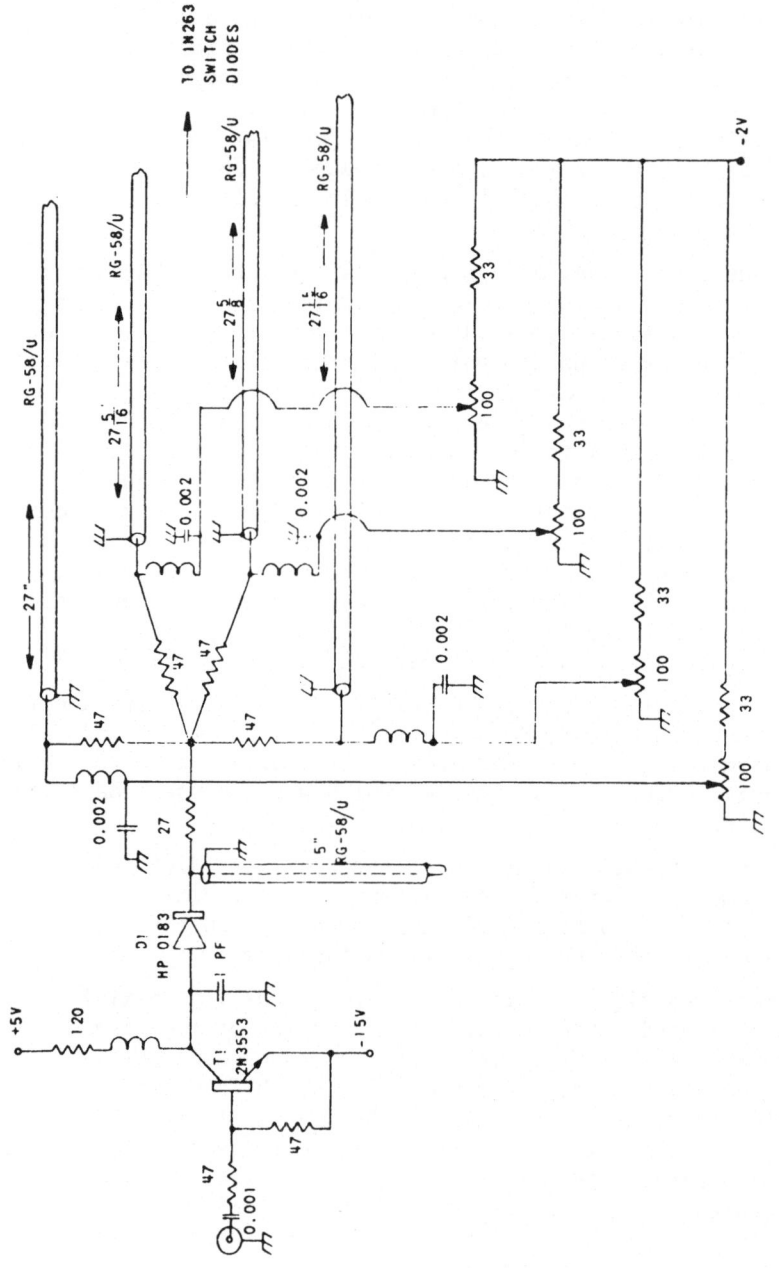

Figure 25   SWITCH DRIVER SCHEMATIC

control the relative time delay between the transmitted pulses and the range gates. The spacing of the individual gates is determined by adjustment of the length of the coaxial cable delay lines between the 2-MHz source and the individual range gates. This relationship remains fixed as the transmitter gate is adjusted.

During radar measurement tests, the transmitter pulse and range gate video drive pulses are displayed on two channels of a sampling oscilloscope. The transmitter pulse is applied to the oscilloscope via a delay line consisting of a length of RG-58/U coaxial cable. With the receiver gate set at the desired position, the delay line is adjusted until both pulses are visible on the scope, with a sweep rate of one nanosecond per centimeter. The relative positions of the two pulses are then held constant throughout the test by small adjustments to the phase shifter. This compensates for equipment drift and insures that the range gate is always positioned within 0.1 nanosecond of the specified range.

The X-band output pulses of the range gates are passed through isolators to bandpass waveguide filters centered at 9.13 GHz. The bandpass characteristic is shown in Figure 26(a). The filter passband is such that the signal associated with the spectral line 12 MHz below 9.2 GHz is attenuated 35 dB, while the signal 12 MHz above 9.2 GHz is attenuated only 5 dB (i.e., the insertion loss of the filter). The filter output is applied to the input port of a synchronous detector. The reference signal for the synchronous detector is derived from the stable 9.2 GHz oscillator via a 10-dB directional coupler. The spectrum of the synchronous detector output is shown in Figure 26(b). The output of the synchronous detector is amplified by a preamplifier which has a center frequency of 11 MHz, a 3-MHz bandwidth, and gain of 38 dB. The passband of the preamplifier includes both the 10- and 12-MHz spectral lines of the pulse (Figure 26(c) and their associated doppler signal sidebands. The signal amplitude channel (log amp) is preceded by a 10-MHz bandpass filter, which rejects all frequencies except the 10-MHz component and its doppler sidebands (Figure 26(d)). This signal is amplified by the logarithmic amplifier (center frequency 10 MHz, bandwidth 1 MHz) and detected. Rejection of the otherwise folded sideband makes the detected output independent of the phase of the target return signal and therefore proportional to signal amplitude. In addition to permitting the measurement of target radar cross section, the signal amplitude channel makes it possible to view the return signals from the test site and take precautions to minimize these returns. It also permits the alignment of the range gate on a stationary calibration sphere located where it is desired to view the moving target.

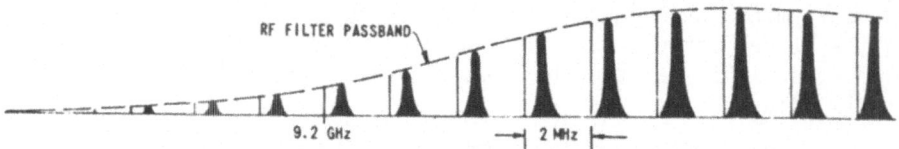

Figure 26(a)   SEGMENT OF RF SPECTRUM OF BANDPASS FILTER OUTPUT

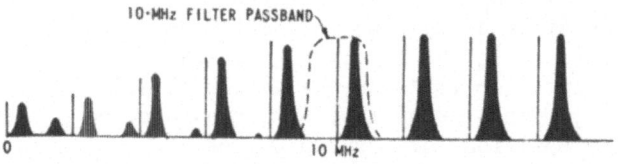

Figure 26(b)   SEGMENT OF SPECTRUM OF SYNCHRONOUS DETECTOR OUTPUT

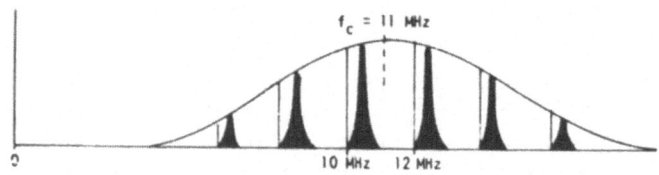

Figure 26(c)   BANDPASS OF PREAMPLIFIER

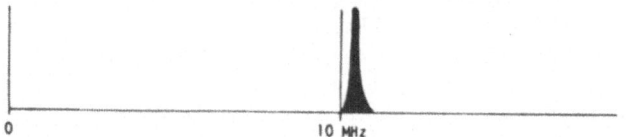

Figure 26(d)   SPECTRUM OF SIGNAL AT 10-MHz BANDPASS FILTER OUTPUT

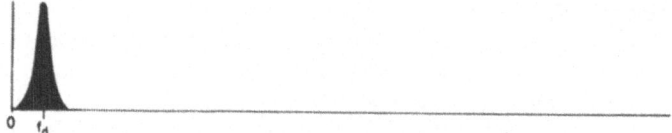

Figure 26(e)   DOPPLER CHANNEL OUTPUT

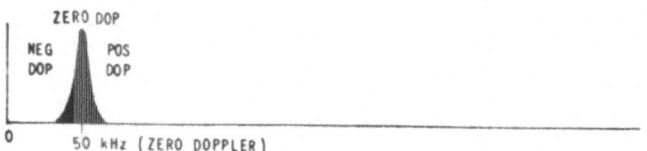

Figure 26(f)   50-kHz CHANNEL OUTPUT

Figure 27 shows the output of the log amplifier as a function of range obtained by sweeping the range gate at a constant rate with a 1-1/2 inch diameter sphere as a target. The width of the main lobe, which is the convolution of the received pulse and the range gate pulse, is approximately 6 inches at the -3 dB level. The range sidelobes following the main lobe are caused primarily by phase and amplitude response characteristics of the traveling wave amplifiers. Better tube response characteristics would reduce the range sidelobe level. In applications thus far, however, range sidelobe effects have not been a problem.

The doppler channel signals are obtained by passing the output of the preamplifier through an amplifier which has two isolated outputs. The doppler channel is fed to a balanced mixer and mixed with a 12-MHz signal derived from the PRF generator. The mixer output is amplified in the doppler amplifier which has a frequency response of 200 Hz to 250 kHz (Figure 26(e). This channel provides a nonoffset or "zero-IF" doppler signal output.

The other output of the isolation amplifier is fed to a second balanced mixer which is referenced to a 11.95-MHz crystal oscillator. The output of this mixer is passed through an amplifier which has a frequency response of 25 to 250 kHz. This channel provides a doppler signal output which is offset in frequency by 50 kHz to avoid spectral folding effects. Thus, when the radar views a stationary target, this channel provides a 50-kHz signal (Figure 26(f)).

It is also possible to provide in-phase and quadrature doppler signals for distinguishing between negative and positive velocities or phase variations in applications where the offset frequency technique cannot be used. This is accomplished by applying the 12-MHz local oscillator signal derived from the PRF generator to both mixers at the isolation amplifier output, with one of the 12-MHz signals shifted 90° in phase relative to the other. The output of one mixer is then of the form $A(t) \cos\phi(t)$, while the other is $A(t) \sin\phi(t)$. The sign of $\phi(t)$ can then be measured and thus velocity sense can be determined unambiguously.

Measurements may be made of the in-phase and quadrature components of the coherent doppler signal time history in a range gate output. The radar also provides a direct measure of doppler-signal amplitude independent of signal phase at the output of one range gate. The forms of the doppler signals, corresponding to the radar output options, are noted below and will be referenced in the next section.

$$E_{n_o}(t) = A(t) \cos \phi(t) \quad \text{(Nonoffset)}$$

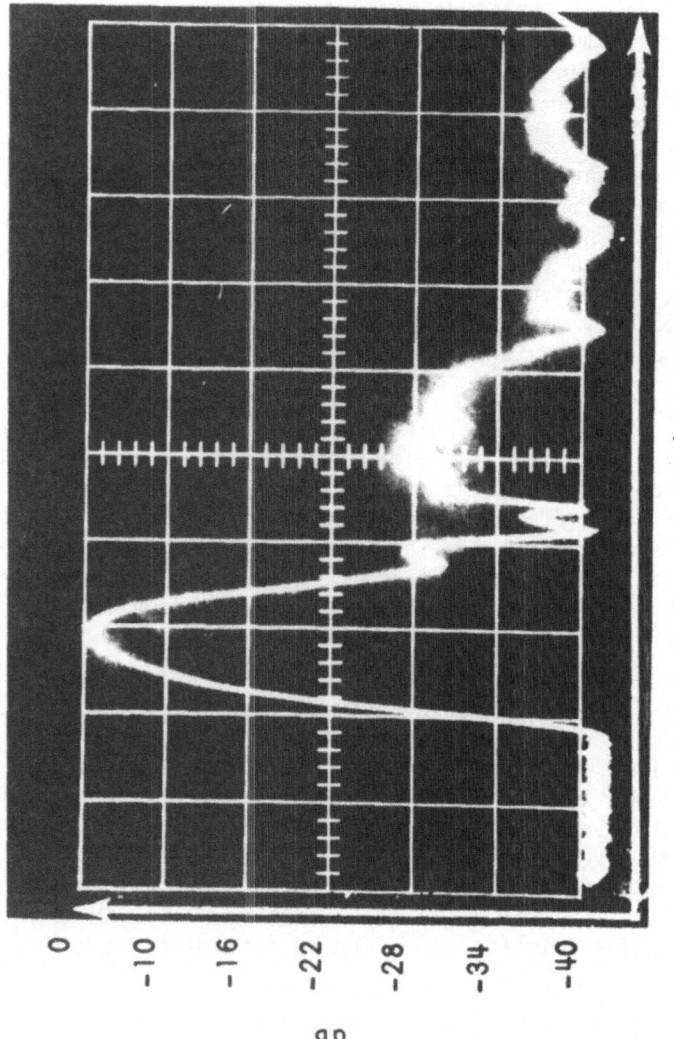

RANGE, 1 FT/DIVISION

Figure 27   RECEIVER OUTPUT AS A FUNCTION OF RANGE GATE
POSITION, TARGET = 1.5-INCH-DIAMETER SPHERE

$$E_o(t) = A(t) \cos [\phi(t) + \omega_o{}^t] \quad \text{(Offset)}$$

$$E_{I,Q}(t) = A(t) \cos \phi(t) \quad \text{(In-Phase/}$$
$$A(t) \sin \phi(t) \quad \text{Quadrature)}$$

where

$A(t)$ = doppler amplitude time history,

(t) = doppler phase time history and may include a constant frequency term corresponding to mean doppler frequency,

ω_o = 2π X offset frequency in hertz (i.e., 50 kHz).

Table 4 summarizes the radar characteristics.

TABLE 4 - SHORT-PULSE DOPPLER RADAR CHARACTERISTICS

FREQUENCY	9.2 GHz
PEAK POWER	1 W
PULSE LENGTH	1 ns (-3 dB)
PULSE REPETITION FREQUENCY	2 MHz
ANTENNA BEAMWIDTH	10° (TWO-WAY)
RECEIVER RANGE GATE DURATION	1 ns (-3 dB)
POSITION	VARIABLE FROM 0 TO 50 FT
RANGE RESOLUTION CELL*	0.5 FT (-3 dB)
NUMBER OF RANGE GATES	3
RECEIVER NOISE FIGURE	9 dB
DOPPLER RECEIVER SENSITIVITY**	-92 dBm (S/N = 1)
RECEIVER DYNAMIC RANGE	40 dB
MINIMUM DETECTABLE*** DOPPLER RCS	$\sigma/\lambda^2 \approx$ -60 dB

*ANTENNA BEAMWIDTH AND RANGE RESOLUTION USED IN APPLICATIONS THUS FAR; OTHER OPTIONS AVAILABLE

**DOPPLER SENSITIVITY FOR EQUIVALENT NOISE BANDWIDTH = 20 kHz

***TARGET RANGE = 43 INCHES, S/N ≈ 1, EQUIVALENT NOISE BANDWIDTH = 20 kHz

3.3 Downward Looking Radar

A third type of high-resolution radar, an all-video radar,

has been developed for the U. S. Army Mobility Equipment
Research & Development Center by the author and associates at
Calspan for very short range downward looking applications.
The radar is man-portable and employs a nanosecond sinusoidal
pulse generator that transmits at MHz repetition rates. System
PRF is generated from a crystal controlled source and is coherent in the sense that the transmitted waveform is identical
from pulse-to-pulse and the time between pulses is always the
same. The pulse-to-pulse coherence permits a wide variety of
signal processing.

A generalized system description of a short-pulse system is
presented in Figure 28. Figure 8 is a sampling oscillograph recording of the short pulse (approximately one nanosecond long
and consisting of one cycle of a sinusoidal signal) which is
transmitted to provide the high range resolution required for
this system. Reflections from buried targets illuminated by the
transmitter are received and detected using a sampling technique.
The sampled output is displayed and provides an A-scope presen-
tation--signal amplitude as ordinate against time as abscissa--
of the actual signal waveform received over a short time (or
range) interval. The time interval displayed is approximately
10 nanoseconds; individual targets in·this range are visually
detected from the display presentation. In free space (open air),
this 10-nanosecond interval corresponds to approximately 60
inches; in subsurface media, the same 10 nanoseconds represent a
distance ranging from 30 inches for dry sand ($\varepsilon_r \approx 4$) to
approximately 13 inches for very wet soil ($\varepsilon_r \approx 20$).

An early version of a short pulse radar system appears in
Figure 29. This system weighs approximately 30 pounds and consists of three main subassemblies: (1) two antennas, (2) display, transmitter and receiver module, and (3) battery pack and
associated control electronics module. The use of separate receiving and transmitting antennas is called "bistatic" in radar
terminology; if a single antenna is used for both purposes, the
radar is called a "monostatic" radar as shown in the block diagram, Figure 30. A monostatic man-portable high-resolution radar
is shown in operation in Figure 31. Both systems employ a soil
lock circuit which presents a stablized display despite antenna
height variations resulting from operator motion. For a downward looking radar to perform effectively, it must overcome the
effects of the geometric loss incurred in accordance to the inverse square law of radiated fields or R^{-4} for a backscattered
signal and the absorptive loss of the media.

$$P_r = \frac{P_T \, G_T \, G_R \, \lambda_m^2 \, \sigma}{(4\pi)^3 \, R^4} \, e^{-4\alpha R} \qquad (20)$$

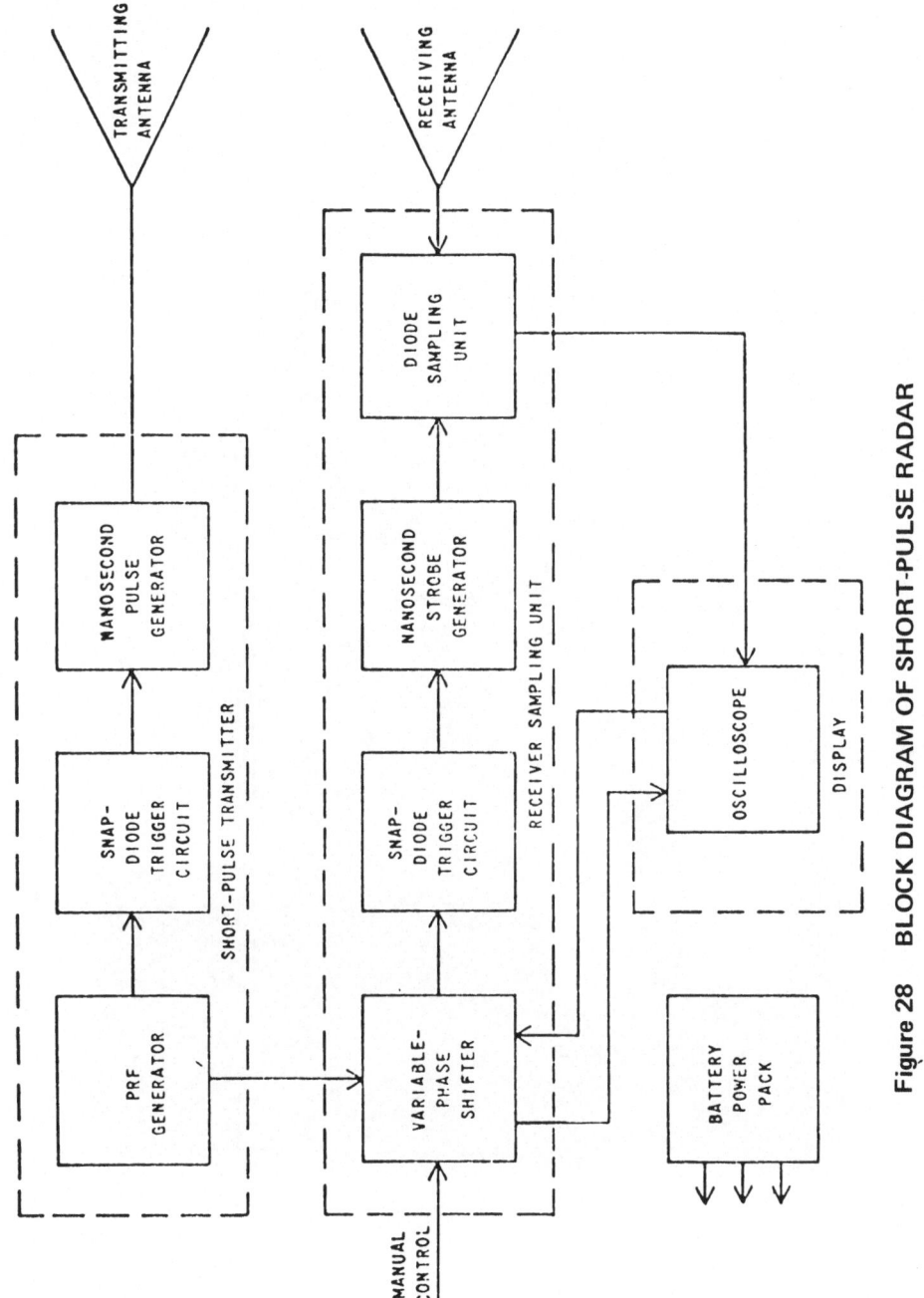

Figure 28 BLOCK DIAGRAM OF SHORT-PULSE RADAR

Figure 29 - EXPERIMENTAL BISTATIC SHORT-PULSE RADAR

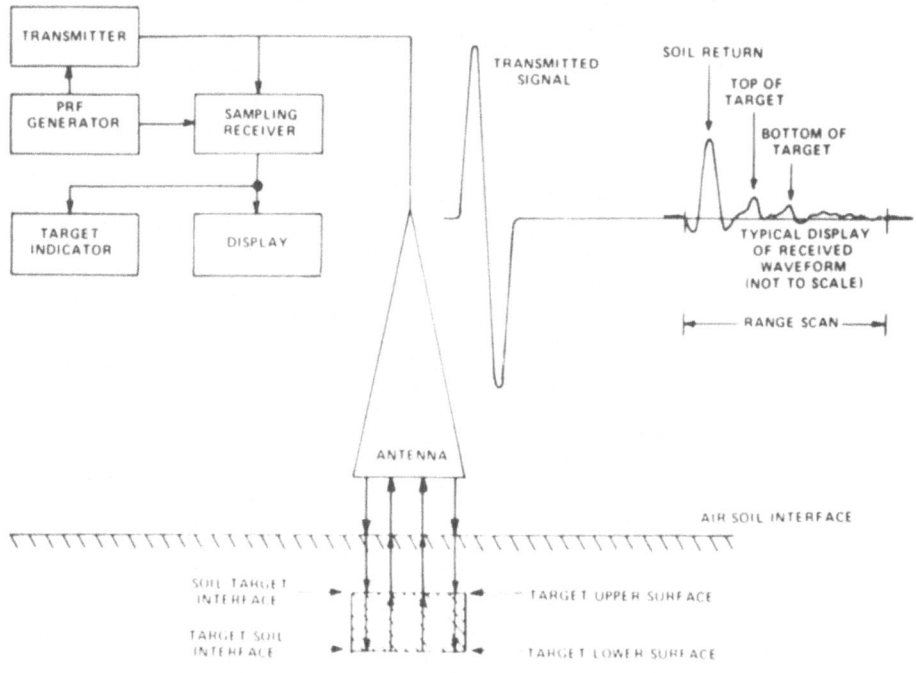

Figure 30   BASIC SHORT PULSE MONOSTATIC RADAR OPERATION

Figure 31   EXPERIMENTAL MONOSTATIC SHORT-PULSE RADAR

where P_R = received power,

             R = distance to the target,
             P_T = transmitted power,

             G_T = transmitting antenna gain,

             G_R = receiving antenna gain,

             λ_m = wavelength in the medium which is the free-
                        space wavelength shorted by the refractivity
                        of the medium,
             α = coefficient for attenuation in the medium,
             σ = target radar cross section

Von Hippel [19] defines the attenuation coefficient in terms of
the loss tangent

$$\frac{1}{\alpha} = \frac{\lambda_o}{2\pi} \left[\frac{2}{\epsilon_{\hbar} \sqrt{1 + tan^2 \delta} - 1} \right]^{1/2} \tag{21}$$

where $tan\ \delta$ = $\frac{18\beta}{f\epsilon_{\hbar}}$

             $\epsilon_{\hbar}$ = relative permittivity
             β = conductivity of the medium (millimho/m)
             f = frequency (MHz)

The absorptive losses for soil dictate the use of low frequencies whereas high resolution requires wide bandwidths. This im-
poses design problems, principally on the antenna as indicated
earlier in this paper. A TEM antenna designed to give a smooth
impedance match from 50-ohm coaxial cable to free-space was
found to have acceptable levels of pulse distortion. The antenna response is that of a high-pass filter network, while the
pulse stretch is approximately 20%. Internal reflections at
the feed input to the antenna is 46 dB below the applied signal
and 30 dB below the applied signal at the end as shown in
Figure 32. Figure 33 shows the transmitted pulse and the received waveform.

4.   EXPERIMENTAL RESULTS

     Selected representative high-resolution radar data are pre-
sented to illustrate the performance of the various high-resolution radars discussed here. Figure 34 gives an example of diagnostics high-resolution in which the principal scatterers of an
aircraft are shown. The data is from the Micronetics short
pulse range with the display sweep from left to right at 2 nanoseconds per centimeter. The model is a 6-foot-long scale model
of an F-102 fighter aircraft. The skin of the model is fiberglass, and therefore the dominant contributors to the radar cross
section are metal structures inside the aircraft. Reading from
left to right on the display, the first pulse is from a battery
inside the radome, the second and third pulses are from structural

5 ns/cm
BALUN: −46 dB
END: −30 dB

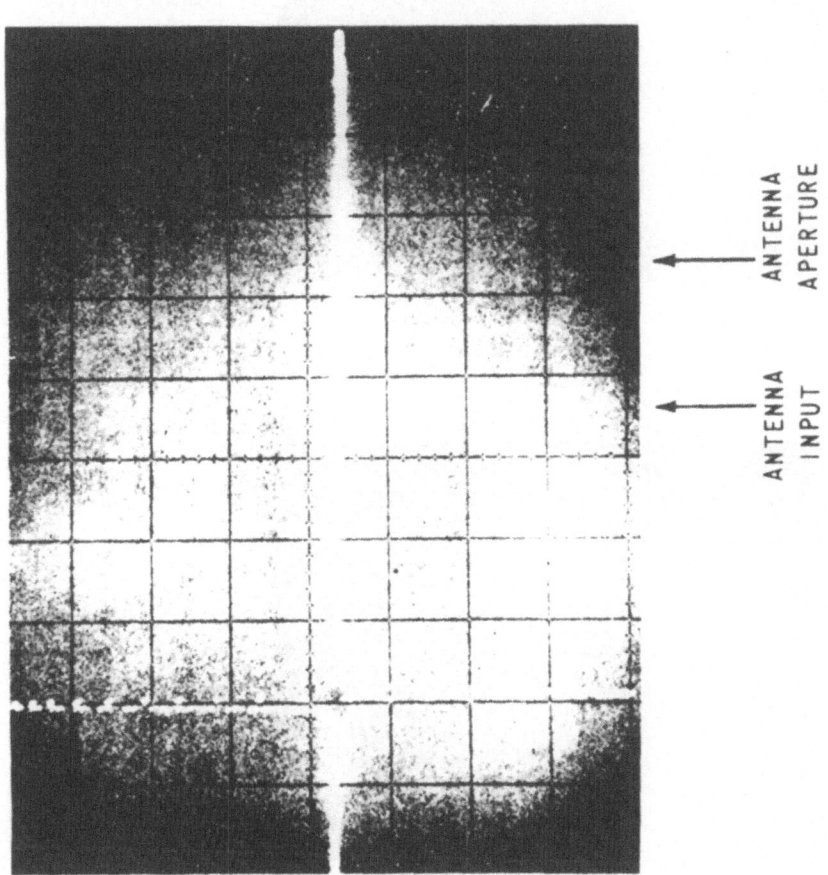

ANTENNA
APERTURE

ANTENNA
INPUT

Figure 32   TRANSMITTER PULSE AND ANTENNA REFLECTIONS

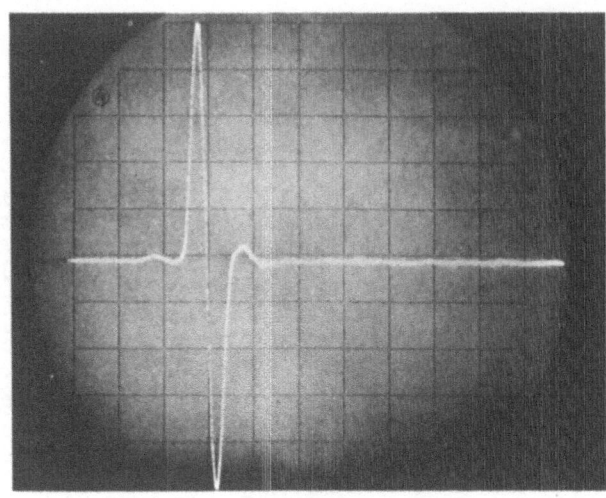

TRANSMITTED PULSE
1 ns/cm
12V PEAK TO PEAK

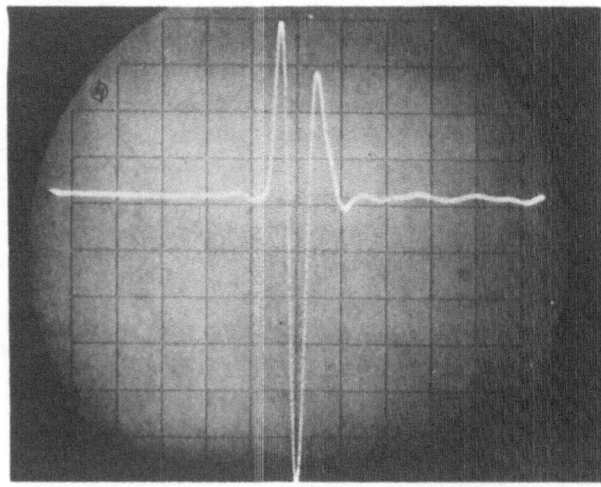

RECEIVED PULSE
1 ns/cm

Figure 33 - 40-INCH CONSTANT-FLARE, VARIABLE-WIDTH ANTENNA,
10 INCHES WIDE

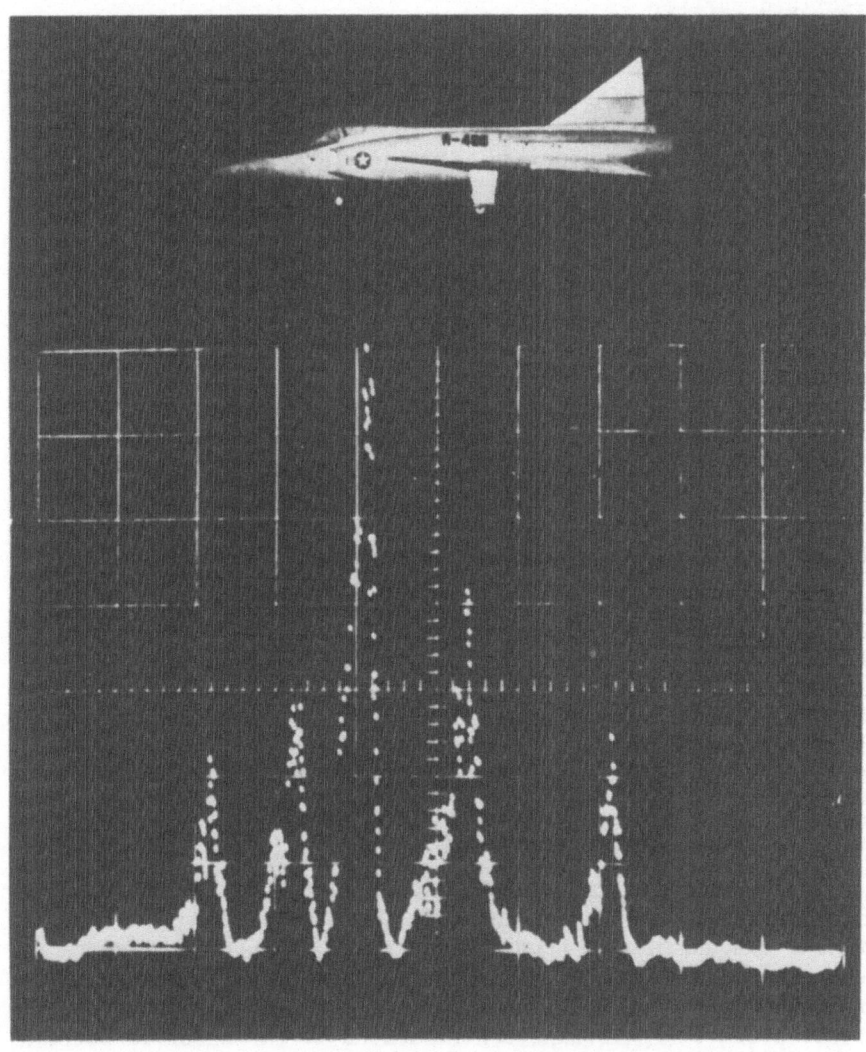

Figure 34   COMPOSITE PHOTOGRAPH SHOWING FIBERGLASS AIRCRAFT
ABOVE THE RADAR DISPLAY OF ITS SHORT PULSE SIGNATURE

cross braces inside the fuselage, the fourth pulse is from the
main landing gear struts, and the fifty pulse is from a control
cable mechanism inside the fuselage. Figures 35 through 39 present data taken with the X-band FM/CW high-resolution radar
described earlier.

4.1 X-Band FM/CW Experimental Results

The standard reference target was a thin, flat disk four
inches in diameter, supported by 5/8 lb. nylon string passed
through three small holes 120° apart. The backscattering cross
section of this flat plate is $10^3 \lambda^2$ or 0.900 square meters at
10 GHz. A representative calibration record is shown in Figure
35.

In all measurements of this series, we have recorded the
response to the reference flat plate target along with records
of the particular target reported to permit determining whether
structure of small amplitude seen on the target records may
properly be ascribed to that target, or, instead, to radar limitations. Obviously, when a large return is observed, one must
expect a corresponding range-sidelobe level at least as great
as that indicated by the flat plate measurements. Thus, the
reference flat plate data assist in analyzing and interpreting
the records for each scattering center of the target.

The possibility that scattering from the supporting strings
might introduce extraneous returns was explored by measuring
the samples of various strings separately, over the range of
aspect angles encountered in practice. In all aspects at which
targets presented small radar cross sections, the measured values
of strings were found to lie below $10^{-3} \lambda^2$. We have seen no
recognizable evidence of support interference in any of the
measurements.

4.1.1 Two Separated Spheres

To demonstrate the resolution capability of the radar and to
establish an indication of the effect of two simple scatterers
at nearly equal ranges from the radar, two 3-inch diameter
spheres separated by 18 inches between centers were used. Since
previous experience had indicated that supporting structure could
itself produce significant contributions to the observed pattern,
a minimum of such structure was used in this experiment. In this
case, the spheres were separated by a 15-inch length of 2-pound-
test nylon monofilament which was fastened to a supporting cage
made of the same material. This cage consisted of three orthogonal loops placed along great circles of the spheres. Thin
supporting lines were led from each sphere upwards to a common
yoke about six feet above the common level of the two spheres.
Five-eighths-pound nylon line fastened to the bottom of each
cage was pulled taut to hold the spheres at the 18 inch center-to-
center distance permitted by the horizontal yoking string. The

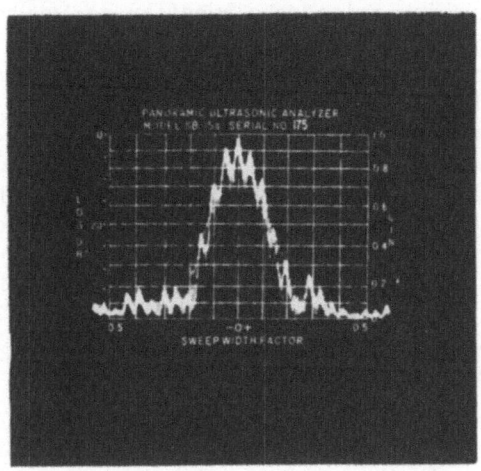

EACH SMALL PEAK DENOTES A SEPARATE SPECTRAL LINE. INTERLINE SPACES REPRESENT 1.47" WITH POINT NEAREST THE RADAR ON THE LEFT. THE ABSCISSA IS LOGARITHMIC WITH 0db (AT THE TOP) CORRESPONDING TO $10^3 \lambda^2$.

Figure 35 - SPECTRUM ANALYZER RECORD OF RADAR CROSS SECTION FOR A 4" DIAMETER FLAT DISC AT NORMAL INCIDENCE

resulting target is shown in Figure 36.

The cross section of each of the two spheres at the band center frequency is very nearly $5\lambda^2$ (45 square centimeters, or 0.0045 square meters). The close proximity of the two spheres in this configuration raises the question as to whether the field scattered from the nearer of the two spheres toward the more remote one will appreciably affect the illuminating field for the second sphere. Since forward scatter from the first sphere (for which the bistatic cross section is 0.3 square meters) is involved, an estimate of the magnitude of this perturbation was made. The estimate indicated that the field scattered from the first sphere toward the second one when the two spheres lay with their common axis pointing toward the radar was 36 per cent in amplitude of that of the incident field. The 3-dB width of this secondary scattered field was 23°, in the far zone, and the second sphere lay beyond the normal far zone boundary of the first sphere.

The data on the two-sphere target are presented in Figure 37 for intervals of 10° from the 0° position (sphere centers on common line through the radar boresight) up through 90°. The last 20° are covered at intervals of 4° to demonstrate the effects of interference between two objects closely spaced in range. The center-to-center spacing of the spheres was 18 inches. This spacing, properly reduced to account for inclination at the higher aspect angles, is marked by the two arrows which appear on the individual frames of the figure. The scattering cross sections of the two spheres, as shown in Figure 36, are equal to the 0° point and are approximately 5 dB above λ^2. The calculated cross section was 7 dB above λ^2. Interference of the forward-scattered wave from the closer sphere modified the return from the second one at 10° and again at 30°, although it apparently did not at 0°. The scattering from the two spheres appears quite independent until an angle of 70° is reached; at this point the centerline separation projected on the radar line of sight has been reduced to 6.2 inches. Successive steps from 70° to 90° in 4° increments carry the spheres through the entire range of interferences, and some evidence of the attendant lengthening effect is seen at 82°. The pronounced shape distortion observed on the target spectra at the higher angles is characteristic of the effects of target interference. A spectrum for the flat plate used for calibration is included for comparison.

4.1.2 Sphere-Capped Cone

The sphere-capped cone is an object of strong interest because of its low radar cross section. The sphere-capped cone used for these measurements consisted of a sharp-pointed conical nose of 15° semi-conical angle joined smoothly to a sphere 8-1/4 inches in diameter to yield a body of overall length about 20

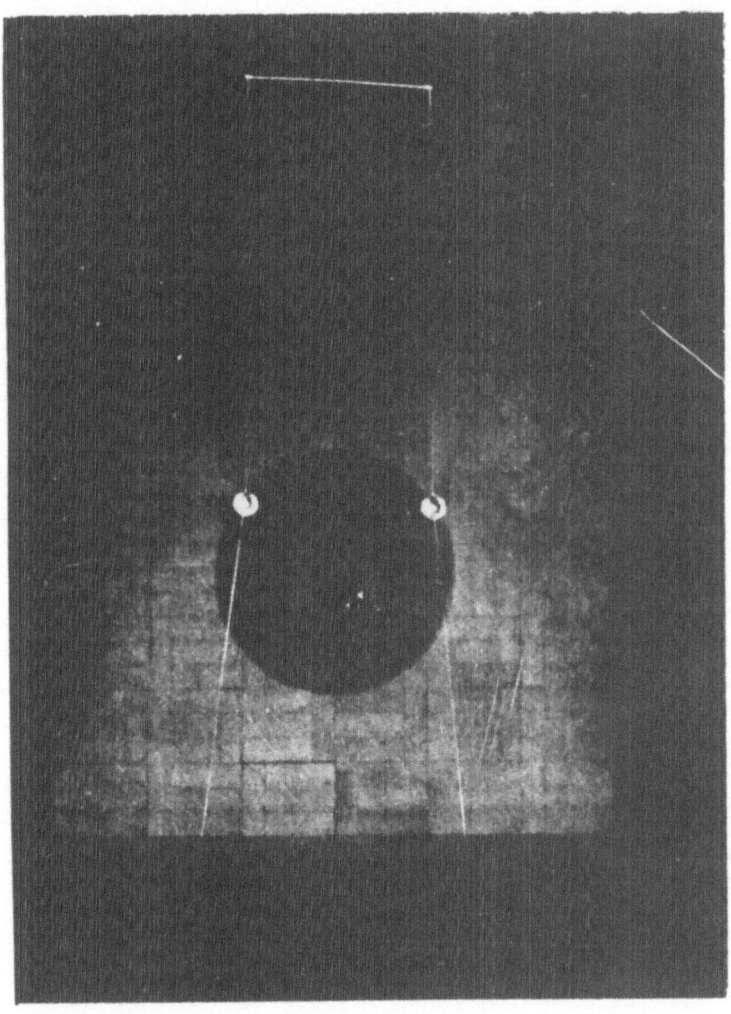

Figure 36 - TWO 3" SPHERES, SHOWING STRING SUSPENSION (THE
         SPREADER BAR IS ABOVE THE UPPER NULL OF THE
         RADAR BEAM)

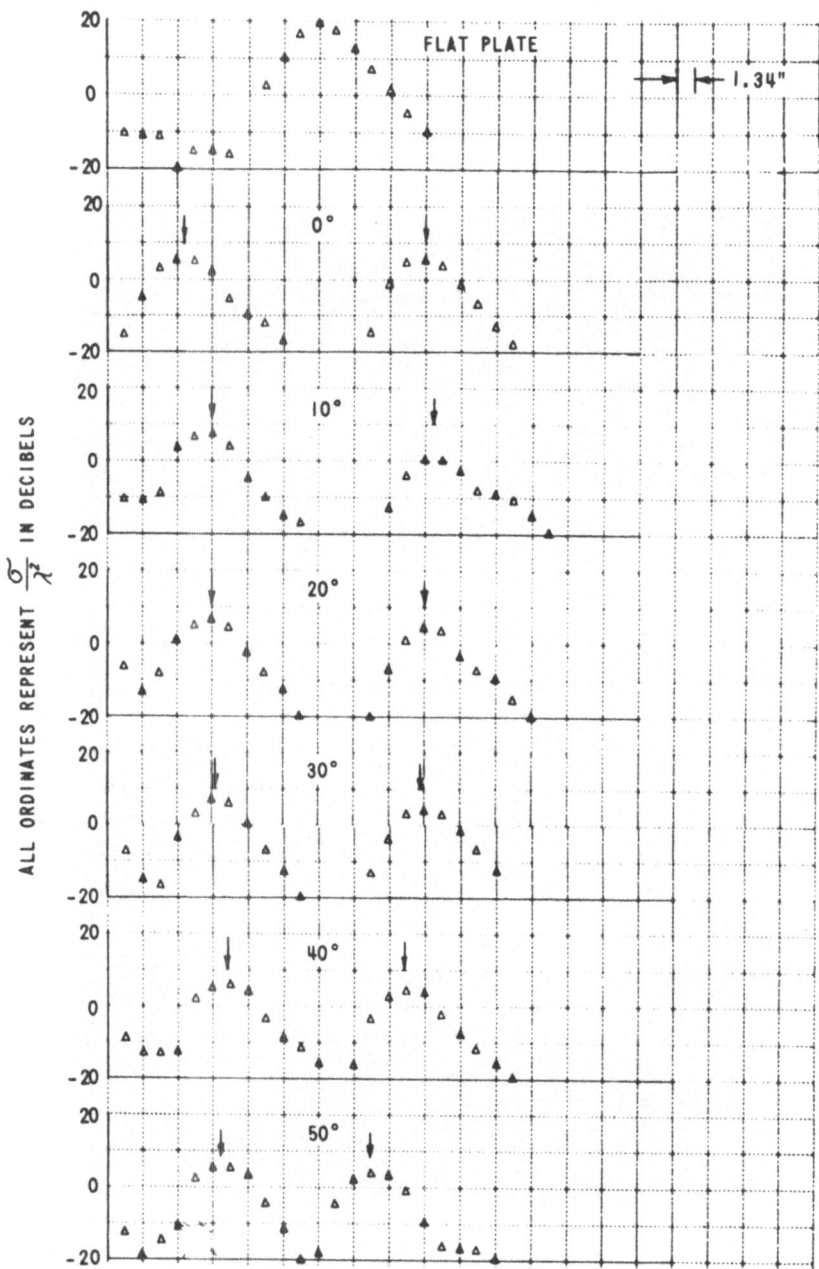

Figure 37(a)   RADAR CROSS SECTION OF TWO 3″ SPHERES SEPARATED
18″ AT VARIOUS ASPECT ANGLES

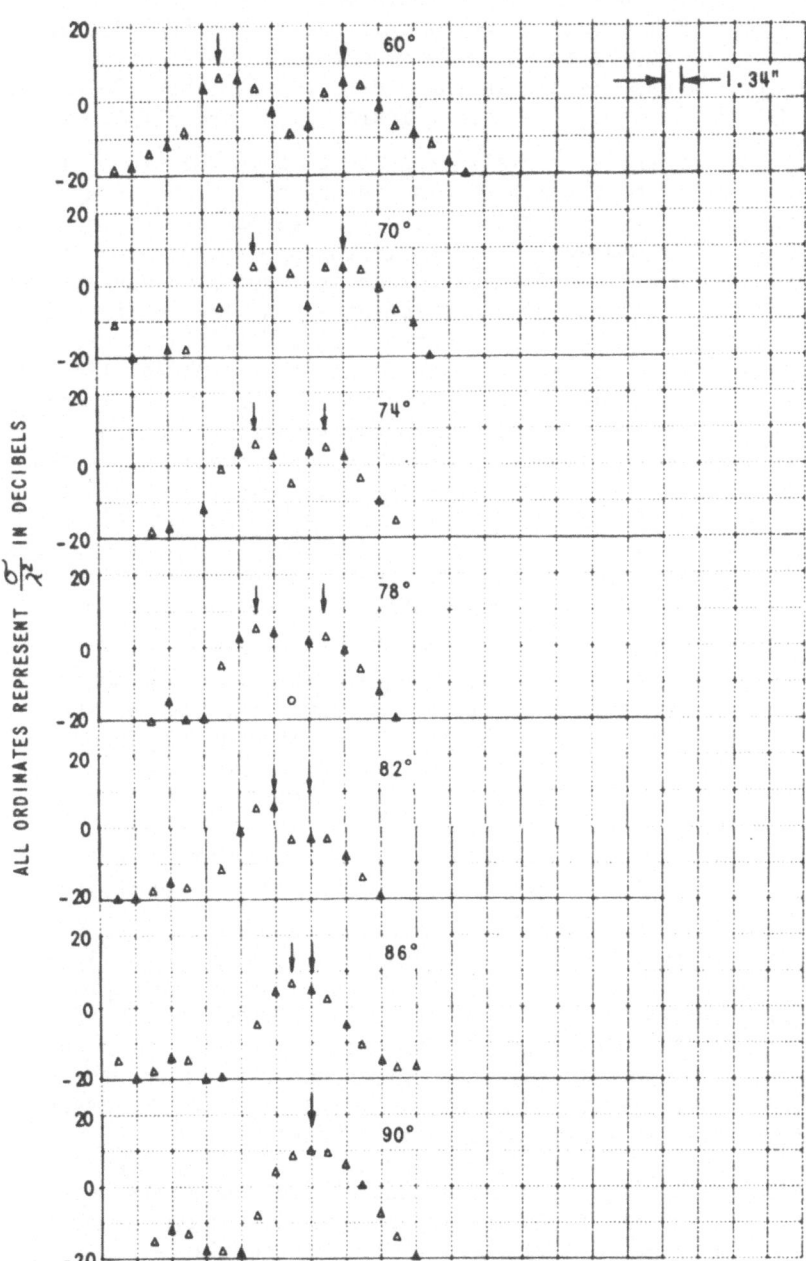

Figure 37(b)   RADAR CROSS SECTION OF TWO 3'' SPHERES SEPARATED
18'' AT VARIOUS ASPECT ANGLES

inches (see Figure 38). Data for it are shown in Figure 39. In the course of making these measurements, low-frequency noise (usually out of the range gate of this radar, but included in the present measurements which were made with the wideband spectrum analyzer) was visible in the record and lay about 30 dB below λ^2 at its worst position (Figures 37(a) and 37(b).

Because of the very small cross section of the tip of the capped cone and low-frequency noise in the system at a range corresponding to the location of the tip, the radar return from the tip is not immediately evident in the spectrum. A 0° (nose-on incidence) one line of the spectrum is about 5 dB higher than the corresponding lines of the spectrum which were obtained when the antenna was blocked off. By means of a flat plate serving as a range marker, it was determined that this spectral line results from scattering at a range corresponding to the location of the tip of the body. Calculation of the target return as the interfering signal necessary to perturb the original signal by 5 dB + 1 dB, demonstrates that the tip cross section must lie between -40 and -30 dB relative to λ^2. Since the tip is rounded slightly rather than being a perfectly sharp point, thus having a somewhat higher cross section than the sharp point would have, this magnitude is in reasonable agreement with the computed value (-40 dB) for the sharp point.

The two principal radar returns are from the cone-sphere junction and from the wave diffracted around the rear of the body. At 0° these returns should be separated by 7.6 inches or 4.9 spectral lines: for nose-on incidence the two peaks are separated by about 4 lines. Interaction between the radar return from the two scatterers can slightly shift the locations of the peaks (a problem inherent in radar operation with limited resolution and not peculiar to the Delta radar), so precise agreement is not to be expected.

The amplitudes of the two major returns have an easily understood behavior and are as predicted by modern concepts of electromagnetic scattering, because of the discontinuity in second derivative of the generating curve at that point. On the basis of asymptotic physical optics one can predict the radar return from such a discontinuity. At nose-on incidence the entire discontinuity forms a ring, scattering from all points of which is coherent. The predicted cross section is -16 dB relative to λ^2; the measured value of the cross section was -19 dB. The discrepancy results in part from measurement error, in part from interference with the other large return nearby, and in part from inadequacies of the theory. At a large angle off-nose-on the return is from only one point on the ring because of interference between returns from different elements on the ring that are at different distances from a constant-phase plane. This return can also be determined by means of the asymptotic-physical-optics approximation; for aspect angle

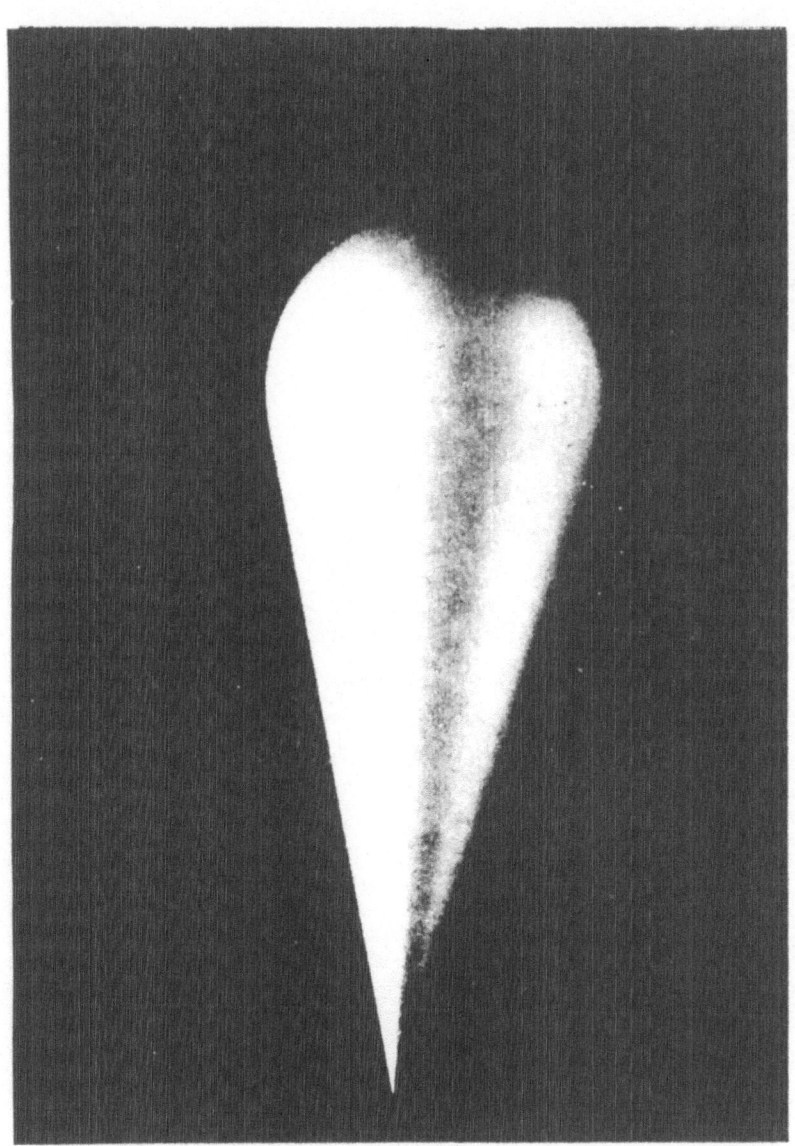

Figure 38 — SPHERE-CAPPED CONE WITH STRING SUSPENSION

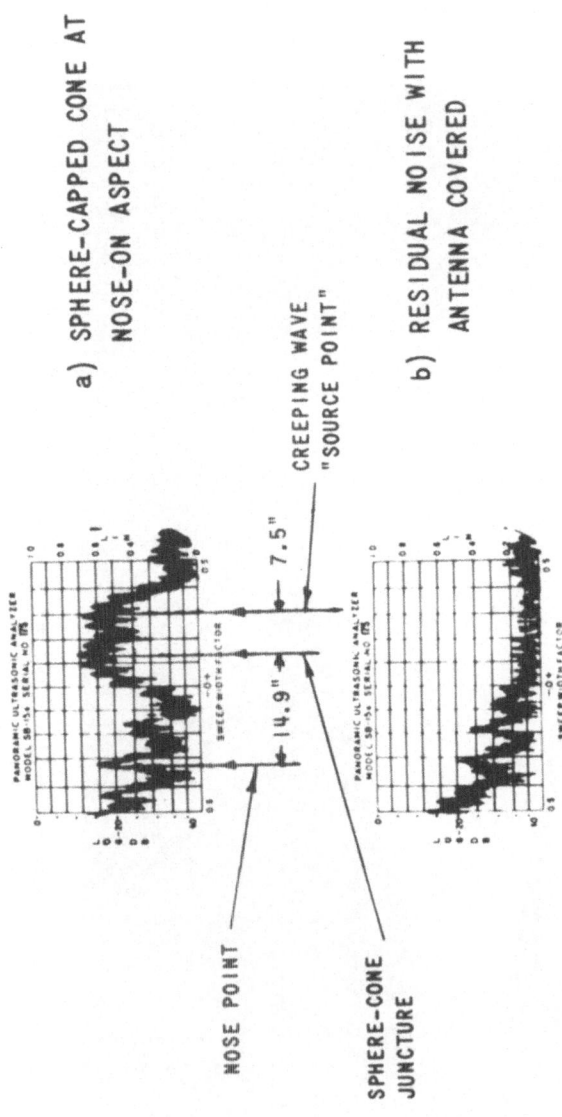

a) SPHERE-CAPPED CONE AT NOSE-ON ASPECT

b) RESIDUAL NOISE WITH ANTENNA COVERED

(EACH SMALL PEAK DENOTES A SEPARATE SPECTRAL LINE. INTERLINE SPACES REPRESENT 1.56", WITH THE POINT NEAREST THE RADAR ON THE ORDINATE IS LOGARITHMIC WITH 0 db (at top) CORRESPONDING to $1.6 \times 10^{-1} \lambda^2$)

Figure 39   SPECTRUM ANALYZER RECORDS OF RADAR CROSS SECTION

from 10 to 40 degrees the results are given in Table 5.

TABLE 5 - COMPARISON OF THEORY AND EXPERIMENT FOR RADAR
SCATTERING FROM A CONE-SPHERE JUNCTION

ASPECT ANGLE (degrees)	THEORETICAL $\frac{\sigma}{\lambda^2}$ (dB)	MEASURED $\frac{\sigma}{\lambda^2}$ (dB)
10	-31	-31
20	-31	-25
30	-29	-28
40	-25	-30

Discrepancies arise because of the aforementioned limitations
upon measurement and analysis.

The second major radar return results from diffraction of
energy behind the body. From the scattering properties of a
sphere we deduce that this return should, for nose-on incidence,
correspond to σ / λ^2 -18 dB; the measured value is -20 dB.
At an aspect angle of 10 degrees the illumination on the rear of
the body is essentially unchanged, to a first order, at least,
because the tip of the cone does not yet shadow the spherical
capping. Consequently, one would again expect a normalized cross
section of -18 dB; the measurement yielded a value of -20 dB. At
larger aspect angles the illumination of the rear of the body is
affected by the shadow of the tip. As would be expected, the
creeping-wave return becomes smaller (-25 dB at 20° and -22 dB
at 30°) with increasing aspect angle.

At angles greater than 40° the effective scatterers are too
close together to permit interpretation of the measurements in
terms of isolated scatterers. For angles greater than 75° the
return is essentially that from a sphere.

Downward Looking Radar, Experimental Results

Examples of data taken with the downward looking radar are
shown in Figures 40 through 43. Figure 40 shows the radar response of the surface alone. The radar response obtained from
a six-inch-square flat metallic plate buried at depths of four
and eight inches in clay soil of approximately 4 per cent moisture content are shown in Figures 41 and 42. These measurements
were performed with the early bistatic radar in a soil of known
constituency. The fain vertical pulse located on the 2-cm line
from the left edge of the graticule is the soil return, and
the signal return from the buried metallic plate appears displaced to the right of the surface return. The time displacement increases for increased plate depth. Table 6 presents
the radar-measured time delay in nanosecond as a function of
plate depth and the calculated time delay based on knowledge
of the relative dielectric constant for clay (ε_r = 2.8) at 4

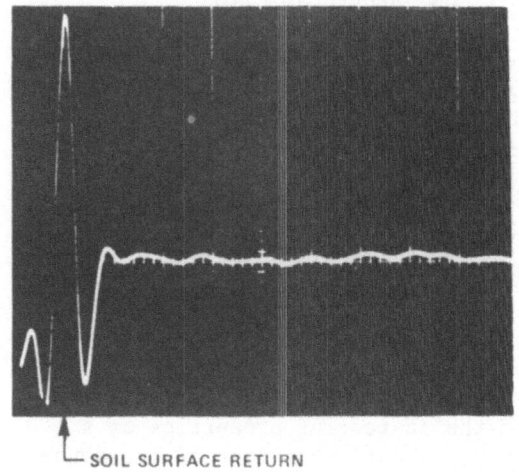

VERTICAL SENSITIVITY: 100 mV/cm
HORIZONTAL SENSITIVITY: 1 ns/cm

SOIL SURFACE RETURN

Figure 40   SOIL SURFACE RADAR RETURN SIGNAL

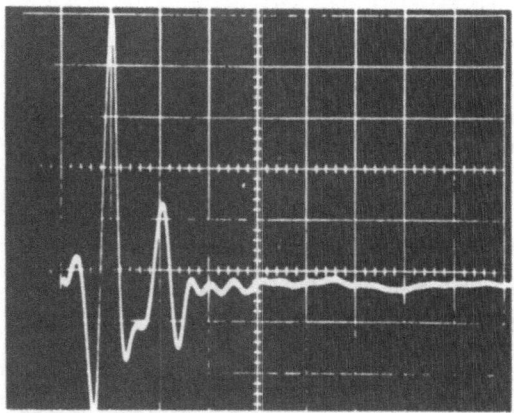

Figure 41   SIGNAL RETURN FROM FLAT PLATE AT DEPTH OF 4 INCHES

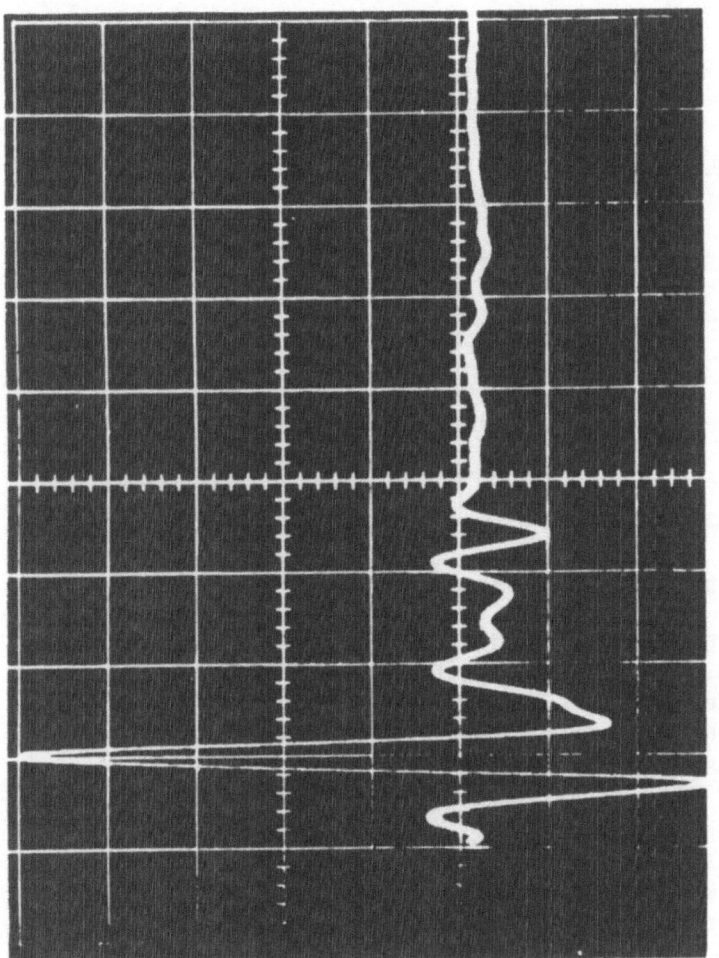

Figure 42   SIGNAL RETURN FROM FLAT PLATE AT DEPTH OF 8 INCHES

per cent moisture. The velocity in the soil is given by:

$$V_s = c/\sqrt{\epsilon_r} \qquad = 0.178 \text{ meters/nanosecond} \qquad (21)$$

where c = velocity of light in meters/second; thus, the time
delay is given by

$$t = 2d \left(\frac{1}{V_s}\right) \qquad (22)$$

where d = depth in meters. It can be appreciated that a slight
variation in ϵ_r would shift the values correspondingly.

TABLE 6 - TIME DELAY vs. PLATE DEPTH
(DRY CLAY: 4% MOISTURE CONTENT)

PLATE DEPTH (in)	MEASURED TIME DELAY (ns)	CALCULATED TIME DELAY (ns)
2	0.5	0.566
4	1.1	1.132
6	1.8	1.7
8	2.1	2.26

The measured values of time delays are visually determined
by noting the position of the positive peak value of the signal
return from the buried plate and measuring the distance to the
center of the positive peak signal representing the surface
return (located 1 cm from the left edge of the graticule). Note
that the width of the oscilloscope trace is on the order of
one-tenth nanosecond. With this limit to reading accuracy,
there is very close agreement between calculated and measured
time delays.

Figure 43 shows the radar response for a buried styrofoam
block, which was 9" x 12" x 6" and was buried 7 inches deep,
with the upper face horizontal. This response illustrates the
capability of a radar to detect holes in the ground. Styrofoam,
for all practical purposes, approximates the dielectric constant of air. Observe that the polarity of the backscatter
waveform has reversed since styrofoam has a much lower ϵ_r than
the surrounding medium as compared to the metallic target
case. Finally, Figure 44 shows a radar image of the subsurface
taken with the downward looking radar.

5.   SUMMARY AND CONCLUSIONS

It is hoped that this presentation may indicate to the

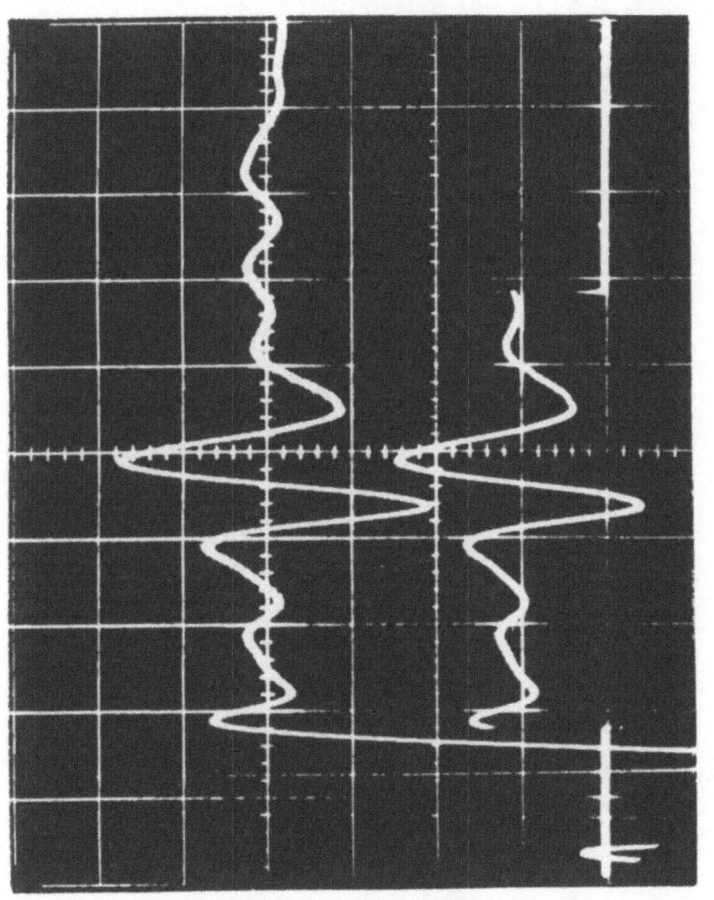

Figure 43 RADAR RETURN FROM STYROFOAM BLOCK 7 INCHES DEEP

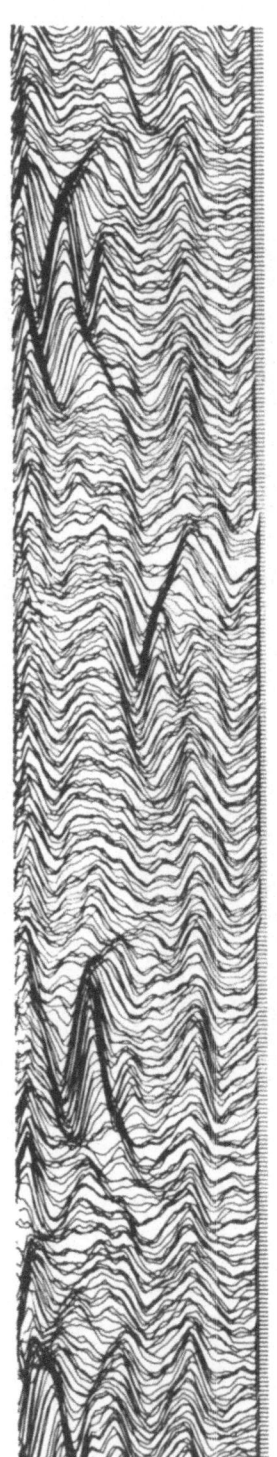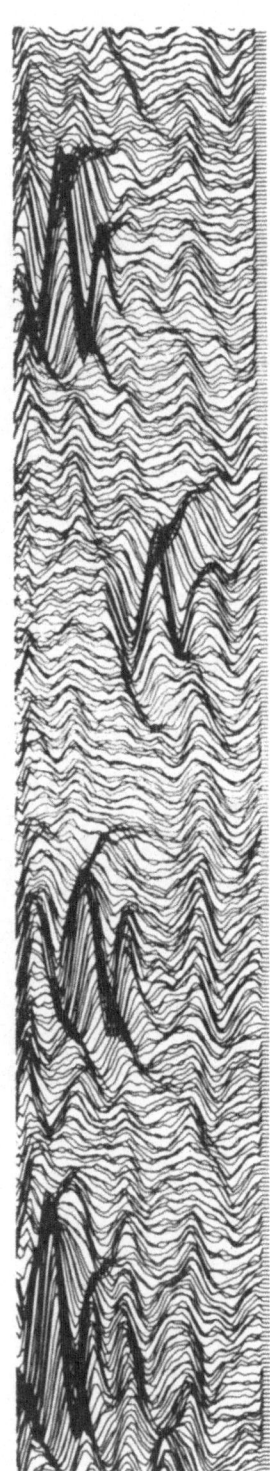

Figure 44   RADAR MAP OF SUBSURFACE PROFILE

to the reader some of the problems common to high-resolution radar and some of the approaches which may be taken to solve those problems. Three different types of high-resolution radars have been described which employ different forms of wide bandwidth waveform generation. We have shown experimental data that demonstrates high-resolution and its applications to target imaging.

In conclusion, several techniques currently exist for the design and implementation of short-pulse radar or of equivalent wide bandwidth radar. Although the radars described in this presentation are special purpose instruments, it is believed that the techniques demonstrated by them may be applied toward the design of target imaging radars.

I would like to acknowledge the contributions of the numerous present and past associates at Calspan who have worked with me in the high-resolution radar area; these include R. E. Kell, D. J. Newton, M. E. Bechtel, R. J. Wohlers, R. V. Gallagher, B. R. Tripp, M. C. Young, and H. R. Prinsen.

REFERENCES

1. M. I. Skolnik, Introduction to Radar Systems, McGraw-Hill Book Co., Inc., 1962, pp. 8-13.
2. R. R. Hively, "Sub-Nanosecond Pulse Methods of Radar Cross Section Measurement," Radar Reflectivity Measurements Symposium, Rome Air Development Center Technical Report No. RADC-TDR-25, April 1964.
3. A. C. Beck and G. D. Mandeville, "Microwave Traveling-Wave Tube Millimicrosecond Pulse Generators," IRE Transactions on Microwave Theory and Techniques," Vol. MTT-3, p. 48, December 1955.
4. R. V. Garver, E. G. Spencer and M. A. Harper, "Microwave Semiconductor Switching Techniques," IRE Transactions on Microwave Theory and Techniques," Vol. MTT-6, p. 378, October 1958.
5. Sylvania Microwave Newsletter, Sylvania Electric Products, Inc., Vol. 1, No. 1, Copyright 1956.
6. A. V. Alongi, R. E. Kell, and D. J. Newton, "A High-Resolution X-Band FM/CW Radar for RCS Measurements," Proc. IEEE, vol. 53, pp. 1072-1076 (1965).
7. M. C. Young, B. R. Tripp, "Nanosecond Pulse-Doppler Radar for the Investigation of Turbulent Wakes in Shock Tunnels and Ballistic Missile Ranges," IEEE Transactions on Aerospace and Electronic Systems," ICIASF'69 Record, 69-C-19-AES.
8. A. V. Alongi, G. C. Vorie, "An Experimental Investigation and Evaluation of a Doppler-Radar Technique for Measuring Projectile Velocities of a Rapid-Fire Machine Gun," Calspan Report No. GH-3074-E-1.

REFERENCES (Cont.)

9. A. V. Alongi, "A Short-Pulse High-Resolution Radar for
    Cadaver Detection," First International Conference on
    Electronic Crime Countermeasures, Edinburg University,
    Scotland, 18-20 July 1973.
10. H. S. Rothman, H. Guthart, T. Morita, "Nanosecond Pulse
    Scattering from Extended Laboratory Targets," Radar
    Reflectivity Measurements Symposium, Rome Air Development
    Center Technical Report No. RADC-TDR-64-25.
11. R. S. Berkowitz, Modern Radar, John Wiley & Sons, pp. 195-
    215, 1965.
12. M. P. Forrer, "Analysis of Millimicroseconds R-F Pulse
    Transmission," Proc. IRE, Vol. 46, pp. 1830-1835, November
    1958.
13. P. R. McIsaac and I. Itzkan, "A New Class of Switching
m, Devices and Logic Elements," Proc. IRE, Vol. 48, pp. 1264-
    1271, July 1960.
14. G. C. Bachman, "A Study of Nanosecond Pulse Techniques
    in Radar Transmission," Cornell Aeronautical Report No.
    UB-1426-P-1, February 1961.
15. G. Piefke, "Die Exponentialleitung und ihre Wellseablosong,
    Teil I," Archiv der Electrischen Ubertrogung, Vol. 7, 1953.
16. G. Piefke, "Die Exponentialleitung und ihre Wellseablosong,
    Teil II," Archiv der Electrischen Ubertrogung, Vol. 7, 1953,
    pp. 274-280.
17. "Atmosphere Limitations to Radar Resolution," Cornell
    Aeronautical Laboratory Report No. UB-1363-P-3, March 1960.
18. P. Van Etten, Private Communication, Rome Air Development
    Center, Griffis Air Force Base, New York.
19. A. R. Von Hippel, "Dielectric Materials and Applications,"
    MIT Press, Cambridge, Mass., 1954.

SIDE-LOOKING RADAR

K. G. Corless

RRE Malvern, UK

1   INTRODUCTION

If the man-in-the-street were to be stopped and asked what he
understood by the term RADAR, then the most probable sensible
reply would be that it is a method for detecting objects such as
aircraft, vehicles or ships; and to that extent he would be
correct. He might even add that other echoes from sea, clouds,
rain or land was something called clutter to be eliminated as
effectively as possible.

However, one man's clutter is another man's signal and a
ground mapping radar is designed to map "clutter" from land and
sea according to its spatial, plan-position, location. It also
exploits the fact that such a radar map bears a close relation to
a topographic map of the same area. Airborne ground-mapping
radar has been used for over 30 years as a means of navigation
and/or reconnaissance when the ground is invisible to the eye
because of cloud cover or darkness.

2   EARLY TYPES OF MAPPING RADAR

Fig 3.2 shows an example of the display of one of the early
types of mapping radar. The radar uses a radially-scanning
antenna with a beam which is fan-shaped in the vertical plane.

In the example shown, the X-band (10 GHz) radar used an
antenna with a horizontal aperture of about 2 metres which was
(and still is) the maximum length of antenna which can be rotated
about a vertical axis within the confines of the fuselage of an
aircraft.

Jeske (ed.), Atmospheric Effects on Radar Target Identification and Imaging, 157–178.
All Rights Reserved. This article Copyright © 1976 by Controller HMSO, London.

Tangential resolution is produced by the antenna beam width in the horizontal plane, and radial resolution is produced by the radar pulse length.

The principal limitations of this radar so far as quality of mapping was concerned, were

a   the relatively poor ground resolution (approx 300 metres at 20 Km) due to antenna beamwidth,

b   the poor dynamic range of the map presentation on the cathode ray tube (CRT), due to the need for a long afterglow phosphor.

3   SIDE-LOOKING RADAR

Two developments followed fairly quickly with beneficial results so far as ground mapping was concerned:

1   locking the antenna in the position which produced an antenna beam directed sideways,

2   introducing a photographic display with rapid chemical development which was intrinsically capable both of improved grey scale and of improved signal-integrating properties compared with the cathode ray tube.

Instead of the display showing a radial intensity modulated line time-base which rotated on the CRT in synchronism with the antenna, the photographic paper was exposed to the now fixed line time-base. The paper was in the form of a roll which was pulled past the CRT at a speed proportional to the forward speed of the aircraft. Fig 3.1 shows the radar scanning arrangement diagrammatically. Note that the antenna horizontal beamwidth determines the along-track resolution of the map, and the transmitter pulse-width determines the across-track resolution.

The system advantages of this arrangement over the radially-scanning radar are that, while a target is in the beam, it remains within a range resolution cell so that signal integration can be more easily performed. This results in an improved signal-to-noise ratio (SNR), and hence also of contrast, on a display which can also exploit that improvement. Fig 3.4 is an example of this improved imagery and was derived from a side-looking version of the same radar as that which produced Fig 3.2. Both Figs 3.2 and 3.4 show the estuary of the river Thames east of London, of which Fig 3.3 is a topographic map.

Fig 3.5 shows a montage of England and Wales composed of this improved imagery.

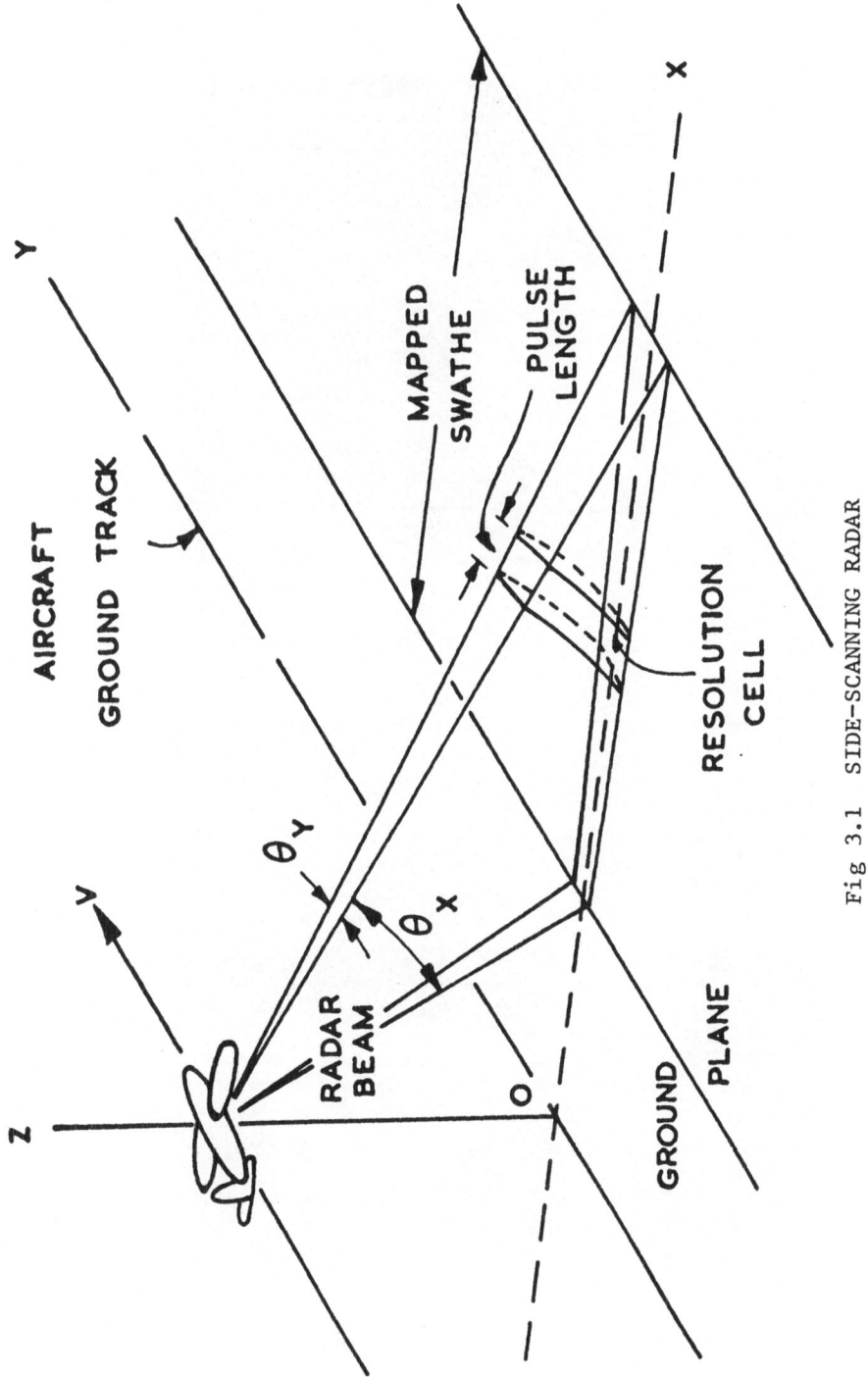

Fig 3.1 SIDE-SCANNING RADAR

Fig 3.2

Fig 3.3

Fig 3.4

Fig 3.2   RADIAL-SCAN RADAR IMAGERY OF THAMES ESTUARY

Fig 3.3   TOPOGRAPHIC MAP OF THAMES ESTUARY

Fig 3.4   SIDE-SCAN RADAR IMAGERY OF THAMES ESTUARY

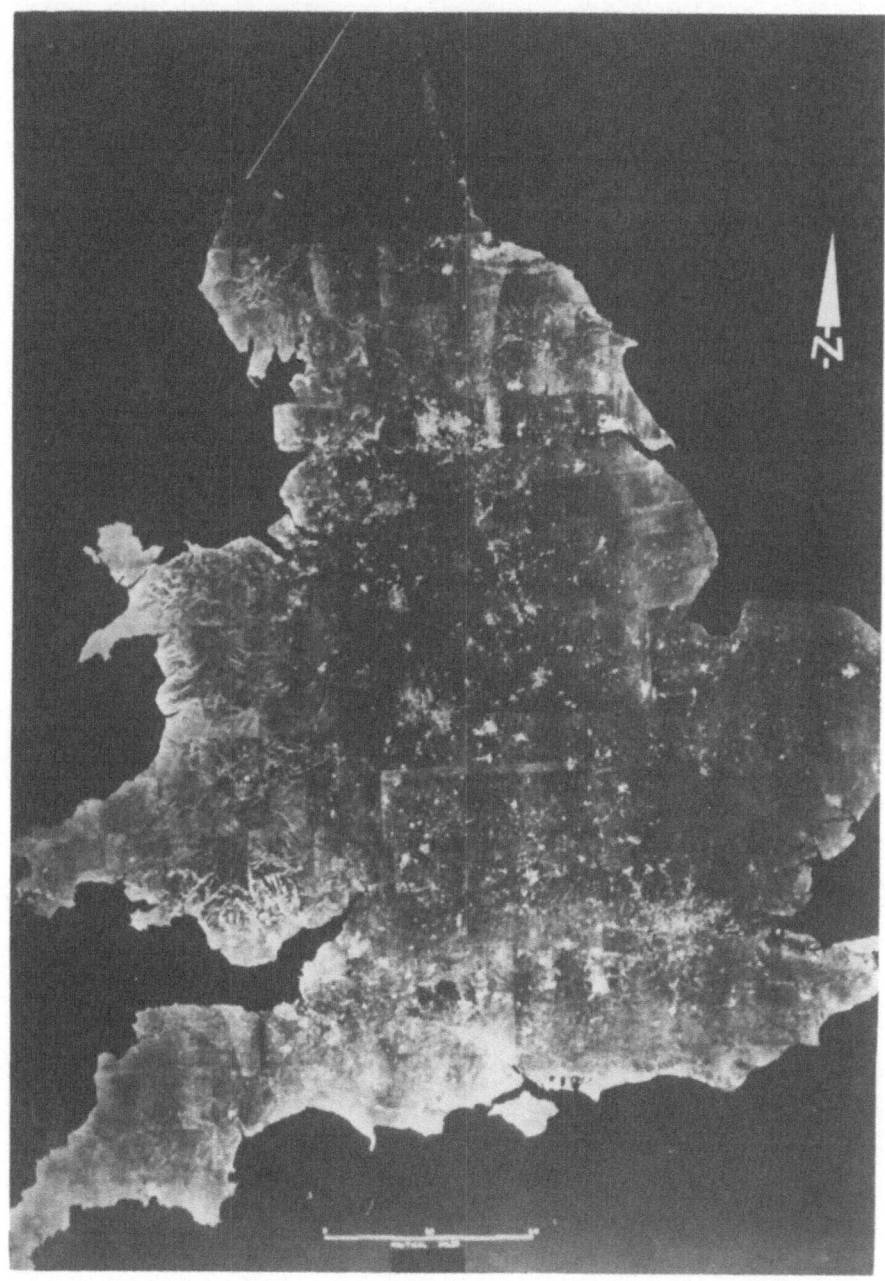

Fig 3.5 SIDE-LOOKING RADAR IMAGERY OF ENGLAND AND WALES

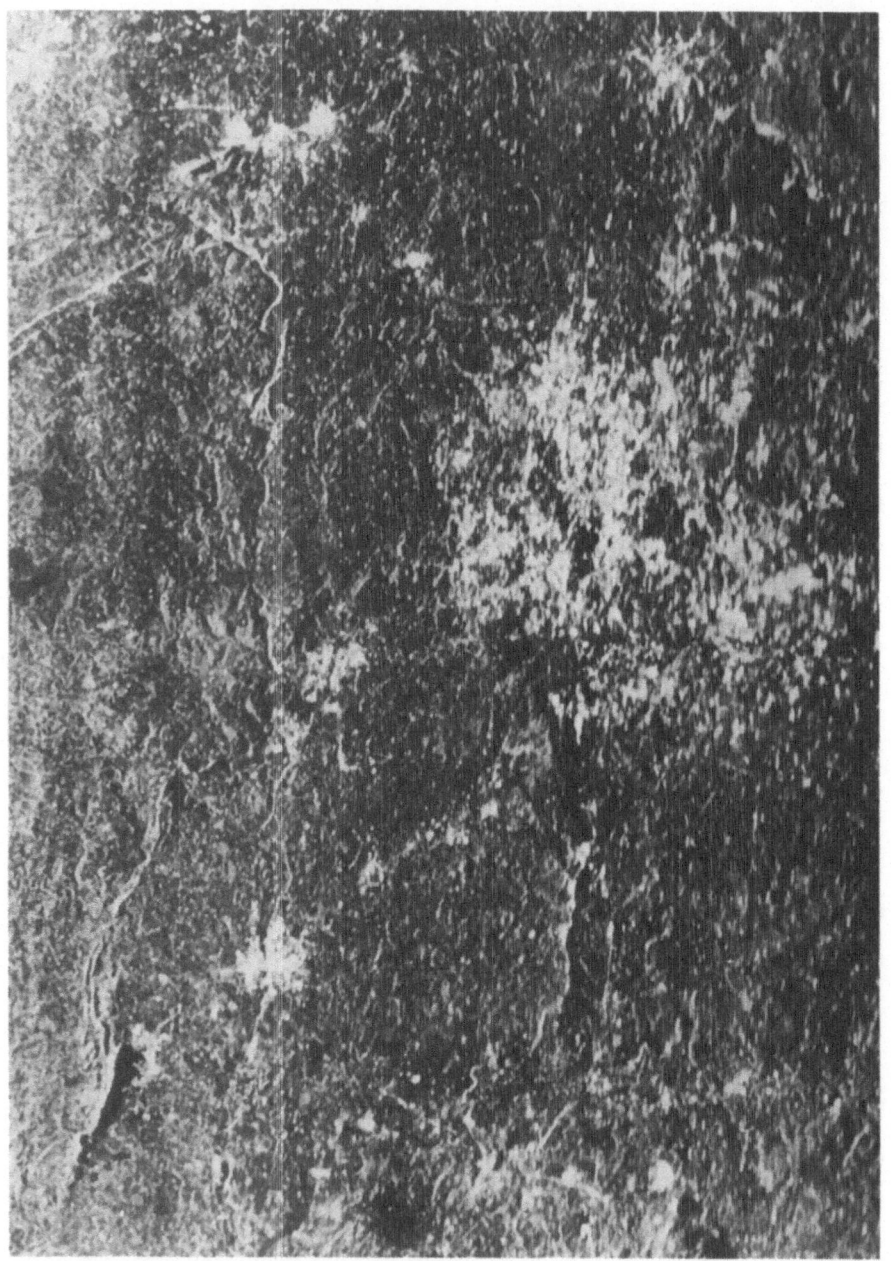

Fig 3.6   SIDE-LOOKING RADAR IMAGERY OF WEST MIDLANDS.
MAP WIDTH APPROX. 50 Km. X-BAND, 5 m REAL APERTURE.

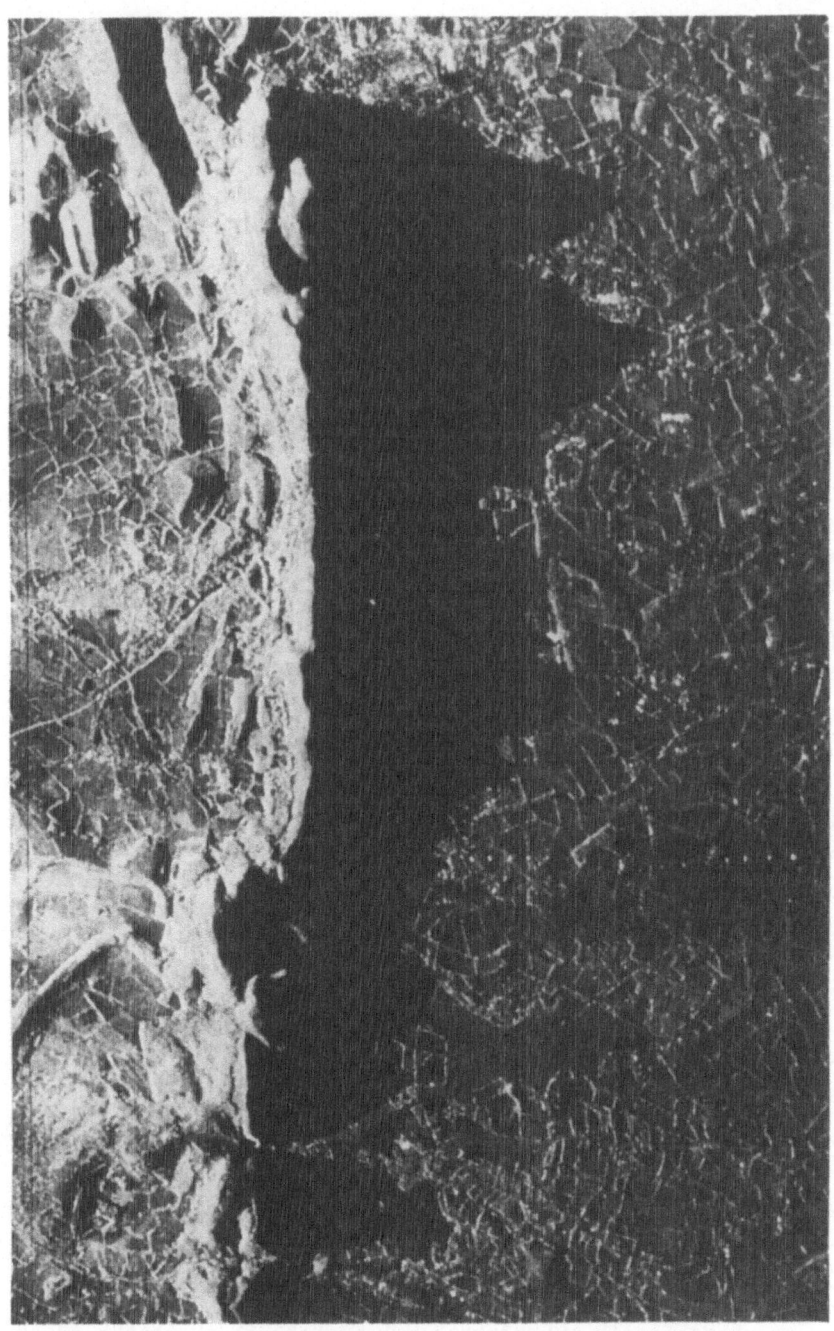

Fig 3.7 MALVERN HILLS. Q-BAND, 4 m REAL APERTURE

Having locked the antenna sideways the next obvious step was to increase its length and thereby improve the azimuthal resolution. Fig 3.6 is a sample of imagery from such a system with a 5 metre X-band antenna which resulted in an azimuthal resolution of about 100 metres at 20 Km range (circa 1955).

The area shown is that of the Midlands with the city of Birmingham prominent upper right of centre and the line of the Malvern Hills, with their radar shadow, in the lower left corner.

By 1958 the side-looking technique had been further extended by reducing the wavelength from 3 cm to 8 mm, retaining an aperture size of 4 metres, and by using a high resolution photographic film recorder which further enhanced the mapping grey scale. System resolution was about 20 metres at 10 Km. Fig 3.7 is an example of imagery from an 8 mm wavelength radar. It is of the Malvern area and should be compared with Fig 3.6.

All radar aerials are designed to produce diffraction-limited angular resolution in at least one plane. (Slight variations on this are usually invoked in order to control angular sidelobes but these do not fundamentally affect the argument.) The Rayleigh criterion of angular resolution for diffraction-limited apertures relates the far-field angular resolution (θ) to aperture dimensions (D) and operating wavelength (λ) by

$$\theta = 1.2 \frac{\lambda}{D} \qquad\qquad 3.1$$

θ is usually defined as the angular separation of the half-power points on the main lobe of the response.

It is this equation which has been the traditional spur behind the historical evolution of radar from metric through centimetric to millimetric wavelengths.

However further improvements in REAL APERTURE resolution became increasingly hampered by limitations in

a   millimetric radar technology where it becomes difficult to acquire components for engineered radars at wavelengths less than 8 mm

b   mechanical engineering tolerances on antenna manufacture. Even at 8 mm the demanded tolerances are approximately 0.1 mm over a 5 metre aperture and these must not only be achieved in manufacture but maintained in flight in the face of temperature and vibration.

So it was that attention was drawn to the technique of
APERTURE SYNTHESIS as a means of deriving an improvement in
azimuthal resolution amounting almost to two orders of magnitude
in potential.

4   SYNTHETIC APERTURE RADAR

The technique of Aperture Synthesis has found application in
such fields as radio and radar astronomy and in ground and airborne radar. Fundamentally the technique synthesises an aerial
aperture of large dimensions by using a much smaller aperture as
a probe to scan the spatial electromagnetic field incident upon
the synthesised aperture. Subsequently this field distribution
is processed to form an aerial beam of width corresponding to the
large synthesised aperture. It is in all respects a microwave
analogue of the possibly better known technique of optical holography, and both single - and two - dimensional apertures can be
synthesised.

The particular application to be described is an airborne,
side-looking ground mapping radar where a single-dimensional
horizontal aperture is synthesised in order to enhance the radar
azimuthal resolution.[1,2]

4.1 The Synthetic Aperture Principle

In contrast to the real aperture radar which simply maps the
intensity of signal power in the echoes from the ground, the
synthetic aperture radar employs a phase-coherent pulsed transmitter (with a range resolution commensurate with that obtained
in azimuth) with a side-looking aerial to illuminate a strip of
ground to one side of the aircraft track. The radar is caused to
scan the ground by the forward movement of the aircraft (Figure
3.1). The phase-sensitive receiver measures the changes in
carrier-frequency amplitude and phase which occur from discrete
ground echoes as these echoes pass through the real beam in a
direction parallel to track. These phase and amplitude measurements are recorded at video frequency and at that stage they
constitute a map of the incident electromagnetic field across the
aperture to be synthesised. This recording then forms the input
to the synthetic-aperture beam-forming processer from which the
final radar map is produced.

The range-resolving capability of the pulsed transmitter is
used to establish many contiguous range gates, the several outputs
of which constitute parallel and independent channels to be
processed separately and recombined when the final radar map is
formed, as in the case of a television raster.

The function of the processer is to perform a weighted
vectorial summation of the amplitude and phase histories, in each
of many hundreds of range cells, along the flight path, in such a
way as to synthesise sets of overlapping synthetic apertures, one
set for each range cell. Each synthetic aperture is focussed by
the processer at the range of its range cell. The synthesised
aperture can be many hundreds of metres in length.

4.2 The Synthetic Aperture Process

The enhancement of azimuthal resolution is achieved by an
observation of the phase changes which occur in the received
signal from a point reflecting object as the object passes through
the radar beam. (Fig 4.1a). Simple geometry shows that these
phase changes are a quadratic function of along-track distance,
centred about the beam axis (assuming the latter is normal to
aircraft track) (see Fig 4.1b, and Section 4.3). A phase function
which is quadratic is equivalent to a frequency which changes
linearly with time, (Fig 4.1c) just as in CHIRP pulse compression.
The analogy to CHIRP is close, the principal difference being the
order of magnitude of frequency and bandwidth. The CHIRP waveform
is centred at IF, say 300 MHz, and has·a bandwidth of say 100 MHz.
The doppler return from the target as it passes through the SAR
beam is centred on zero frequency and has a bandwidth of say
100 Hz. Otherwise the processes for the enhancement of resolution
are identical and amount to matched filtering.

The following table indicates the basic similarity between
CHIRP and SAR, using order-of-magnitude values:

CHIRP		SAR
10^{-6} second	Signal duration (T)	1 second
10^8 Hz	Bandwidth (B)	10^2 Hz
100	T x B	100
10^{-8} sec	Temporal resolution (waveform decorrelation time)	10^{-2} sec
5 x 10^8 ft/sec	"Velocity of propagation"	10^3 ft/sec
5 ft	Spatial resolution	10 ft

Note that the "velocity of propagation" for CHIRP range compression is $\frac{1}{2}$c (c = velocity of light) and for SAR azimuth
compression it is the aircraft track-velocity.

One significant practical difference in the processers for
CHIRP and SAR follows from the fact that, in CHIRP, the linear FM
is imparted to the transmission and is sensibly unaffected by the

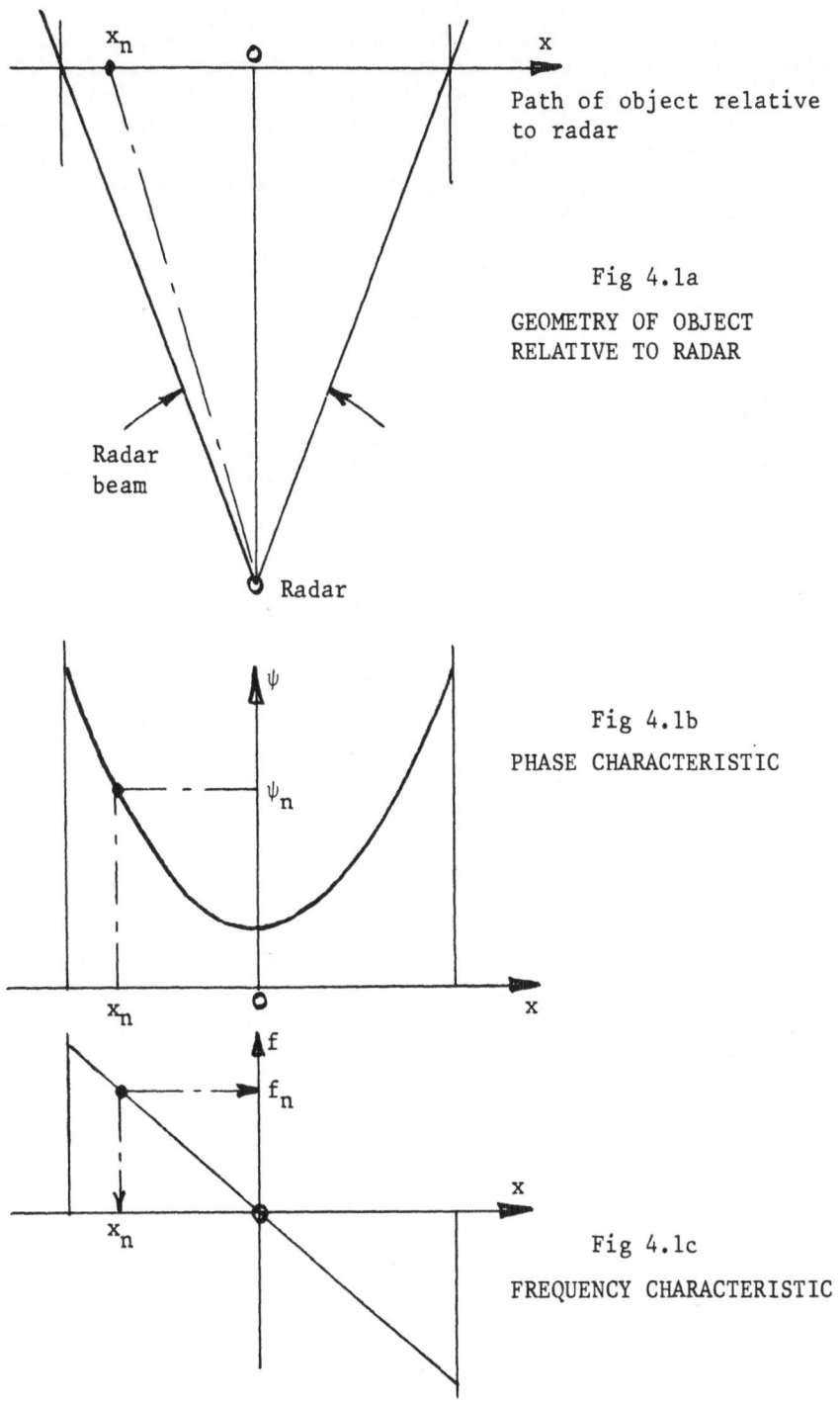

Path of object relative
to radar

Fig 4.1a

GEOMETRY OF OBJECT
RELATIVE TO RADAR

Radar
beam

Radar

Fig 4.1b

PHASE CHARACTERISTIC

Fig 4.1c

FREQUENCY CHARACTERISTIC

target motion whereas in SAR the modulation of the received waveform is due entirely to the relative motion of the target across
the beam. The effect of this is that the frequency deviation is
constant for all ranges (and hence the azimuthal resolution is
constant for all ranges) whereas the time to pass through the beam
is proportional to range. This results in a linear FM the slope
of which is inversely proportional to range, and the processer
must cope with this. This results in FOCUSSED processing.

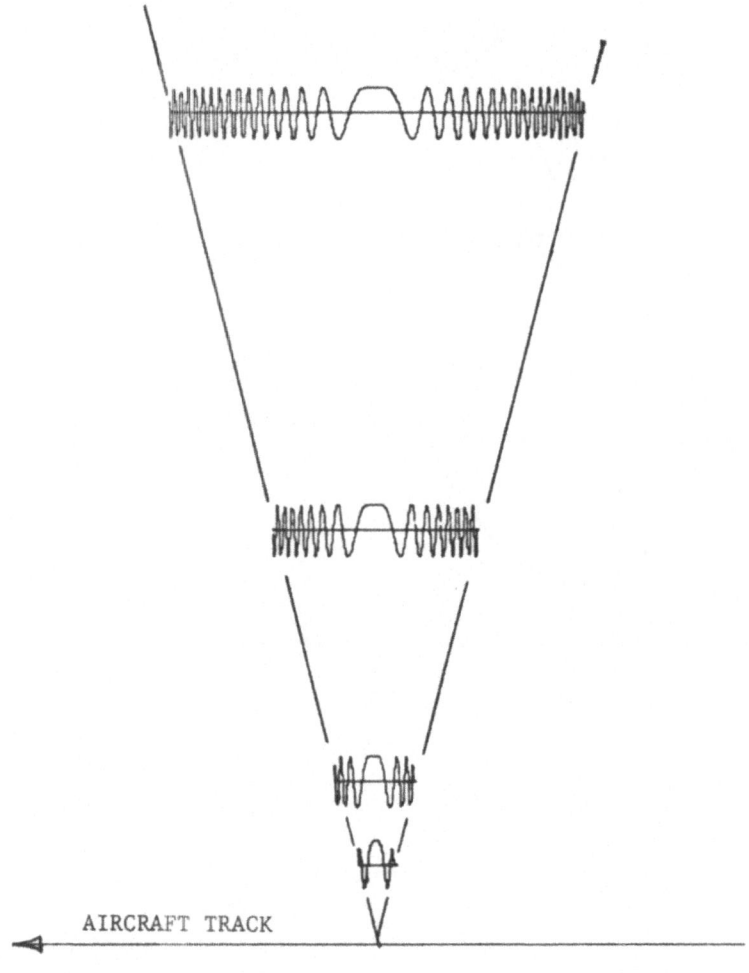

AIRCRAFT TRACK

Fig 4.2   SYNTHETIC APERTURE DOPPLER WAVEFORMS

Fig 4.2 shows the SAR doppler waveforms diagrammatically and indicates in particular that the maximum values of doppler frequency, which occur at the edges of the beam, are independent of range but that the slope of the linear frequency characteristic is a function of range as is the TB product of the waveforms.

There is a simplified form of SAR which only processes over the central stationary-phase region of these doppler waveforms using a simple integrator in each range gate. As can be seen from Fig 4.2 this results in a useful increase in azimuthal resolution but the resolution does degrade with range though not as quickly as does that of a real aperture radar. This simplified form is known as UNFOCUSSED SAR and is discussed in more detail later.

4.3   Target-Radar Geometry and Absolute Doppler Phase Characteristic

The geometry is shown in the diagram below:-

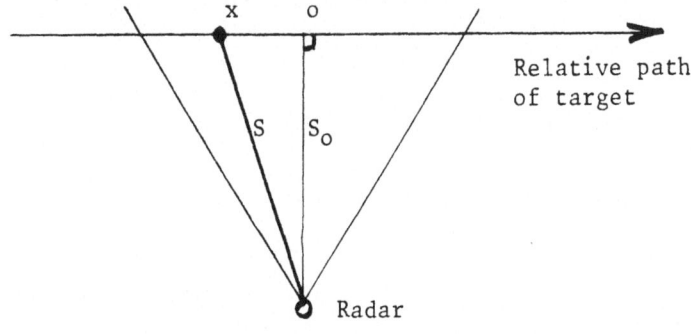

Fig 4.3

where the geometry is simplified by fixing the radar in space and causing the (stationary) target to move at aircraft velocity V. S is the instantaneous range to the target at time t, x is the displacement of the target along its path from the origin which is taken as the point of closest approach. S_o is the minimum range.

For a constant velocity V

$$x = Vt \tag{4.1}$$

The instantaneous range to the target is

$$S = (S_o^2 + x^2)^{\frac{1}{2}}$$

ie $S(t) \; \widehat{=} \; S_o + \dfrac{V^2}{2S_o} t^2$ (for x << S_o)   4.2

Converting S from linear measure to absolute radar phase ψ

$\psi(t) \;\; = \;\; 2 \cdot \dfrac{2\pi}{\lambda} S(t)$

$\qquad\quad = \;\; \dfrac{4\pi}{\lambda} \cdot S_o + \dfrac{4\pi}{\lambda} \cdot \dfrac{V^2}{2S_o} t^2$ 4.3

when t = 0

$\psi(0) \;\; = \;\; \dfrac{4\pi}{\lambda} S_o$

$\qquad\quad \equiv \;\; 2n\pi + \gamma$

this is the minimum phase at closest approach, and γ is the phase modulo 2π. Obviously γ depends on the precise range of a target within a range resolution cell.

Consequently Equation 4.3 becomes:

$\psi(t) \;\; = \;\; 2n\pi + \gamma + \dfrac{2\pi V^2}{\lambda S_o} t^2$ 4.4

The essential features of $\psi(t)$ are preserved by dropping the $2n\pi$ term and re-writing Equation 4.4 as follows:

$\theta(t) \;\; = \;\; \gamma + \alpha t^2$ 4.5

where $\alpha \;\; = \;\; \dfrac{2\pi V^2}{\lambda S_o}$ 4.6

This phase characteristic is parabolic in form over a range of t for which Vt << S_o as mentioned in Equation 4.2 and is shown in Fig 4.4 where both ψ and θ scales are shown.

Note that over the range of t for which the phase character-istic is parabolic the rate of change of phase, or doppler frequency, is linear, with a slope of 2α or $4\pi V^2/\lambda S_o$.

The waveform is identical in form to that used in linear FM pulse compression; the difference lies in the slope of the

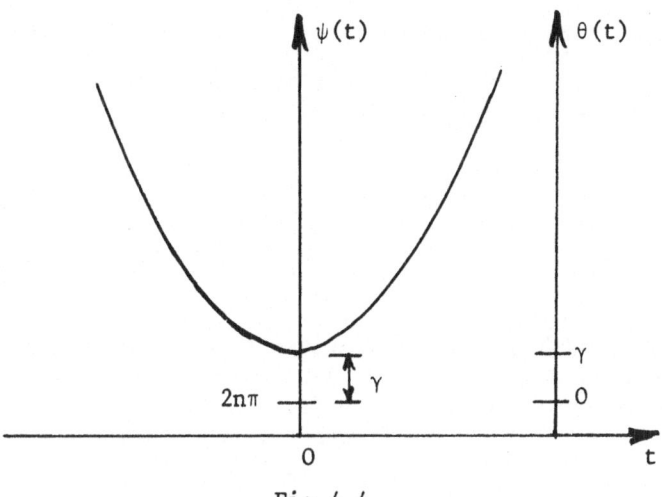

Fig 4.4

frequency characteristics, that for PC being six orders of magnitude greater in bandwidth and occupying six orders of magnitude
less in time, as mentioned in Section 4.2.

4.4   Focussed Azimuth Resolution

The resolving power of a waveform can be estimated by the
degree to which it differs from its shifted self.[3] The autocorrelation function of the waveform is therefore a guide to its
resolving power.

However, for reasons which will become evident later, a
better feel for resolution, in this instance, can be achieved by
forming the cross-correlation function of the target waveform with
one identical to it but shifted in frequency (to represent moving
targets with a finite radial velocity) and shifted in phase at
the point of stationary phase (to represent unknown target range
measured in fractions of a wavelength). Such a correlation
function is known as the cross-ambiguity function.

If the transmitted waveform is

$$\mu(t) \quad = \quad \exp jw_c t \qquad\qquad w_c = \text{carrier frequency}$$

then the return signal is obtained by modulating u(t) in phase
by the target phase characteristic given by Equation 4.5, ie, the
received signal is

$$u(t) \quad = \quad \exp j(w_c t + \alpha t^2 + \gamma) \qquad\qquad\qquad 4.7$$

Mixing down to zero IF effectively sets $w_c = 0$ in Equation 4.7 and may also add another constant phase shift term which can be absorbed in γ, so we obtain an expression for the coherent video waveform

$$v(t) = \exp j(\alpha t^2 + \gamma) \qquad 4.8$$

and it is the resolving power of this waveform with which we are concerned.

Following Woodward[3] the modulus is formed of the complex cross-correlation function $R(T,f,\phi)$ between the received waveform (Equation 4.8) and a reference waveform which is identical to $v(t)$ but in error by f c/s in frequency, and ϕ radians in phase at the point of stationary phase. Let this reference waveform be

$$x(t) = \exp j(\alpha t^2 + \gamma) \cdot \exp j(2\pi ft + \phi) \qquad 4.9$$

Then

$$R_{vx}(\tau,f,\phi) = \int_{-T/2}^{T/2} v(t + \tau) \cdot x^*(t,f,\phi)\, dt \qquad 4.10$$

where the interval T seconds is the integration time and related to the synthetic aperture size x by x = VT. Note that T can be either the total time the target appears in the beam, or less than that.

Combining Equations 4.10, 4.8 and 4.9

$$R_{vx}(\tau,f,\phi) = \int_{-T/2}^{T/2} \exp j[\alpha(t+\tau)^2 + \gamma]\, \exp -j(\alpha t^2 + \gamma) \, \exp -j(2\pi ft + \phi)\, dt$$

which reduces to

$$R_{vx}(\tau,f,\phi) = T.\exp j(\alpha\tau^2 - \phi) \cdot \mathrm{sinc}(\alpha\tau - \pi f)\, T$$

R is the unnormalised cross-correlation function (CCF). The more useful normalised CCF, ρ, can be simply derived by dividing R by the value at $\tau = f = \phi = 0$

ie $$\rho_{vx}(\tau,f,\phi) = \frac{R_{vx}(\tau,f,\phi)}{R_{vx}(0,0,0)}$$

$$\therefore \quad \rho_{vx}(\tau,f,\phi) \quad = \quad \exp j(\alpha\tau^2 - \phi) \ \text{sinc}(\alpha\tau - \pi f) \ T \qquad\qquad 4.11$$

This expression consists of the familiar sinc function modulated by a complex periodic function of $(\alpha\tau^2 - \phi)$. This indicates that the CCF cannot be reliably formed by other than a complex reference function, that is phase and quadrature correlation channels with appropriate combination of outputs, must be used otherwise the CCF would vanish for particular values of α, τ and ϕ.

Note that for a particular value of ϕ the coefficient of the sinc function is more slowly varying with τ than is the sinc function itself since the sinc function varies as the product τT ($T \gg \tau$) whereas the coefficient varies as τ^2. This means that in either the phase or quadrature channel the coefficient produces near-uniform amplitude weighting of the sinc function.

To determine the resolution of the process, it can be seen from Equation 4.11 that the nulls in the sinc function occur for

$$(\alpha\tau - \pi f) \ T \quad = \quad \pm \ \pi$$

ie for $\qquad\qquad \tau \ = \ \dfrac{1}{\alpha} \ (\pi f \pm \dfrac{\pi}{T})$

giving an interval between first nulls of $2\pi/\alpha T$. If the resolution of the sinc function is defined as the separation of the -3 dB points, then the resolution is approximately half that between first nulls. The temporal resolution can therefore be defined as

$$\Delta T \ = \ \frac{\pi}{\alpha T} \qquad\qquad\qquad 4.12$$

This can be related to linear resolution on the ground, as is done for optical processers, by considering the following diagram of the SLAR geometry.

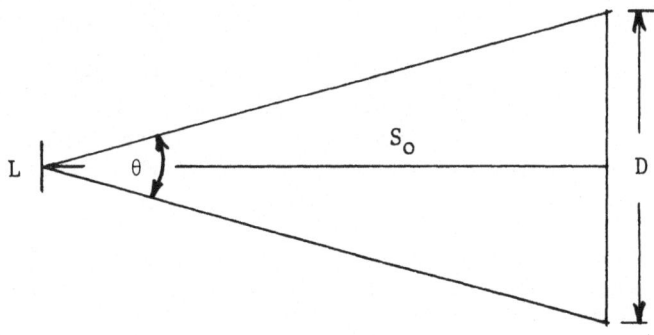

Fig 4.5

L is the aperture size of the physical aerial, generating a
beamwidth θ which gives a linear beamstrike D at range S_o.

$$\text{Then} \quad \theta = \frac{\lambda}{L} ; \quad D = S_o\theta = VT \qquad\qquad 4.13$$

Assuming that all the available target energy is coherently
integrated, ie over a complete beamwidth, then

$$T = \frac{S_o\lambda}{VL} \qquad\qquad 4.14$$

From Equations 4.12, 4.6 and 4.14

$$\Delta T = \frac{L}{2V}$$

Hence the linear spatial resolution on the ground is given by:

$$\Delta D = \Delta T.V = \tfrac{1}{2}L \qquad\qquad 4.15$$

which is the familiar result for optical processers, and not
surprising since the two processers are mathematically identical.

Fig 4.6a indicates this process diagrammatically for a single
simulated target. The degree of enhancement of azimuthal resolution is given by the degree to which the dispersed FM waveform
is compressed. For an unadulterated linear FM this compression
factor is equal to the time-bandwidth product of the LFM waveform.
Fig 4.6b indicates the resolving power of the process for two
just-resolvable simulated targets.

A SAR signal processer, whether optical or electronic,
effectively performs a cross correlation function, in each of
many independent range-gates, between the doppler waveform at the
output of the receiver and a stored replica of the predicted waveform. Such a processer not only improves resolution but is a
coherent integrator which also improves signal-to-noise ratio.

The result of Equation 4.15 suggests that the linear
azimuthal resolution on the ground is independent of such factors
as wavelength, range and aircraft velocity. In the ideal case
this is true but there are practical limits to the extent to
which this can be exploited, for example:

WAVELENGTH - clearly the longer the wavelength the broader
is the radar beamwidth θ (Fig 4.5) and the greater the
azimuth compression which the processer must achieve. As
technology progresses, processers can be made more powerful

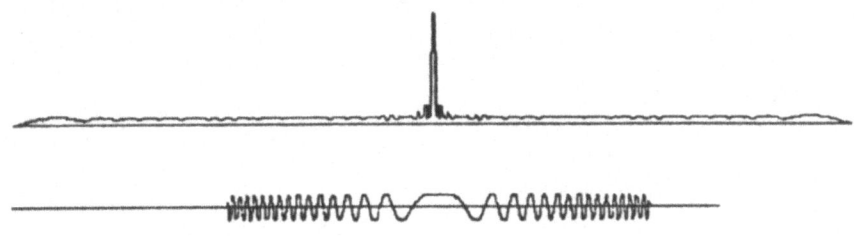

4.6a   Single target

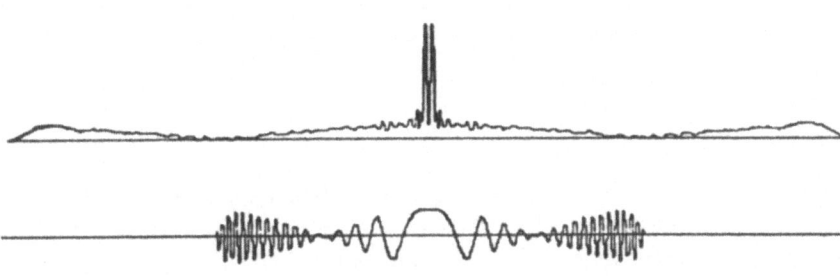

4.6b   Two targets

Fig 4.6   SYNTHETIC APERTURE FOCUSSED PROCESSING

but there is still much to be said for keeping θ reasonably small, say less than a few degrees. So far it has been assumed that the processed doppler bandwidth corresponds to that for the whole beamwidth. This does not have to be so. An antenna shorter than the theoretical "L" can be used and only part of the doppler waveform be processed. More will be said later in the lecture by Dr Brooks.

RANGE - as range increases so does the TB product of the doppler waveform to be processed, which places more demands on processing power if the ½L resolution is to be maintained.

VEHICLE - it has been tacitly assumed that the aircraft flight path is a perfectly straight line. This will not be the case

in turbulent weather conditions and this will cause corruption of the LFM phase characteristic which will degrade
azimuth sidelobe levels or, in extreme cases, degrade the
resolution itself. More will be said about this in Section
5.

4.5 Depth of Focus

From equations 4.13 and 4.14 the length of the synthesised
aperture can be expressed as

$$D = \frac{S_o \lambda}{L} \equiv \frac{S_o}{(L/\lambda)} \qquad\qquad 4.16$$

In optical terminology this is an expression of aperture size (D)
to focal length (S_o). For example, if L = 20 ft (theoretical)
and λ = 0.1 ft (X-band) then

$$D = \frac{S_o}{200}$$

which is a "well stopped-down aperture" having a good depth of
focus. In particular this focussed processing is quite inadequate
to provide range resolution – a pulsed radar must be used for
that.

4.6 Unfocussed SAR Azimuthal Resolution

For any antenna focussed at infinity – that is with a uniform
distribution of phase across the aperture – there is a transition
region of range beyond which the radiating phase fronts can be
assumed to be parallel and normal to the direction of propagation
and within which they are curved and centred on the antenna
radiating elements (Huygen's principle).

At the transition range the antenna beamwidth is approximately
equal in width to the antenna aperture.

In an unfocussed SAR, the processer becomes an integrator
which synthesises a uniform-phase aperture of the required length
to achieve a transition region at the particular range being
processed.

If d is the length of unfocussed synthetic aperture then, at
range R, the linear azimuth resolution is given by $\lambda R/d$ which is
also equal to d as explained above. That is

$$\frac{\lambda R}{d} = d = \text{azimuth resolution}$$

whence azimuth resolution = $\sqrt{\lambda R}$

This technique offers a useful improvement in azimuth resolution with a much simplified processer. Unfortunately the processing gain is much less than that of a focussed SAR because the processer rejects all the energy in the doppler waveform except that at the central region of stationary phase (see Fig 4.2).

5   EFFECT OF DOPPLER PHASE ERRORS ON SAR

In Section 4.4 mention was made of the adverse effects on SAR resolution and sidelobe level of the corruption of doppler phase by uneven motion of the aircraft. The particular motion to which the SAR is sensitive is that along the line joining the radar antenna to the object on the ground being imaged. To evaluate the effect of such phase corruption on the sidelobe level and resolution, it is necessary to add a corresponding phase-corrupting term into the bracketed term of equation 4.8 and proceed from there.

Before doing that it must be appreciated that there are many other potential sources of phase error in the SAR process, notably in the

RADAR - due to instabilities in transmitter timing and in spectral impurities in the local oscillator.

TARGET - the analysis of sections 4.3 and 4.4 assumes a stationary target. Any movement of a target with respect to terra firma will give rise to phase errors and resultant imagery distortion. A ship rolling, pitching and yawing on the sea is an extreme example of this.

PROPAGATION - given a perfect radar, smooth flight and a stationary target, any spatial and/or temporal variations in the refractive index of the propagation medium are also potential sources of phase error across the synthesised aperture.

Theoretically the imagery degradation due to any or all such sources of phase error can be evaluated in terms of synthetic aperture SIDELOBE LEVEL, RESOLUTION and POINTING ACCURACY once the errors can be specified quantitatively for inclusion in equation 4.8. In this approach it is usual to attempt to separate the phase errors across the synthetic aperture into two kinds:

PERIODIC effects, including any bias or "tilt" term,

RANDOM effects.

The periodic effects result in the accentuation of particular sidelobes (on the Paired Echoes principle) or, in the case of bias, a pointing error due to a displacement of the main lobe of the synthesised beam.

The random effects cause a uniform rise in sidelobe level according to their magnitude.

In fact, once the phase errors can be specified, their effect can be found by reference to standard works on antenna design where the relation between increased sidelobe level and aperture phase errors are usually graphed for easy reference[4].

The main problem becomes that of identifying the nature and magnitude of the phase errors, and in the case of propagation this involves a fairly detailed understanding of the refractive index fluctuations within a volume of atmosphere bounded by the aircraft height, the minimum and maximum ranges of the mapped swathe and the length of the synthesised aperture.

This problem will be examined in detail by Dr Marder in the next lecture.

REFERENCES

1   W.M. Brown and L.J. Porcello, "An introduction to synthetic aperture radar", IEEE Spectrum, Vol. 6, pp 52-62, Sept. 1969.
2   R.O. Harger, Synthetic Aperture Radar Systems, Academic Press, 1970.
3   P.M. Woodward, Probability and Information Theory with Applications to Radar, Pergamon Press, New York, 1953.
4   M.I. Skolnik, Introduction to Radar Systems, McGraw Hill, New York, 1962.

AZIMUTH COMPRESSION PROCESSING OF REAL SLAR DATA

S.R. Brooks

Marconi Research Laboratories,
GEC-Marconi Electronics Ltd., Great Baddow,
Chelmsford, Essex, England.

INTRODUCTION

The real data processing study described is a direct consequence of earlier theoretical studies into digital signal processing by colleagues at Marconi Research Laboratories.   During the
initial phase of this investigation an application of digital
techniques to a Doppler signal processor was considered where radar
return data was cross-correlated with a bank of single-tone
sinusoidal filters.   In a practical system the operation would be
repeated over data in many parallel range gates.   It was of particular interest to study the effects of extreme quantization or hard
limiting where only the sign of the sampled bipolar radar data is
retained because of the simplicity of the arithmetic operations in
the processing.   The results of this part of the study have been
published ((1) and (2)).

The scope of this work was later widened to include the application to the azimuth compression of side-looking airborne radar (SLAR)
returns to produce aperture synthesis.   Although similar in basic
design to the Doppler processor, this application was sufficiently
different in purpose to require some alteration in the method of
performance assessment previously used.   In addition, multi-bit
processing was included in the investigation.   On the basis of
results derived from this theoretical work recommendations were made
on a broad design for a digital SLAR processor.   A hardware feasibility study showed the acceptability of the design in all respects
and development of a prototype system commenced soon after.

The complete system may be regarded as comprising three units –

Jeske (ed.), Atmospheric Effects on Radar Target Identification and Imaging, 179–191.
All Rights Reserved. Copyright © 1976 by D. Reidel Publishing Company, Dordrecht-Holland.

    a) the radar interface and bandwidth compressor
    b) the processor
    c) display.

The development schedule has enabled the interface to be completed
before the remaining units, and the availability of real data
before the operational hardware processor provided an opportunity
to process real data on a general purpose computer. The aims of
such an exercise were to provide some validation of the digital
design and to explore alternative processing and display options
for use in future systems.

The organisation of the hardware processor is outlined,
followed by a description of the computing facilities used and
some examples of processed terrain 'maps' obtained from the real
data are presented.

SYNTHETIC APERTURE RADAR

In concept the synthetic aperture principle is straightforward (3). The real aperture carried on the airborne platform
points broadside to the aircraft track and is scanned by the platform motion. A longer aperture having enhanced resolution is
synthesised by coherently combining successive radar returns.
If the wavelength of transmission is λ and the real aperture
length is d then the unprocessed resolution ρ_{REAL} (the antenna
beamwidth) at a range R is approximately

$$\rho_{REAL} = R\lambda/d.$$

If the platform moves with velocity V and returns are combined
over a time τ then an aperture of length $V\tau$ is formed with
resolution ρ_{SAR} where

$$\rho_{SAR} = \frac{R}{2} \left(\frac{\lambda}{V\tau}\right)$$

the factor of two appearing because both out and back paths are
utilised. The length of the synthetic aperture that can be
formed is limited by the real beamwidth and this leads to the
maximum attainable resolution being equal to half the real
aperture or d/2.

The standard technique for carrying out the coherent integration of the returns has been to record the radar return on film
with phase information. The film is then transported through a
lens configuration which focuses the recorded phase front onto a
second moving film. The disadvantages of this optical technique,
such as the limited dynamic range and non-linearity of the film,
and the difficulties associated with the implementation of a real
time system, have led to the consideration of alternatives. We
will consider the design of the digitally processed system.

THE DIGITAL PROCESSOR

A simplified block diagram of a possible system is shown in Figure 1. After phase detection, bipolar radar video is sampled (range-gated) at intervals of the compressed pulse length and quantized. Some pre-processing may be carried out on this data before it is either stored for on-line processing or recorded for off-line work. The processor store will contain a large array of immediately previous pulse returns over all range gates within a chosen swath. The size of the store will determine the bounds on the swath width and integration length (and hence resolution) in any particular radar situation.

Processing involves parallel operations over all range gates which may be represented as a cross-correlation of the stored radar return data with a stored and similarly quantized reference waveform which provides the phase adjustment needed to focus the synthetic antenna. The reference waveforms for a broadside antenna will be a set of linearly frequency modulated waveforms centred about a frequency of 0 Hz and having l.f.m. 'rates' appropriate to the focusing over the ranges within the swath. In practice some tolerance may be allowed in this operation so that each reference may span several consecutive range gates usually with insignificant performance loss. The reference waveform store may therefore be a small fraction of the radar return storage area.

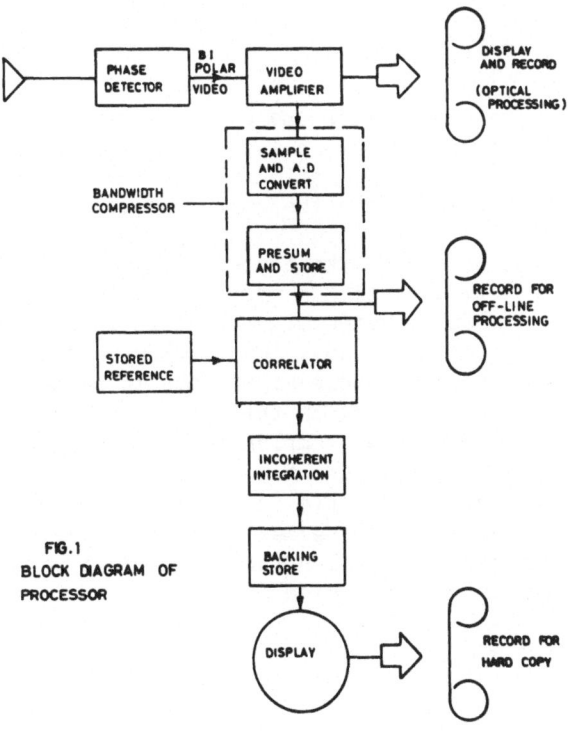

FIG. 1
BLOCK DIAGRAM OF
PROCESSOR

The processing within one range gate may be represented as forming

$$U_k = \sum_{n=1}^{N} \left\{ A_n X_{n+k} + B_n Y_{n+k} \right\}$$

$$V_k = \sum_{n=1}^{N} \left\{ A_n Y_{n+k} - B_n X_{n+k} \right\}$$

where the pulse returns are

in phase channel $\quad \left\{ X_m \right\}$

quadrature channel $\quad \left\{ Y_m \right\}$ for $-\infty < m < \infty$

and the 'matched' reference is

in phase channel $\quad \left\{ A_n \right\}$

quadrature channel $\quad \left\{ B_n \right\}$ for $1 \leqslant n \leqslant N$.

The quantities U_k and V_k are then subjected to square law detection to give an output Q_k where

$$Q_k = \left\{ U_k^2 + V_k^2 \right\}^{\frac{1}{2}}.$$

In a processing cycle a similar operation is carried out over all range gates and the outputs are passed to the display store after, perhaps, some incoherent integration. New radar return data can then replace older data within the store before the next cycle.

This simplest of frameworks conceals some considerable variation that may be permitted in the system, but still the processing task is hugh. A modest system might require input over 1000 range gates with integration over, say, 100 samples. Depending on platform velocity a processing cycle might have to be carried out in 0.01 second. The requirement in this situation is for a storage area greater than 10^5 complex samples and a processing rate of 10^7 complex multiplications per second. In the radar interface the A/D converter will need to sample at 40 nsec or less. It follows that it is very necessary in the design of a real-time in-flight processor to consider the bit lengths of the data very closely in order not to overdesign the system.

QUANTIZATION

These comments on quantization will be necessarily brief although the subject has formed a large study. The aim was to produce an optimized processor design as a function of data sample lengths which would be feasible to construct.

The first consideration was sensitivity. Transmitter power should not have to be uprated just because the system is digital instead of optical. Using computer simulation as our assessment tool, the situation where a single idealized target is detected in a noise background was studied to determine a 'processing loss' for the digital system over the equivalent sampled analogue system. Results showed that a hard limited system (where only the sign of the sample is stored) would incur a processing loss of 3-4 dBs. This loss declines quite rapidly when more bits are used in both the signal and reference quantization schemes. In the case of a two-bit system it is between 1.5 - 2 dBs while for three bits the loss is around 1 dB, valid over a wide range of signal-to-noise ratios.

However, other criteria must also be considered in a complete system design; such properties as linearity, fidelity and dynamic range are important. In order to assess their dependence on quantization, various multiple target situations were simulated. In particular one situation involved a small target in the presence of a larger target where measurements were made on the distortion introduced by the processing. A measure of system dynamic range was defined as the greatest difference between the signal strengths at which the smaller target was reliable detected. The dynamic range of a hard limited system turns out to be around 10 dBs for correlation over lengths of 100 samples. This is due to a pronounced 'capture effect' where on output from the processor the strength of the larger target is enhanced at the expense of the second target.

Good synthetic aperture design requires that for sidelobe reduction of the synthetic beam an amplitude taper should be applied across the aperture. In the case of a hard-limited reference waveform, amplitude control is completely absent, but effective tapers can be introduced as more bits are used. Amplitude tapers employed have usually been approximations to the cosine on a pedestal type of which the Hamming weighting function is an example. The result of using such a taper with multi-bit reference waveforms is greatly reduced sidelobes but with a broadening of the main lobe of the beam. This implied loss of resolution can be recovered by increasing the incoherent integration length as long as there is sufficient real beamwidth available. The addition of the taper increases the dynamic range of a three bit processing scheme to 20 dBs.

It might be added that the dynamic range of the system measured by the range of signal strengths over which a linear relationship between signal-to-noise on input and output holds is dependent on integration length but is greater than 30 dBs for a 3-bit system correlating over 100 samples. This may be increased at a rate of about 6 dBs per additional bit.

COMPUTING FACILITIES

Digital data recorded in the air on magnetic tape in a high density format is converted to standard 9 track magnetic tape using a PDP configuration. In this form it can be handled by an ICL System 4-70 at Marconi Research Laboratories. This computer is suitable for the bulk of the processing task. The airborne program is directed by the Royal Radar Establishment, Malvern, England.

The display of the compressed SLAR imagery is carried out using a Marconi Myriad computer linked to an X2000 Display. Interface with the System 4-70 is by disc. The display is organized in a scan mode to show an array of up to 1024 x 1024 cells, each with intensity modulation to 6 bits (64 levels on a grey scale). The brightness law can be controlled at run time. The scan time for an array of the maximum size is approximately 60 seconds. This speed and flexibility allows the user to display and photographically record, with some interactive control, many stored processed maps. The display still exhibits some significant sources of image degradation, but it has the advantages of being readily available on-site, using a proven interface, and having acceptable scan time. We are considering the use of alternative display techniques at the present time.

An interactive data analysis system having graphical display facilities has been developed using the same Myriad configuration for the handling of large volumes of data. This system has been employed for quick-look analysis of the raw SLAR data prior to processing.

DATA PROCESSING SOFTWARE

The data processing task is characterised by large data volumes (collected in a short time period) and long computer runs because of the unsuitability of the processing to serial computing devices. The computer software is described with reference to the block diagram of Figure 2.

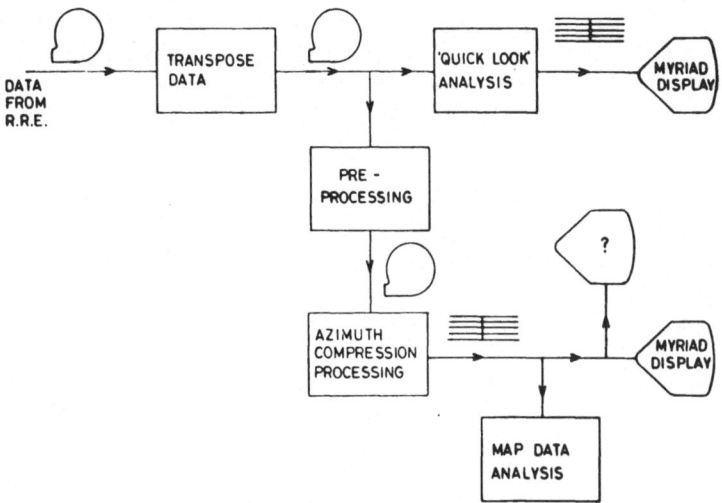

Figure 2 : Data Processing Scheme

The organization of the data at real-time recording is incon-
venient for efficient handling. A data recording consists of an
array of complex data samples with dimensions equal to the number
of pulses by the number of range gates recorded with range the
more rapidly varying. Input to processing software is more
easily organized if data is ordered with pulse varying more rapidly
so that data can be input range gate by range gate. A first
requirement is to transpose the original data array and store but
for large arrays needing storage areas many times greater than
computer core store this is not a trivial operation. The operation
is carried out in two stages, some reordering of the data over
sub-arrays being carried out during the initial PDP data transfer,
to the limit of the PDP store. The transposition can then be
efficiently completed on the System 4-70. Since this operation
need be carried out only once per data recording it is allocated
the sole function of one program.

Quick-look analysis can be performed on the raw data at this
stage using the in-house data analysis facility called MIDAS
(Myriad Interactive Data Analysis System).

In situations where the radar returns are oversampled (pulse
repetition frequency too high) with respect to the desired final
resolution of the processed image there exists an opportunity for
bandwidth reduction which will result in a time saving at the
azimuth compression stage. A separate software module provides
the means to pre-process the data, usually involving presummation
over a number of consecutive samples within a range gate with,
possibly, a re-quantization of the result. Other uses may be

found for the pre-processor such as a crude removal of dc offset
and frequency shifting of the return.   The output is to tape and
again the operation is isolated from the processing since it may
be intended to derive several pre-processed tapes from a single
section of raw data with each tape giving rise to many processing
runs.

The software processor accepts input from tapes created by
the pre-processor.   Many options are provided in the processing
including the more important study techniques such as :-

- variation of bit lengths used at all stages of the processing,

- sub-aperturing, where more than one synthetic aperture is
  formed within the real beamwidth, the images derived being
  added incoherently,

- full quadrature or non-quadrature processing,

- processing with frequency offset references.

The output from the program is stored on disc for interface with
the Myriad display facility.

Having derived a processed map, it is very necessary to be
able to obtain qualitative measurements of image characteristics
since 'quality' is often difficult to assess in purely visual
terms.   A small analysis program is in a progressive state of
development to fill this role.   It is proving particularly valuable in a comparitive mode for comparing the results of similarly
processed areas of the same raw data by some of the alternative
schemes mentioned above.

REAL DATA RECORDING

The initial recordings of radar data have been made under
restricted operating conditions.   In the ideal SLAR system the
radar aperture is gyro-stabilised in all planes with stabilisation
in the yaw plane being particularly important.   However, during
the first series of equipment flight trials and data recording the
radar aperture was fixed to the airframe.   Therefore an accurate
system relied on the pilot maintaining a straight and consistent
heading in still air conditions.   The precise pointing direction
of the aperture was imperfectly known.   No cross-track motion
correction for the additional aircraft induced Doppler frequency
was applied to the data.

This lack of fine adjustment has invalidated some of the data
recorded where yaw was excessive or rapidly varying.   The task of

processing such data is much less straightforward than that where
the data has been obtained in the ideal situation.

The data which has given rise to the digitally processed
imagery of the following section is X-band radar data.   The radar
pulse has a 25 MHz bandwidth and the compressed return has been
sampled at intervals of 40 nsec, or 6 m. in range.   Polarization
is horizontal.   The pulse repetition frequency is tied to the
aircraft velocity and pulses are transmitted at 0.3 m. intervals
along track.   The radar aperture is 1.8 m. in length.

Quantized radar returns in both channels are recorded to
3-bit accuracy over approximately 960 range gates covering a
swath 5760 m. wide (slant range) from a minimum range of 1900 m.
The aircraft was flying at a height of approximately 300 m. above
the terrain during recording.

COMPUTER PROCESSING

A section of the data of length 4096 samples representing
an along track translation of 1250 m. was chosen for computer
processing.   A final resolution in the along track dimension of
6 m. was considered to be an appropriate processing goal since
there would then be a close correspondence between this and the
range resolution.   The display cell was chosen to have the same
dimensions as a resolution cell.   Quantization is carried out to
three-bit accuracy.

Prior to processing samples were presummed in pairs to give
an effective 0.6 m. sample interval into the processor.

Figure 3 shows the processed map over an array of about 200
cells along track by 960 cells in range.   One aperture has been
employed in the processing centred about zero Doppler frequency.
The brightness characteristic in the final map is linear in amplitude.   The result is a grainy image.   For comparison, Figure 4
shows the same data processed using two sub-apertures to form the
final image.   A noticeable improvement in the image is evident
due to the smoothing effect of the incoherent integration.

The following comments are confined to Figure 4.   Large scale
image degradation is immediately noticed.   The very prominent
dark band at the left of the image is caused by aircraft yaw or
cross-track motion which has the effect of squinting the real beam
out of the azimuthal angle within which the synthetic apertures are
formed.   Yaw motion alone will result in loss of gain whereas
cross-track motion which is uncorrected will cause, in addition,
some image distortion consisting of alternate expansion and contrac-
tion of sections of the map.   Less prominent banding visible on the
image will have originated from the same source but of lesser degree.

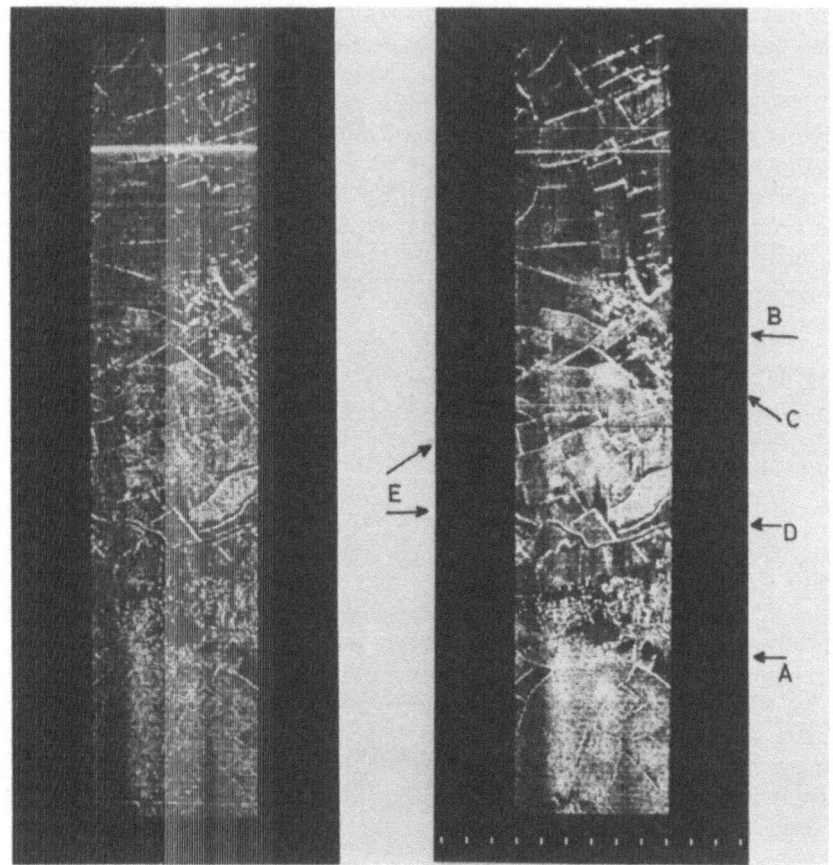

Figure 3   Figure 4

Some of the degradation caused by the display is also immediately evident. This includes occasional brightening of scan lines
which occurs irregularly. The broad, bright band of scan lines
occupying a group of range gates towards the end of the swath is
caused by the processing of added test data. Less prominent
display instabilities cause a slight 'washboard' effect along the
swath. There is slight misplacement of scan lines caused by
rounding errors in the display. All the faults can be tolerated
in the role for which the display is intended. The missing line
denotes the centre of the display.

Two further examples obtained by processing data from this
same area are shown in Figures 5a and 5b. The range covered is
precisely range gates 200 to 680 of the processed image in Figures
3 and 4. In both cases, the processing carried out synthesizes a
squinted beam with the remaining parameters unchanged. The
squint in Figure 5a is +0.29° and in Figure 5b -0.29°. The extent
of the aircraft motion is now more clearly shown, and the two images

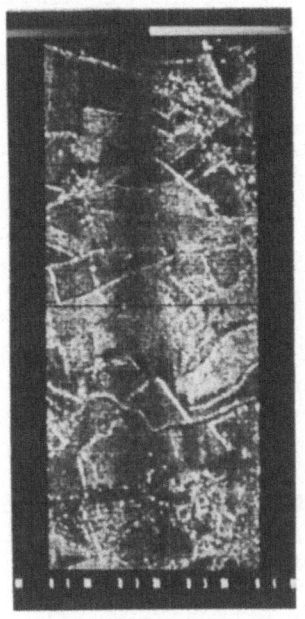
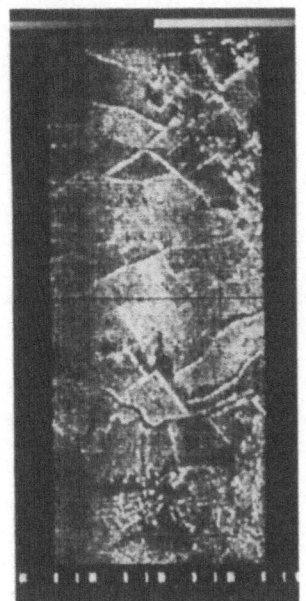

Figure 5a   Figure 5b

are complementary in that a dark yaw band in one corresponds to a
better illuminated region in the other.   The distortion introduced
by the squinted aperture has not been corrected in the image.
Where sub-aperturing is employed this correction must be applied
prior to the incoherent integration stage.   The images of Figures
5a and 5b are similar to those obtained from each of the two sub-
apertures of Figure 4 although in this case the aperture squints
were approximately ± 0.22°.

IMAGE INTERPRETATION

     The terrain region processed and displayed in Figures 3-5 is
a predominantly agricultural region near Pershore, Worcs.   The
data was gathered in winter at a time when ground conditions were
very wet.   The region is generally flat but two hills peak at
about 30 m above the terrain level at A and B.   Hill A causes
considerable shadowing of the built up area down range, and the
mottled return is caused by rooftops etc. at grazing incidence to
the radar.   Some structure can be determined within this area.
The hill at B with its gentler slopes has less effect on the image
although trees growing on its crest cause considerable shadowing.

     A village extends into the swath at C and the return is again
coarsely mottled.   The village includes several farms with typical
buildings.

The River Avon crosses the swath at D and the far bank sets up an almost continuous bright return. The bank may be differentiated from hedges with their accompanying shadows. The river is determined by an area of low radar return.

At a resolution of 6 m. it is possible to distinguish single hedges from hedge-flanked tracks since in the latter feature the hedge image is doubled. Two such tracks are evident (E) in both cases the tracks being about 10 m. wide with hedges about 1 m. high. The road which cuts the swath at top-right also demonstrates a similar but more distinct doubled return. Hedge separation here is about 30 m.

Shadows caused by trees and buildings are very clearly defined at mid-ranges. Measurements made on shadow lengths lead to accurate predictions of tree height.

The informal ground truth survey that provided the basis for the above comments was not carried out close enough to the time of radar recording to enable firm statements to be made concerning the correspondence between crop type and image. The brightest field returns are however caused by beet and cabbages while ploughed fields return weakly.

SUMMARY

An on-going study into the digital processing of side looking airborne radar returns has been described with comments on hardware and associated computer software. First recordings of data, obtained under restricted operational conditions, indicate the potential of the digital processing technique.

An initial recording described in this contribution has been processed in a variety of ways generally to 6 m. resolution. A ground truth survey has indicated the usefulness of the information derived from the processed image. The flexibility offered by the general purpose computer has been found particularly useful in evaluating performance under different processing systems.

ACKNOWLEDGEMENTS

This work was carried out with the support of the Procurement Executive of the Ministry of Defence. The author would like to thank Dr. K. Corless and Mr. L. Saunders for their cooperation in the supply of data and the Technical Director of GEC-Marconi Electronics Limited for permission to publish this paper.

REFERENCES

1. J.H. BLYTHE, MRS. K. EDGCOMBE, A. RYLEY, W.D. WORTHY,
   "Signal Filtering using Hard-Limited Digital Processing
   Part I - General Description and Performance with White-Noise
   Background", Proceedings IEE, Vol. 177, pp. 1768-1776,
   September 1970.

2. J.H. BLYTHE, MRS. K. EDGCOMBE, A. RYLEY, W.D. WORTHY,
   "Signal Filtering using Hard-Limited Digital Processing
   Part II - Performance with a Single Target in a Coloured
   Noise Background", Proceedings IEE, Vol. 180, pp. 825-832,
   August 1973.

3. W.M. BROWN, "Synthetic Aperture Radar", IEEE Trans. Aerosp.
   Electron. Sys. Vol. AES-3, pp. 217-229, March 1967.

SYNTHETIC APERTURE RADAR

S. Marder

Environmental Research Institute of Michigan,
Arlington, Virginia, U.S.A.

ABSTRACT. The previous discussions of this A.S.I. of target
scattering and effects of the atmosphere upon microwave propagation
are extended to the problem of imaging by synthetic aperture radar
(SAR). The foundation for discussion of SAR system performance
has been given by Corless [1]. A SAR system generates an image of
the reflectivity distribution of the surface being observed, as it
appears at the wavelength employed by the radar. Differences in
gross reflectivity, shadowing, texture, and shape of the illuminated terrain features are analyzed to yield information required
for specific applications. These may include mapping, agricultural
land use, ice surveillance, and measurement of ocean waves. The
implications of terrain scattering for imaging radar system performance and initial applications of a dual-wavelength, dual-polarization SAR are discussed. The influence of atmospheric turbulence
on SAR is then examined.

1.   IMPLICATIONS OF TARGET SCATTERING

The advantages of radar with respect to aerial photography in
its independence from an uncontrolled and variable source of illumination and its ability to penetrate adverse weather have often
been called to attention. Associated with the very factors which
produce these advantages are off-setting disadvantages: the re-
placement of the sun is costly and complex; penetration of the atmosphere requires the use of long wavelengths which limit the resolution attainable. Additionally, the use of both coherent illumination and long wavelengths entail target scattering effects which
often are unwholesome. Some implications of these effects for target imaging are now addressed.

Jeske (ed.), Atmospheric Effects on Radar Target Identification and Imaging, 193–217.
All Rights Reserved. Copyright © 1976 by D. Reidel Publishing Company, Dordrecht-Holland.

Target classification is capable of infinite refinement depending upon the level of detail of information which is of interest. Here only a few major categories are distinguished. These are: Mobile man-made objects, Terrain (including man-made structures), Urban areas, and Oceans. Only the first two will be considered explicitly.

The vehicles, aircraft, ships, weapons, farming and constuction equipment which constitute the first class are composed of surfaces which are quite smooth at microwave frequencies and of a relatively small number of major localized scattering centers due to discontinuities and surface intersections. When illuminated by coherent, directed, microwave radiation, the scattering is highly specular. As an example, the scattering patterns of aircraft show intensity variations of 20 dB over angles as small as $1/3^{\circ}$. Depending upon orientation, visually prominent surfaces such as the fuselage or wings may produce no discernible signal, whereas an antenna or wheel well may make a major contribution to the backscatter. The dynamic range of the signals is extremely high, and variations in radar flight path or target orientation angle can cause significant variations in the target signature. As a result, the image is said to be unstable.

The case of terrain is quite different. Though some terrain features have large cross sections, and man-made structures such as buildings, bridges, power lines, and fences can be highly specular, the general case is otherwise. The average reflectivity per projected unit area typically lies in the range from -10 to -30 dB, though it can fall to -40 dB for especially smooth terrain surfaces. Each resolution cell tends to be composed of large numbers of independent scatterers, none of which predominate. For an individual resolution cell, the vector sum of these returns can vary significantly on successive flight passes even with controlled flight conditions, though on a macroscopic scale the imagery of areas the order of 100 resolution cells or more in size appears stable. The variability of the return from an individual cell is commonly attributed to the sensitivity of the cross section to variations in aspect and depression angle which alter the phase relations of the component scatterers.

Considering only these two quite general target classes, one then finds that SAR systems must cope with three kinds of image problems: adequate resolution, wide dynamic range, and reproducibility of target returns (i.e., image stability).

Table 1 illustrates the problem of radar dynamic range and contrasts it with that of photography. The examples for radar in the lower two lines apply to a flat plate of the indicated size at normal incidence compared to a horizontal terrain patch of grass or concrete. Whereas 17 dB is as much dynamic range as

TABLE 1.   DYNAMIC RANGE REQUIREMENTS

SYSTEM	CONDITIONS	REQUIRED DYNAMIC RANGE (DB)[*]
Photographic	Sunlit Snow/Asphalt in Shadow Ground Level	(>30)
Photographic	Cities and Towns/Average Haze 4000 ft. Altitude Ground	10 (17)
Photographic	Satellite Altitudes	7-10
Radar	Flat Plate/Grass Contrast 1 ft. Resolution (X-Band)	54
Radar	Flat Plate/Concrete Contrast 10 ft. Resolution (X-Band)	88

[*] Values of photographic contrast are taken from Photographic Considerations For Aerospace, edited by Hall and Howell, Itek Corp., 1965 (Numbers in parentheses are maximum values.)

aerial photography is called upon to handle, the radar dynamic range could exceed this by 70 dB. Since photography and SAR both utilize film recording, which typically accommodates a dynamic range of 24 dB, this is quite satisfactory for photography, but grossly inadequate for radar. The last two lines of the table illustrate two effects. Primary is the potential reduction of dynamic range requirements if resolution could be improved. For the scattering situation considered, a factor of ten improvement in linear resolution would provide 20 dB of dynamic range improvement. The second effect is the reduction in dynamic range obtained by moderating the demands placed upon the system, as one raises the minimum signal requirement from smooth concrete to grass, a difference of 14 dB. The two effects combine to provide a reduction in required dynamic range of 34 dB.

To alleviate the problems of both dynamic range and target instability, it has been suggested that diversity be added to imaging radars. By diversity is meant that the radar obtains more than one independent view of each resolution cell. The proposed methods of providing diversity to a SAR are shown in Table 2. These are through the use of either wider bandwidth or aspect angle than required for the desired resolution or through the use of additional polarization channels. In Table 2, Δf and $\Delta \theta$ refer to the bandwidth and aspect angle employed, while W and β are the bandwidth and aspect angle required for the design resolution in range and azimuth.

TABLE 2. FORMS OF DIVERSITY

Frequency

   a) "Small" Diversity: $\Delta f \approx W$

   b) "Large" Diversity: $\Delta f \gg W$

Aspect Angle

   a) "Small" Diversity: $\Delta\theta \approx \beta = \dfrac{\lambda}{2\rho_x}$

   b) "Large" Diversity: $\Delta\theta \gg \dfrac{\lambda}{2\rho_x}$

Polarization

   a) Transmit Linear, Receive Parallel and Cross

   b) Transmit Circular, Receive Left- and Right-Circular

The utilization of diversity information through the noncoherent addition of images obtained at different frequencies or at different target aspects has been discussed by Mitchell [2]. He has performed a mathematical simulation of the effect of noncoherent addition using a simple statistical model composed of about 5000 scatterers uniformly spaced over the shape of a cross, each arm of which was 32 scatterers wide. A random phase was generated for each scatterer, and several target amplitude distributions with equal ensemble means were investigated, assuming a Gaussian response function for the radar (i.e., ignoring sidelobes). Without the diversity of noncoherent averaging, resolution of about one-tenth the shortest dimension of the target was needed for reliable recognition, and this number was highly sensitive to the detail of information that was desired about the target. When independent images were noncoherently summed, the summation of four images qualitatively compensated for a degradation by a factor of two in resolution over variations in resolution from one-sixteenth to one-half the shortest target dimension.

In summary, the use of coherent, microwave radiation creates imaging problems due to resolution limitations, dynamic range requirements, and image fluctuations. To some degree, the latter problems can be alleviated by diversity, but this requires either additional bandwidth, aspect angle, or polarization channels. The requirements for atmospheric homogeneity imposed by the addition of diversity to a SAR system still need consideration. One way of utilizing diversity (i.e., through noncoherent averaging)

has been discussed. An alternative will be considered in the
next section.

2. SYNTHETIC APERTURE RADAR APPLICATIONS

Among the world's population of radars, synthetic aperture
systems are a rare and exotic species. Among the myriad
of air-traffic control, weather, navigation, Doppler, police, and
intrusion detection radars in use, there is--to my knowledge--
only one commercial synthetic aperture system. This is an X-band
mapping radar built by Goodyear Aerospace Corporation. Since
experience with SAR imagery is limited outside a small technical
community, a short presentation of SAR applications may lend a
greater air of reality to subsequent discussions of atmospheric
effects. In discussing these applications, imagery obtained with
an Environmental Research Institute of Michigan (ERIM) dual-
frequency, dual-polarization radar will be relied upon. This
imagery has the additional virtue of illustrating some effects
of frequency and polarization diversity for the imaging of terrain.

The parameters of this system are shown in Table 3. Of primary interest are the wavelengths (3 and 25 cm), polarization
(either horizontal or vertical, selectable for transmission and
both horizontal and vertical received), resolution (3 x 3 m at

TABLE 3. THE ERIM MULTICHANNEL RADAR PARAMETERS

Parameter	X-Band	L-Band
Center Frequency	9.450 GHz	1.315 GHz
Resolution	3 x 3 m	6 x 6 m
Transmitter (Peak)	1.2 kw @ 2% duty cycle	6 kw @ 2% duty cycle
Antenna Gain	28 dB	16.5 dB
Antenna Beamwidth	1.1°	7°
Polarization Isolation	>23 dB	>19 dB
Depression Angle	0° to 90°	0° to 90°
Maximum Range	21 Km	21 Km
Number of Spots/Scan	8,000	4,000
Film Capacity	30 m	30 m

3 cm and 6 x 6 m at 25 cm), and range (21 km). The system, which has been operating since April, 1973, is unique in that four channels of data are obtained simultaneously. Processing of the signal histories to obtain images is done at the laboratory after the flight. In most applications, the data have been processed to a resolution of 9.2 m.

Potential remote sensing applications of SAR are listed in Table 4. Additional applications are possible, and only a few on the list will be discussed here. Mapping stands foremost since this use has already attained commercial status with the Goodyear X-band system previously cited. This system has mapped portions of South America and informal reports of the results are quite favorable.

As an example of the mapping capability of SAR, X-band images (3 cm wavelength) of a portion of Florida are shown in Figure 1. The land-water boundaries are very well defined. The following key denotes additional information which can be derived from imagery of this kind.

> 1.1 Residential, mainly single-family homes
> 1.2 Commercial and Services
> 1.3 Industrial
> 1.5 Transportation, Communications and Utilities
> 1.6 Institutional
> 1.7 Strip and Clustered Settlement
> 1.9 Open and Other
> --- Single-track railroad

TABLE 4. REMOTE SENSING APPLICATIONS OF
MICROWAVE IMAGING SYSTEMS

1. Mapping Remote Areas

2. Detection of Geological Features and Minerals

3. Agricultural Survey

4. Classification of Ice

5. Environmental Monitoring

6. Detection of Pollution

7. Measuring Sea State

8. Monitoring Ship Traffic

9. Urban Development

10. Detecting Lunar Subsurface Features

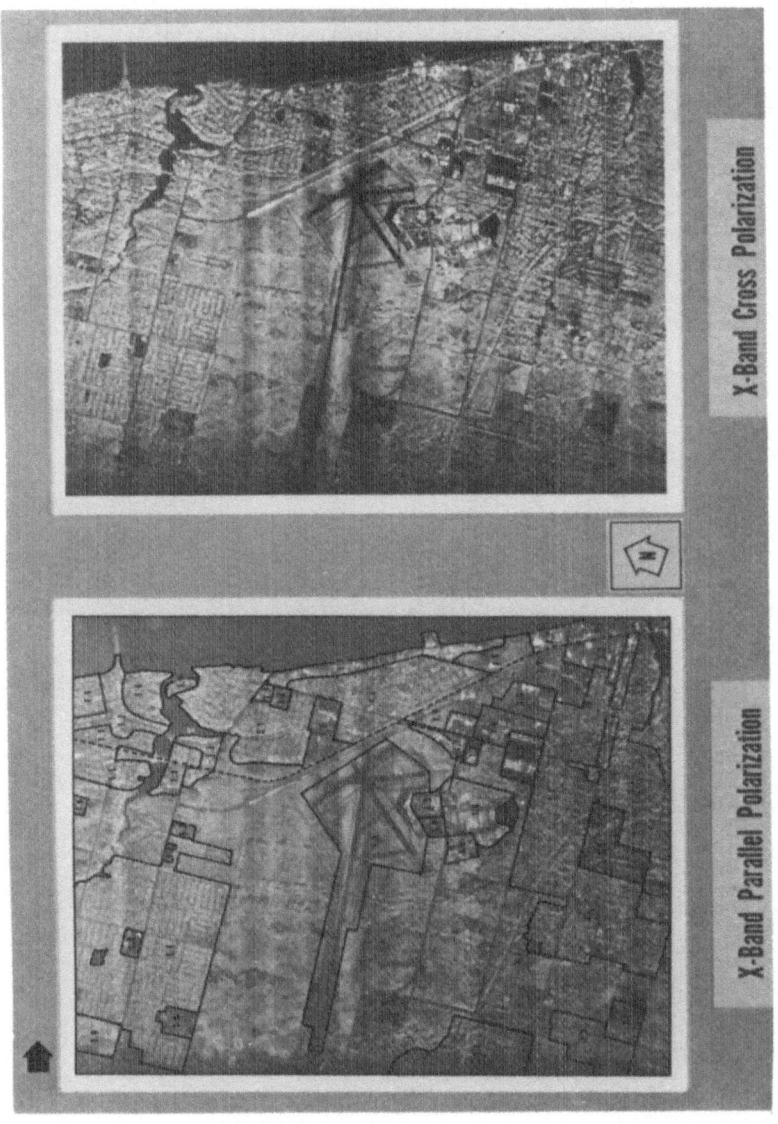

Fig. 1. Dual-polarization imagery of Melbourne, Florida
(Resolution 9.2 x 9.2 m)

An example of SAR imagery for use in agricultural assessment
is shown in Figure 2. The key for this imagery is given in Figure
3. Useful discriminations among the fields can be made from this
imagery [3.] Low corn and soybean stubble can almost always be
distinguished from their standing counterparts. Standing corn
can generally be distinguished from standing soybeans, and winter
wheat can be distinguished from the two. The major differences
in appearance of the different crop types are primarily wavelength
rather than polarization dependent. However, the use of all four
channels is sometimes necessary to differentiate positively between certain crop types.

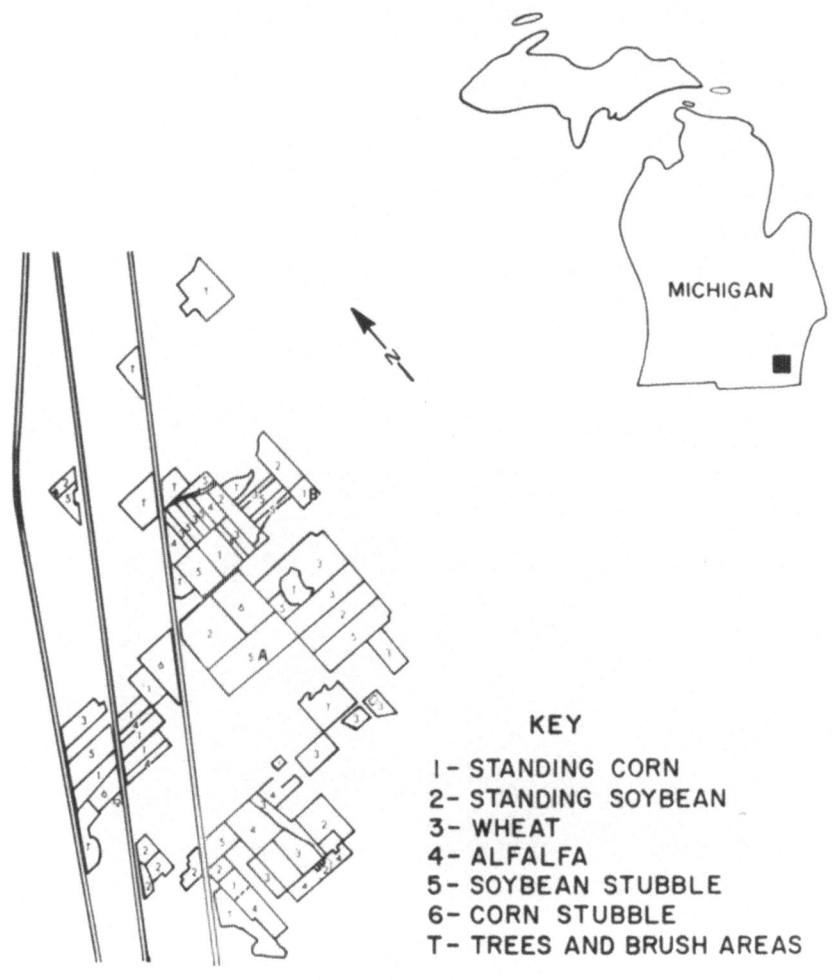

KEY

1- STANDING CORN
2- STANDING SOYBEAN
3- WHEAT
4- ALFALFA
5- SOYBEAN STUBBLE
6- CORN STUBBLE
T- TREES AND BRUSH AREAS

Fig. 3. Key for imagery of Fig. 2

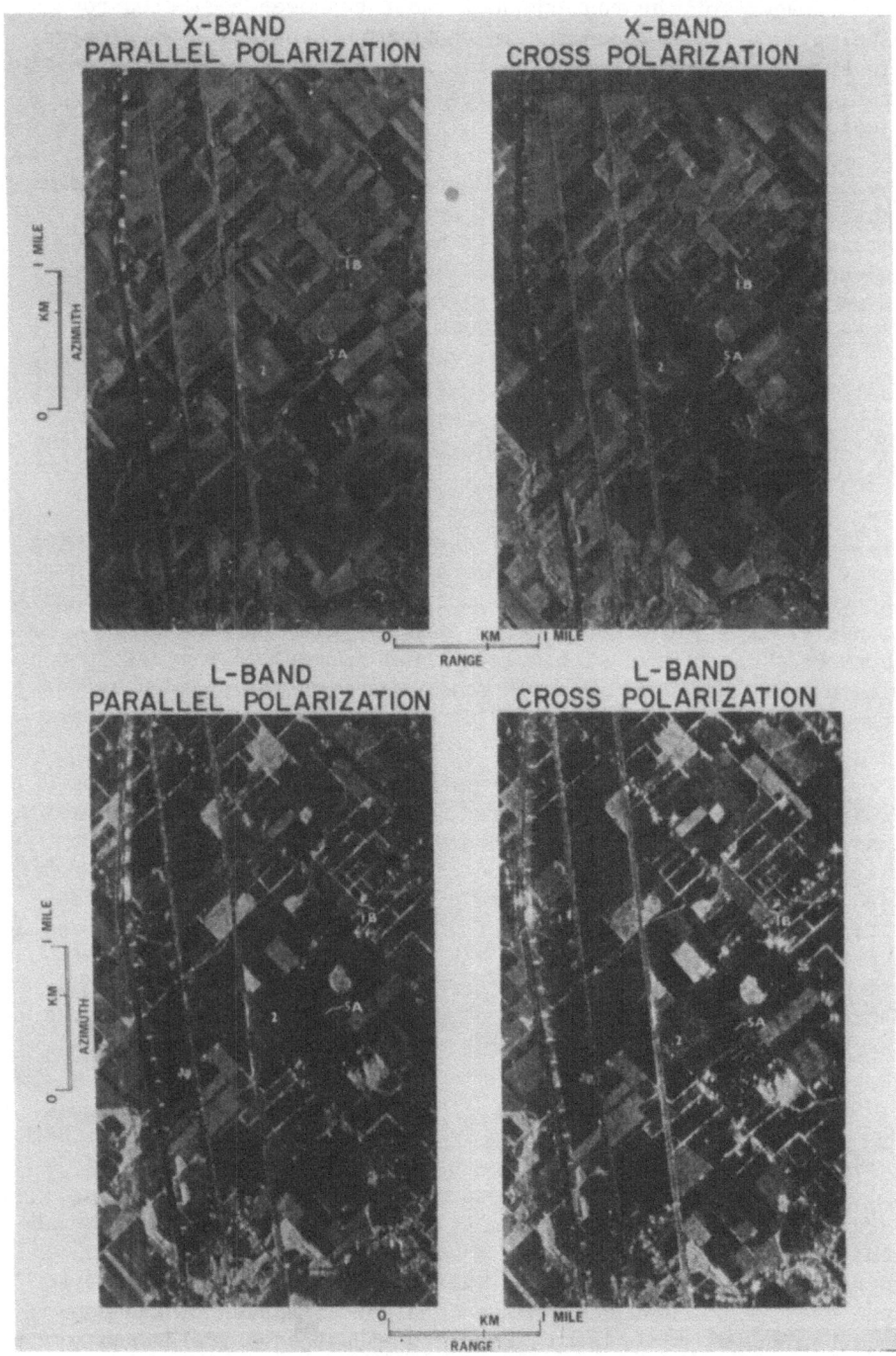

Fig. 2. Multichannel imagery of site in Southeastern Michigan
(Resolution of 3 m at X-band and 6 m at L-band)

Imagery of the four-channel radar has been collected for analysis of fresh water ice in the Great Lakes [4]. The imagery is shown in Figure 4. The analysis is directed toward the utilization of radar for the extension of the Great Lakes shipping season, clearly of importance. In the imagery shown, there is no open water, as the area is entirely ice-covered. A number of features in the image were evident to analysts. These included: the shoreline, Isle Parisienne, areas of new and brash ice, pressure ridges, and tracks left by ships. Attention is being given to the question of whether the depth of ice can be determined from differences in penetration at the two frequencies.

Imagery has been collected to assess the use of SAR for surveillance of strip mines [5]. With the increase in the price of coal, there has been an increase in strip mining. Both the operation and the reclamation of mines come under government regulation in various parts of the United States. The use of SAR in surveillance can insure that many of these regulations are followed. Figure 5 shows four-channel imagery of a strip mining area in Kentucky. Three mining categories were recognized: active mine areas, reclaimed mines, and unreclaimed (or "orphan") mines. Sixty out of 63 mines present within the area were correctly identified with multichannel imagery of 9 m resolution. The primary means of identification was with the unaided eye, though a densitometer was used in the study of reclaimed areas.

For the last example, the use of SAR imagery in detection of soil moisture will be considered. In this experiment, wet fields had more than 25% moisture and dry fields less than 10%. The analysis was more sophisticated since the imagery was digitized and subjected to a multivariate discriminant analysis. In addition, noncoherent averaging of 25 and 100 resolution cells was performed and then analyzed. The table in the upper right of Figure 6 shows that over 90% correct detection can be obtained after averaging. The cluster diagram at bottom right shows the separation of three types of fields obtained with only the two L-band channels.

As a final comment on applications, it should be noted that NASA has included a 22 cm wavelength synthetic aperture radar of 25 m resolution operating at steep depression angle in its program for SEASAT-A. Objectives are (i) the imaging of sea surface and ice, (ii) the detection and measurement of the wavelength and direction of ocean waves, and (iii) detection of slicks, current patterns, and icebergs. This adds a new dimension to the considerations of atmospheric effects on radar propagation and introduces problems of the ionosphere.

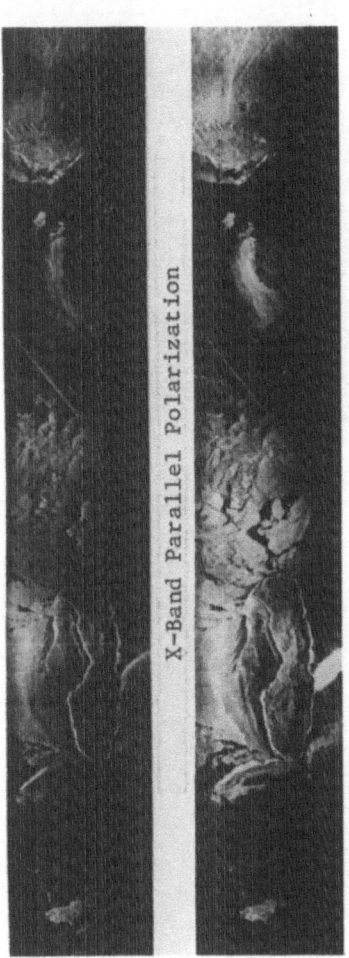

X-Band Parallel Polarization

X-Band Cross Polarization

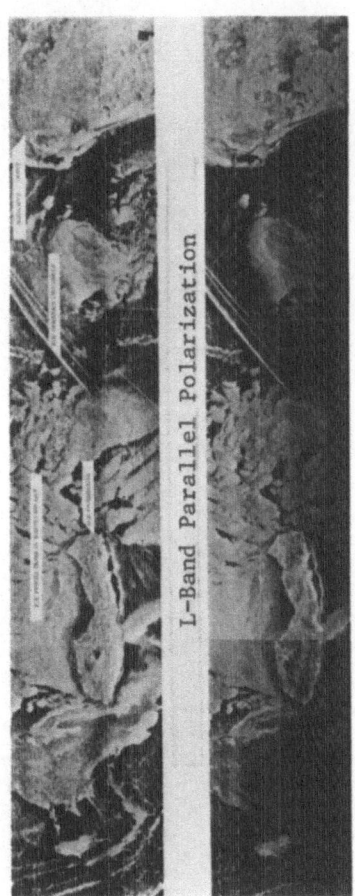

L-Band Parallel Polarization

L-Band Cross Polarization

Fig. 4. Multichannel imagery of Whitefish Bay, Michigan
(Resolution of 9.2 m)

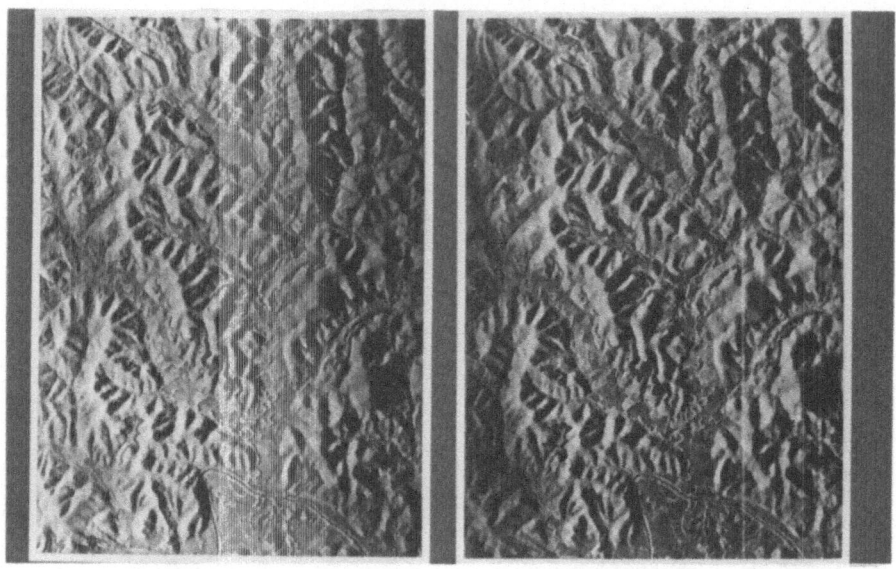

X-Band Parallel Polarization   X-Band Cross Polarization
            3 x 3 M   3 x 3 M

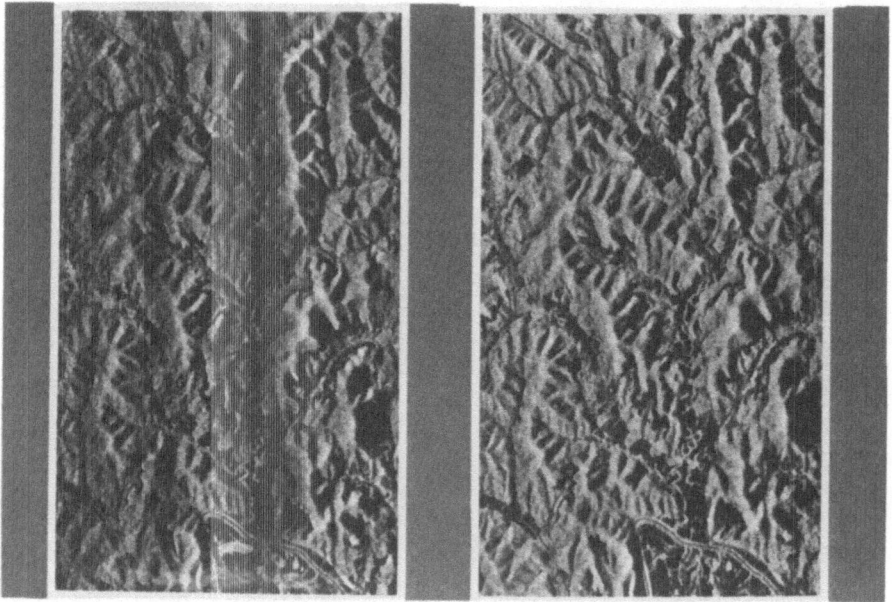

L-Band Parallel Polarization   L-Band Cross Polarization
            6 x 6 M   6 x 6 M
Fig. 5.   Multichannel imagery of strip mining region
                in Southeastern Kentucky

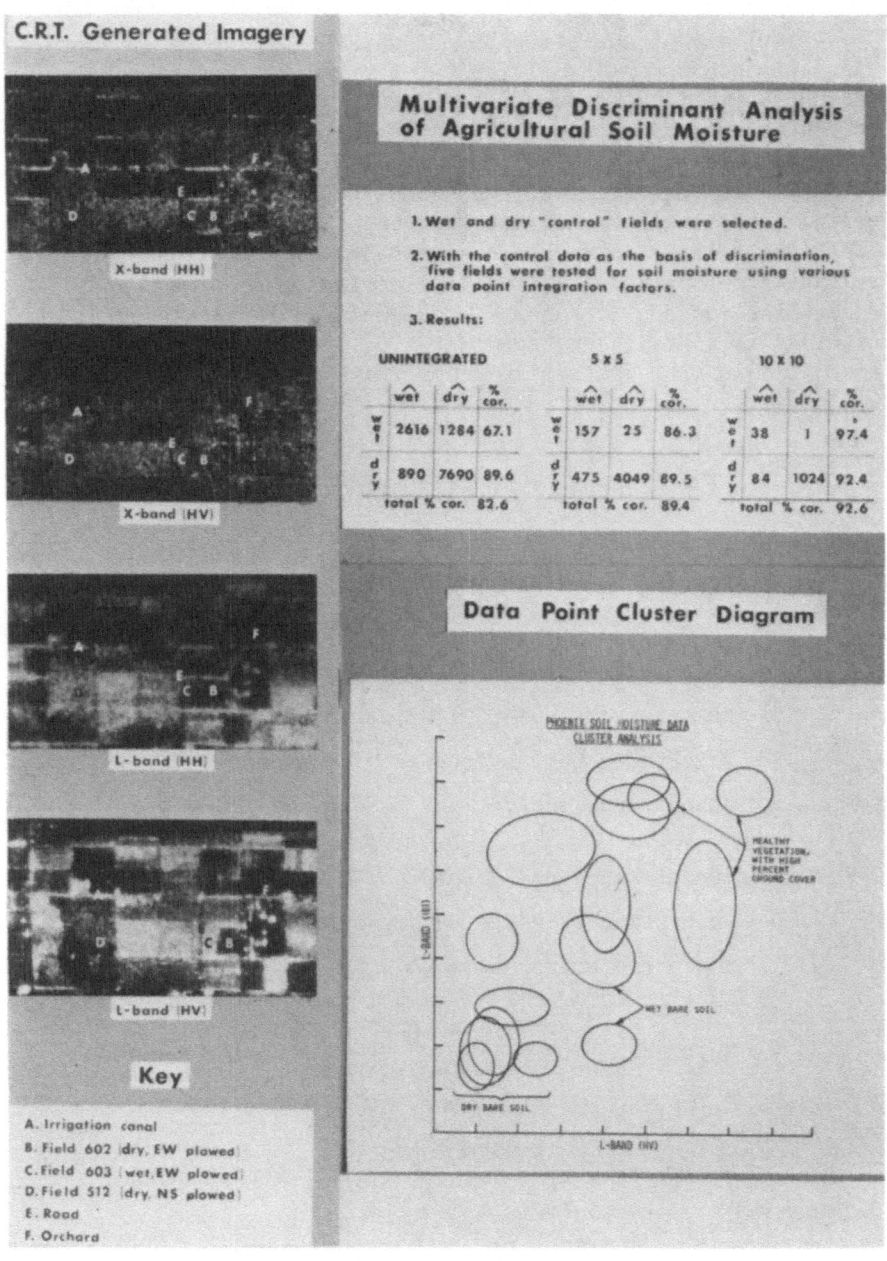

Fig. 6.   Multichannel imagery of fields in the
vicinity of Phoenix, Arizona

3. ATMOSPHERIC EFFECTS ON SAR

The effect of the atmosphere upon SAR imaging is strongly dependent upon the radar altitude. For radars operating below the ionosphere (i.e., roughly below 90 ± 10 km), only tropospheric effects are of importance. Above this altitude the influence of the ionosphere must also be considered. If we restrict our consideration to the effect of the troposphere on microwave systems, we find that the phenomena can be grouped into three categories, according to their influence on the radar signal. These effects are listed in Table 5. Of this collection of problems, one is of peculiar concern to radars which employ a large aperture, of which SAR is a prominent representative. This one is the random fluctuation of the refractive index which is induced by turbulence, and it will be the next topic of discussion.

TABLE 5.   TROPOSPHERIC EFFECTS ON SAR SIGNALS

o Induction of Phase and Time Delay Errors

    -- Phase shift and time delay due to constant n

    -- Refraction due to vertical gradient of n

    -- Phase errors due to fluctuation of n

    -- Dispersion due to frequency dependent
       absorption, scattering, or diffraction

o Induction of Amplitude Errors

    -- Absorption by water vapor, oxygen, and
       condensed water

    -- Diffraction

    -- Multipath

o Addition of Energy

    -- Backscatter by clouds, fog, and precipitation

n = refractive index of troposphere

The principal treatment of the effect of turbulence upon SAR phase errors has been reported in a triad of papers in the September, 1970 issue of the IEEE Transactions on Aerospace and Electronic Systems [6-8]. These papers lead to an analytical estimate of the influence of phase errors on SAR resolution. A number of statistics, other than resolution, which describe the perturbed beam are also of interest, as has been pointed out in earlier papers by Greene and Moller [9] and by Harger and Crimmins [10]. These latter papers, however, assume wide-sense stationarity of the phase error process (known to be invalid), and require Monte Carlo techniques to treat the situation of large variance. Other than noting the statistics discussed by Green and Moller (Table 6), they will not be considered here.

We now turn our attention to the program of the principal references. This is summarized below.

a) Extend the work of Tatarski to obtain a phase structure function of the appropriate form [6].

b) Assuming Gaussian statistics for the phase error and a phase structure function of the form $D_\phi(\mu) = b^n \mu^n$, evaluate the equivalent rectangle resolution and the optimum aperture length for the perturbed SAR response [8].

c) Use data from the Maui experiments [7] to verify the form and determine the magnitude of $D_\phi(\mu)$ [6].

As can be seen, there are three major steps.

TABLE 6.   PERTURBED BEAM STATISTICS

Expected Pattern

RMS Beam Shift

Expected Beam Spreading

Expected Peak Gain Reduction

RMS Azimuth Beam Rotation (as a function of radar platform forward motion)

Expected Ratio of Highest Sidelobe to Main Lobe Gain

Expected Ratio of Main Lobe to Total Sidelobe Power

3.1 Extension of phase structure function to SAR

The extension of Tatarski's work to the SAR problem follows. Turbulence is assumed to be homogeneous and isotropic within some region of size L_o termed the "outer scale". The smallest eddies are of dimension ℓ_o termed the "inner scale," and $\ell_o \ll L_o$. The function $n(\vec{r}, t)$ is a nonstationary random function, but the difference $n(\vec{r}_1, t) - n(\vec{r}_2, t)$ is stationary if $|\vec{r}_2 - \vec{r}_1| < L_o$. The refractive index structure function D_n is defined by

$$D_n(\vec{r}_1, \vec{r}_2) \equiv E\left\{[n(\vec{r}_1) - n(\vec{r}_2)]^2\right\} \tag{3.1}$$

If $r \ll L_o$, where $r \equiv |\vec{r}_1 - \vec{r}_2|$ then $D_n(\vec{r}_1, \vec{r}_2)$ takes the form $D_n(r)$; and for $\ell_o \ll r \ll L_o$, $D_n(r)$ can be shown to be

$$D_n(r) = C_n^2 \, r^{2/3} \tag{3.2}$$

where the refractive index structure constant C_n is given by

$$C_n^2 = C^2 \, L_o^{4/3} \, M^2 \tag{3.3}$$

where

C = constant

M = function of local pressure, temperature, specific humidity (q), potential temperature (H), and dH/dh and dq/dh

For a plane wave propagating a distance R through the turbulent medium, the <u>transverse structure function of the instantaneous phase distribution</u> (or phase structure function) is defined as

$$D_\phi(\mu) \equiv E\left\{[\phi(x + \mu) - \phi(x)]^2\right\} \tag{3.4}$$

where x and x + μ are points in a plane transverse to the direction of propagation.

If $\lambda \ll \ell_o$, then

$$D_\phi(\mu) = a \, k^2 R \, C_n^2 \, \mu^{5/3} \tag{3.5}$$

where

$$k = 2\pi/\lambda$$

$$a = \begin{cases} 2.91 \\ 1.46 \end{cases} \quad \text{for} \quad \begin{array}{l} \mu \gtrsim \sqrt{\lambda R} \\ \ell_o \ll \mu \ll \sqrt{\lambda R} \end{array}$$

If C_n varies slowly along the propagation path, then $D_\phi(\mu)$ may be modified by replacing $C_n^2 R$ by

$$\int_0^R C_n^2(\ell) \, d\ell$$

For SAR, the geometry corresponds to the case of a spherical wave rather than plane (Figure 7). Each region of thickness $(\Delta\ell)_j$ will make a contribution of

$$\left\{ a \, k^2 \, \mu_j^{5/3} \, \left[C_n^2(\ell) \right]_j \, (\Delta\ell_j) \right\} \quad \text{to } D_\phi(\mu).$$

Summing, setting $\mu_j = \theta\ell_j$ and approximating by the integral leads to

$$D_\phi(\mu) = a \, k^2 \, \theta^{5/3} \int_0^R \, C_n^2(\ell) \, \ell^{5/3} \, d\ell \tag{3.6}$$

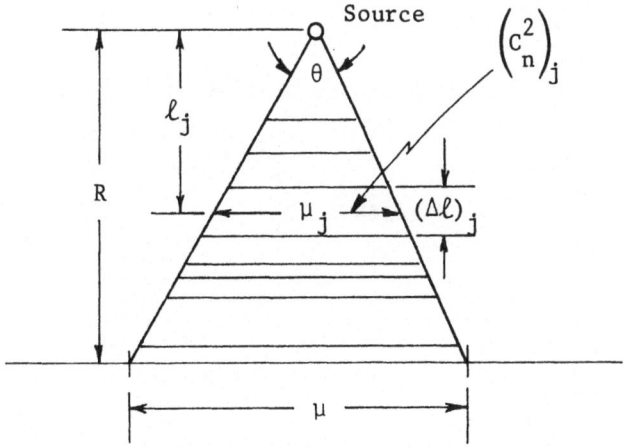

Fig. 7. Geometry for spherical wave case

Setting $\theta = \mu/R$ and $\ell = h \csc \psi$, where ψ is the elevation angle then gives

$$D_\phi(\mu) = a \ k^2 (\mu/R)^{5/3} \ (\csc\psi)^{8/3} \int_0^H c_n^2(h) \ h^{5/3} \ dh \qquad (3.7)$$

It is to be noted that ℓ_o is thought to be of the order of one to ten mm near the surface, which makes $\lambda << \ell_o$ a dubious assumption at microwave frequencies. L_o is thought to range from 100 to 1000 meters away from the surface of the earth. The integral of C_n^2 in Eq. (3.7), which contains the effect of atmospheric parameters is denoted by I. For $\mu << R$, the phase difference $[\phi(x + \mu) - \phi(x)]$ is Gaussian distributed, though $\phi(x)$ is not. For round-trip propagation in the case of radar, the phase structure function is taken to be four times the one-way structure function derived due to the high degree of correlation of the atmosphere for the transmitted and return paths.

3.2   Effects of phase errors on resolution

The formulation of the response of a synthetic aperture radar system has a number of different forms. The form given by Heimiller [11] is most convenient. In Appendix I of Ref. 11, the error-free two-way gain pattern $G(x)$ for a focused synthetic array can be written as

$$G(x) = \left| \sum_{n=-N/2}^{N/2} W_n \ \exp\left(j \ \frac{4\pi V T n x \ \cos\alpha}{\lambda R}\right) \right|^2 \qquad (3.8)$$

The integral approximation can be written as

$$G(x) = \left| \int_{-L/2}^{L/2} f(u) \ \exp(jaux) \ du \right|^2 \qquad (3.9)$$

where

$$a = \left(\frac{4\pi}{\lambda R}\right) \cos\alpha \qquad (\alpha = \text{squint angle})$$

x = azimuth variable at range R

u = distance along array measured from midpoint

By suitably defining f(u), the limits can be extended to ±∞. When the phase of each returned signal is randomly shifted in phase by an amount $\phi(u)$, the response takes the form

$$G(x) = \left| \int_{-\infty}^{\infty} f(u) \exp(jaux) \exp[j\phi(u)] \, du \right|^2 \qquad (3.10)$$

Equation (3.10) is the starting point for the treatment of turbulence-induced phase errors. For notational convenience, s is defined by s = ax and without loss of generality cos α is set equal is defined by s = ax and without loss of generality is set equal to unity. The integral is of the form of a Fourier transform of the product of a deterministic weighting function and a phase function with random argument. The weighting function includes the real beam antenna pattern, the R^4 range dependence, the target cross section, and any weighting imposed by the processor. The case of present interest is that in which the random phase function arises from the effect of turbulence on signal propagation, though other sources of phase errors are of interest in other contexts. The effect of turbulence on the SAR system follows the analysis of Brown and Riordan [8].

The beam tilt is removed by setting $\gamma(u) = \phi(u) - mu$ where m is a random variable selected to center the beam at x = 0 for each sample function of the random process ϕ. The selection of m is made by minimizing the integral of γ^2 over the aperture

$$\int_{-L/2}^{L/2} |\gamma|^2 \, du$$

i.e., by a least squares fit for m. Uniform weighting is assumed for f(u), though this is not strictly necessary. The phase structure function is chosen to be of the form

$$D_\phi(\mu) = b^n \mu^n \qquad (n = 5/3)$$

The "equivalent rectangle" measure of resolution is used to determine the optimum resolution and the associated aperture length. This measure S_R is defined by

$$S_R \equiv \frac{\int |G(s)|^2 \, ds}{|G(o)|^2} \qquad (3.11)$$

S_R is the width of a rectangular beam having the same power gain as $G(s)$ and the same total integrated response. Since $s = 4\pi x/\lambda R$, the value of S_R determines the value of x, or the resolution at the range R. The value of x associated with S_R is denoted by ρ_x.

It is mathematically convenient to work with S_R^{-1}, the inverse of S_R. Maximizing $E(S_R^{-1}) \equiv \rho^{-1}$ then determines the minimum average value of rectangular resolution in the presence of phase errors. Thus,

$$\rho = \frac{4\pi\rho_x}{\lambda R} \qquad \text{or}$$

$$\rho_x = \left(\frac{\lambda R}{4\pi}\right)\rho , \tag{3.12}$$

and ρ_x is the minimum expected ground resolution at range R. Since, in the absence of phase errors, $\rho = 2\pi/L$, the unperturbed rectangular ground resolution ρ_o is $\lambda R/2L$, and

$$\rho_x = \rho_o \left(\frac{L}{2\pi}\right)\rho \tag{3.13}$$

Assuming Gaussian statistics for $[\phi(\mu) - \phi(o)]$, only the calculation of $D_\phi(u - v)$ is required to maximize ρ^{-1}. When this is done for $n = 5/3$, we obtain

$$\rho = 0.985 \, b \tag{3.14}$$

$$L = \frac{13.4}{b} , \quad \text{and} \tag{3.15}$$

$$\rho_x = 2.10 \, \rho_o \tag{3.16}$$

Thus, when the turbulence can be described by a phase structure function of the form $D_\phi(\mu) = b^n \mu^n$, simple expressions can be obtained for both the minimum expected ground resolution, ρ_x, and the associated synthetic aperture length, L. As turbulence (i. e., the value of b) increases, the optimum aperture length decreases and the value of ρ_x grows. For each optimum aperture length, the expected ground resolution in turbulence is just 2.10 times the unperturbed resolution which could be obtained with the same aperture. This can be restated that the resolution is not degraded by turbulence until b is sufficiently large that the optimum aperture length is reduced to twice the value for which designed.

Three additional comments are now in order:

(i)   Since $b^n \sim \beta^{-5}$, $\rho \sim \lambda b \sim \lambda \lambda^{-2/n}$; for $n = 5/3$,
$\rho_x \sim 1/\lambda^{-1/5}$. Thus, the resolution slowly varies
with λ and longer wavelengths are favored.

(ii) If the beam tilt wanders during the formation of
the aperture, then removal of the tilt is in-
appropriate and the resolution is additionally
degraded.

(iii) When Brown and Riordan relaxed the assumption
of uniform weighting and a tapered weighting
function of the form $f(x) = 1 - gx^2$ was used,
the optimum resolution was improved by only
1% for optimum choice of g.

3.3   Determination of the phase structure function

The round-trip propagation phase structure function was determined to be

$$D_\phi(\mu) = 4a \ k^2\mu^{5/3} \ (\csc \psi)^{8/3} \ IR^{5/3} = b^{5/3} \ \mu^{5/3} \qquad (3.17)$$

Thus,

$$b = \frac{(4aI)^{3/5} \ k^{6/5} \ (\csc \psi)^{8/5}}{R} \qquad (3.18)$$

where

$$I \equiv \int_0^H \ c_n^2(h) \ h^{5/3} \ dh \qquad (3.19)$$

Knowledge of the behavior of c_n^2 as a function of altitude would
then determine b. Though models exist in the optical region,
knowledge of C_n^2 at radar wavelengths is quite limited. In the
absence of a description of C_n^2, Porcello [6] used the Maui
measurements of Thompson and Janes [7] to estimate D_ϕ. Since

$$\Delta\phi = \frac{2\pi}{\lambda} \ \Delta R,$$

then,

$$D_\phi = E\left\{\left[\phi(x + \mu) - \phi(x)\right]^2\right\} = \frac{4\pi}{\lambda^2} \sigma_\Delta^2(\mu) \qquad (3.20)$$

Among the data provided by Thompson and Janes were measurements of σ_Δ^2/v^2 as a function of receiving antenna separation, where v is the wind velocity at the lower site, and σ_Δ^2 is the range-difference variance. Though the correlation of σ_Δ^2 with v^2 was not as strong as one might wish, the data--after separation into night and day components--could be used to estimate D_ϕ. Using the geometric mean values of σ_Δ^2/v^2 and assuming a constant wind speed of 8 m/sec, Porcello plotted $D_\phi = (4\pi^2/\lambda^2)\, 8^2\, (\sigma_\Delta^2/v^2)$ versus μ, and obtained a reasonable fit to the predicted $\mu^{5/3}$ dependence. This is shown in Figure 8 along with estimates of D_ϕ from Eq. (3.17) assuming I = 0.3 or 3 cm^2. The discontinuities in the boundaries are due to the factor of two change in the value of a in the region of $\mu \sim \sqrt{\lambda R}$.

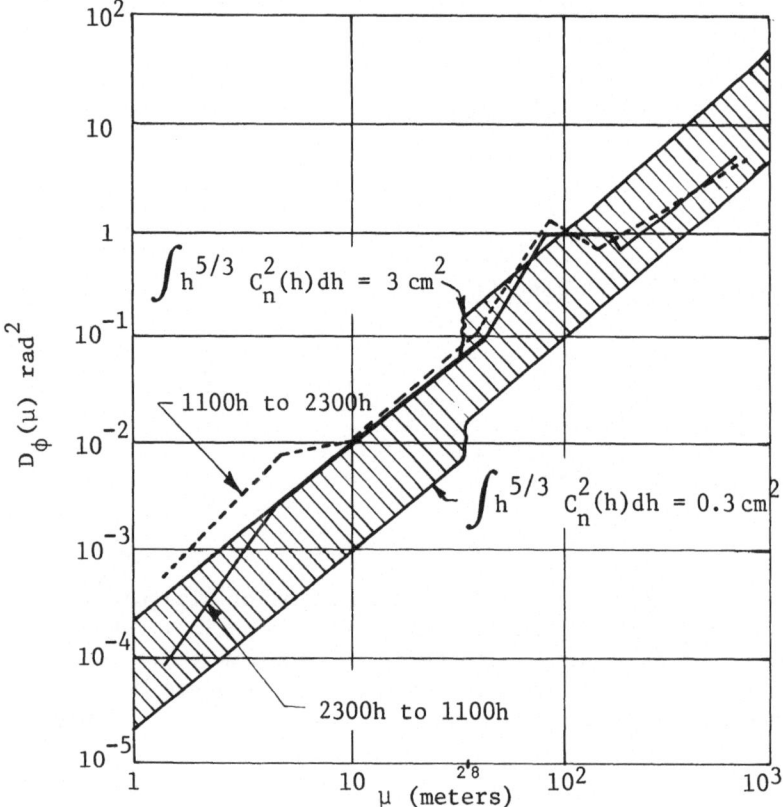

Fig. 8. $D_\phi(\mu)$ Observed Along Maui Path

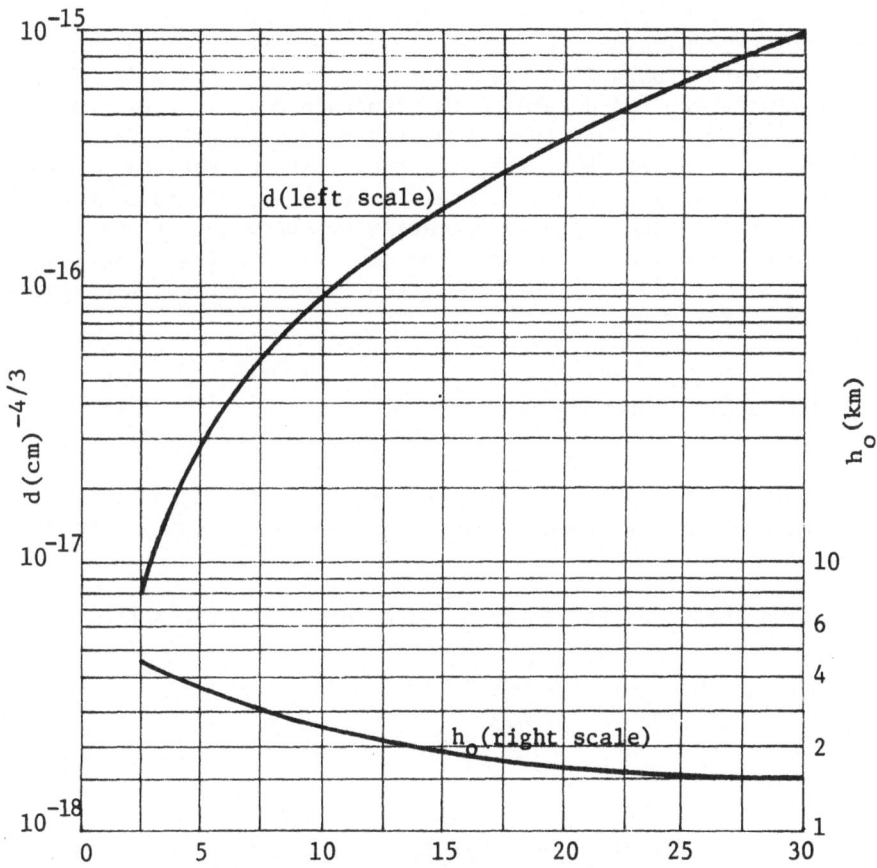

Fig. 9. Empirical Constants versus Sea Level Water Vapor
        Pressure (Due to Sirkis)

3.4 Recent contributions to the effects of phase errors on SAR

Advances in the understanding of the effects of turbulence on microwave propagation have foundered on ignorance of the behavior of C_n^2. Sirkis of Goodyear Aerospace Corporation has used empirical data from a NASA translation of P. N. Tverskoi's "Physics of the Atmosphere" to determine the function M^2 in the expression $C_n^2 = C^2 L_o^{4/3} M^2$. He has found that the variation with altitude of pressure, specific humidity, and the first derivatives of temperature and specific humidity could be stated in terms of their sea level values and depended primarily on the value of sea level water vapor pressure. The variation of temperature with altitude was also determined by its value at sea level. Sirkis then approximated M^2 as follows:

$$M^2 = d/h^{2/3} \exp(-h/h_o)$$

where h is altitude and d and h_o are empirical functions of sea level water vapor pressure. These functions are shown in Figure 9.

Using this expression for M^2, the integral I takes the following form:

$$I = \int_o^H C_n^2 h^{5/3} \, dh = C^2 L_o^{4/3} d \, h_o \, G \qquad (3.21)$$

where $G = 1 - (1 + H/h_o) \exp(-H/h_o)$, and since $C^2 = (2.4)^2$, only a knowledge of L_o is required in order to evaluate I and $D_\phi(\mu)$. The range of validity for this expression for I is still at issue.

REFERENCES

1. K. G. Corless, Side-Looking Radar, NATO Advanced Study Institute of Atmospheric Effects on Radar Target Identification and Imaging.
2. R. L. Mitchell, Models of Extended Targets and Their Coherent Radar Images, Proc. IEEE, 62, 754, 1974.
3. R. A. Shuchman, R. F. Rawson, and B. Drake, A Dual-Frequency and Dual-Polarization Synthetic Aperture Radar System and Experiments in Agriculture Assessment, NAECON '75 Record, 133, 1975
4. M. L. Bryan and R. Larson, Classification of Freshwater Ice Using Multispectral Radar Images, IEEE International Radar Conference Proceedings, 511, Washington, D.C., April 1975.

5. R. A. Shuchman, C. F. Davis, and P. L. Jackson, Contour Strip
    Mine Detection and Identification with Imaging Radar, IEEE
    International Radar Conference Proceedings, 516, Washington,
    D.C., April 1975.
6. L. J. Porcello, Turbulence-Induced Phase Errors in Synthetic
    Aperture Radars, IEEE Trans. Aerospace Electron Systems, 6,
    636, 1970.
7. M. C. Thompson, Jr. and H. B. Janes, Measurements of Phase-
    Front Distortion on an Elevated Line-of-Sight Path, IEEE
    Trans. Aerospace Electron Systems, 6, 645, 1970.
8. W. M. Brown and J. F. Riordan, Resolution Limits with Propa-
    gation Phase Errors, IEEE Trans. Aerospace Electron Systems,
    6, 657, 1970.
9. C. A. Greene and R. T. Moller, The Effect of Normally
    Distributed Random Phase Errors in Synthetic Array Gain
    Patterns, IRE Trans. Military Electronics, 6, 130, 1962.
10. R. O. Harger and T. R. Crimmins, The Effect of Phase Errors
    on Weighted Spectra, IEEE Trans. Military Electronics, 9,
    298, 1965.
11. R. C. Heimiller, Theory and Evaluation of Gain Patterns of
    Synthetic Arrays, IRE Trans. Military Electronics, 6, 122,
    1962.

ON THE EVALUATION OF DOPPLER SPECTRA FOR RADAR CROSS SECTION ANALYSIS AND TARGET RECOGNITION

G. Graf

Deutsche Forschungs- und Versuchsanstalt für Luft- und Raumfahrt E.V., 8031 Oberpfaffenhofen, Germany

ABSTRACT. An analysis of the Doppler frequency spectrum of the radar echo from a moving target is given, showing the relationship between the location of the scattering centers on the target and the spectrum. In addition a scheme is presented, that describes how to get an image of the target from the Doppler spectra.

1. INTRODUCTION

The radar echo from a moving target in general is modulated in a variety of ways. While these modulations in most cases are unliked, they on the other hand contain a lot of information about the target.

There are several origins of the modulations:
In the preceding lectures we learned, that a radar target in general is not a point scatterer, but consists of a variety of scattering centers which are distributed on its surface. The radar cross section of these scattering centers varies with aspect angle, thus modulating the radar return. Scattering centers in the intake of engines, in the vicinity of propellers, on propellers or on wheels etc. often are modulated by shadowing or by their periodic motion. Vibration of parts of the target together with their scattering centers again cause amplitude modulations or at least phase modulations. In general any motion of scattering centers on the target or together with the target at least causes phase variations. Different phase variations of different scattering centers cause amplitude variations of the resulting radar echo, as the waves reflected from these scattering centers then have time varying

phase differences and thus add or subtract as a function of time.

In the following we shall not analyse all of these types of modulation. We only look at the phase modulations caused by an aspect
angle change of a target with several scattering centers and we
shall investigate the relation between these phase modulations and
the location of the scattering centers on the target.

In order to clarify our starting point, without going into details,
we repeat the concept of scattering centers.

A metallic radar target, as seen by a radar, looks very much different from the same target as seen by the eye with natural optical
illumination. There are two reasons for this:

a) Compared to the wavelength of light the surface of radar targets
   in general is rough, so that the light is scattered into all
   directions, whereas the same surface is smooth compared to the
   radar wavelength so that here we only have specular reflections
   and diffractions at discontinuities of the surface.

b) Illumination with a radar in general is with parallel wavefronts
   while natural optical illumination in general is diffuse.

It is easy to simulate how a radar sees a metallic target. We have
to take an object the surface of which is smooth compared to optical
waves and illuminate it with parallel light.

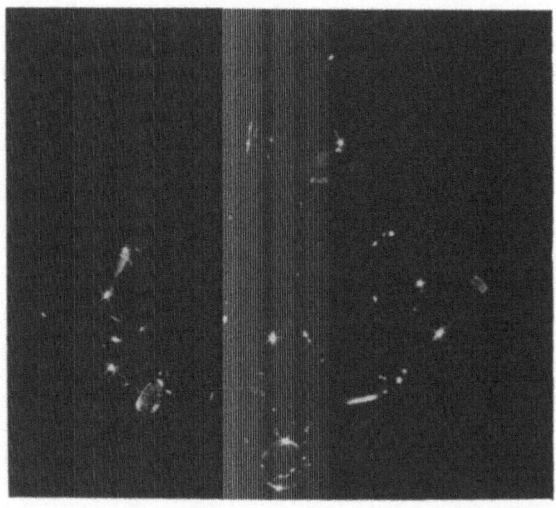

Fig. 1. Water tap with parallel illumination.

A well known object the surface of which is smooth compared to optical waves is an ordinary water tap. Fig. 1 shows a photograph of such a water tap with parallel illumination. From this photograph we get an imagination how a radar sees a metallic target. It does not see the whole target, it only can see the scattering centers on the target.

For our analysis we therefore replace our radar targets by their scattering centers. In general the scattering centers move on the target when the aspect angle is changed. For simplicity however we assume for a moment that the scattering centers were fixed on the target.

Any change of aspect angle in one plane, by an appropriate transformation of coordinates can be replaced by a rotation of the target round an axis with constant speed of rotation $\dot{\varphi} = 1$ and a change of distance. For our analysis we therefore without loss of generality can look at a target which is rotating round an axis.

2. THE PRINCIPLE OF DOPPLER FREQUENCY ANALYSIS

We introduce a coordinate system r, φ, z, the z-axis of which coincides with the axis of rotation (Fig. 2). The angle between the axis of rotation and the direction towards the radar is Θ.

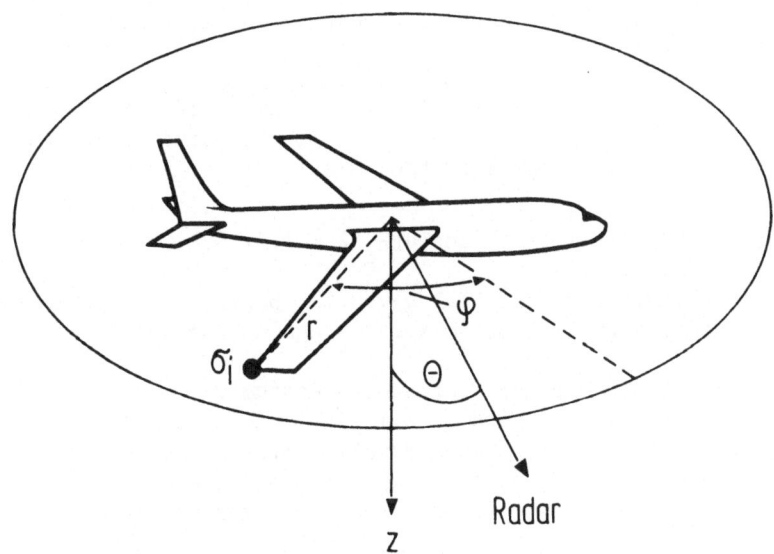

Fig. 2. Coordinate system for the description of the location of scattering centers.

We now describe the location of each scattering center on the target in this cylinder coordinate system by its coordinates r_i, φ_i, z_i and by its radar cross section σ_i.

We transmit a radar wave of the form

$$\underline{E}_0(t) = \frac{E_o}{R} e^{j(\omega t - 2\pi R/\lambda)} , \qquad (1)$$

where E_o is the amplitude, ω is the circular frequency, t is time, R is the distance from the radar and λ is the wavelength.

Part of this wave is reflected by each scattering center on our target. The wave returned from the i-th scattering center is of the form

$$\underline{E}_{ri}(t) = c_1/R^2 \sqrt{\sigma_i} e^{j(\omega t - 4\pi R_i/\lambda + \psi_i)} , \qquad (2)$$

where R_i is the distance of the i-th scattering center from the radar, ψ_i is a constant phase jump that might occur with the reflection and the constant c_1 contains transmit amplitude, antenna gain, attenuation factors and wavelength [1, 2]. (For the amplitude factor we neglect that different scattering centers have different distances from the radar.) The resultant radar return from the target is the sum of all of these waves:

$$\underline{E}_r(t) = c_1/R^2 \cdot \sum_{i=1}^{N} \sqrt{\sigma_i} e^{j(\omega t - 4\pi R_i/\lambda - \psi_i)} , \qquad (3)$$

where N is the number of scattering centers on the target.

As we are interested in the Doppler frequency spectrum of the radar return we mix the radar return (3) with a sample of the transmit signal and thus eliminate the term ωt from the phase of $\underline{E}_r$. At the mixer output we get a signal of the form

$$a(t) = c_2/R^2 \cdot \sum_{i=1}^{N} \sqrt{\sigma_i} \cos(4\pi R_i/\lambda + \psi_i + \beta) , \qquad (4)$$

where β is a constant phase that gives the phase shift of the sample of the transmit signal at the mixer. c_2 again is a constant.

In order to get the Doppler frequency spectrum of this signal we have to take $a(t)$ in a period of time $2\Delta t$ and Fourier transform it. For this we write R_i in (4) as a function of time and express it in our cylinder coordinates r, φ, z, and in the distance R of the origin of our cylinder coordinate system from the radar at $t = 0$.

$$R_i = R + vt - z_i \cos\Theta - r_i \sin\Theta \cos\varphi_i(t) , \qquad (5)$$

where

$$\varphi_i(t) = \varphi_{io} + \dot{\varphi}t \ . \tag{6}$$

φ_{io} is the angular position of the i-th scattering center for $t = 0$, $\dot{\varphi}$ is the angular speed of the rotation and v is the velocity of the coordinate system in the direction from the radar to the target. If we introduce (5) and (6) into (4) the signal at the mixer output becomes:

$$a(t) = \frac{c_2}{(R + vt)^2} \cdot \sum_{i=1}^{N} \sqrt{\sigma_i} \cos \{\frac{4\pi}{\lambda} [R + vt - z_i \cos\Theta - $$

$$- r_i \sin\Theta \cos(\varphi_{io} + \dot{\varphi}t)] + \psi_i + \beta\} \ . \tag{7}$$

As the Fourier transform of this expression is not analytically solvable, we expand the expression $\cos(\varphi_{io} + \dot{\varphi}t)$ into a Taylor series

$$\cos(\varphi_{io} + \dot{\varphi}(t_o + t')) = \cos(\varphi_{io} + \dot{\varphi}t_o) - \dot{\varphi}\sin(\varphi_{io} + \dot{\varphi}t_o) \cdot $$

$$\cdot\ t' + \dots \ , \tag{8}$$

in which we cut off the terms of second and higher order. We replace t' by t and introduce (8) into (7). If Δt is small enough, $c_2/(R+vt)^2$ is nearly constant in the time interval $2\Delta t$ and we can put for simplicity $c_2/(R+vt)^2 = 1$.

$$a(t, t_o) = \sum_{i=1}^{N} \sqrt{\sigma_i} \cos \{\frac{4\pi}{\lambda} [R + vt_o + vt - z_i \cos\Theta - $$

$$(9) \qquad\qquad - r_i \sin\Theta \cos(\varphi_{io} + \dot{\varphi}t_o) + r_i \sin\Theta \cdot $$

$$\cdot \dot{\varphi}\sin(\varphi_{io} + \dot{\varphi}t_o)\ t] + \psi_i + \beta\} \ . $$

In a time window of width $2\Delta t$ centered on t_o, $a(t, t_o)$ is approximately equal to $a(t)$ if Δt is small enough. The frequency spectrum of (9) in that time window is given by

$$A(u, t_o, \Delta t) = \int_{t_o - \Delta t}^{t_o + \Delta t} a(t, t_o)\ e^{2\pi jut}\ dt \ . \tag{10}$$

In order to expand the integration from $-\infty$ to $+\infty$ we multiply $a(t, t_o)$ by a window function $w(t)$ which is unity in the time interval $2\Delta t$ and zero outside. We then have

$$A(u, t_o, \Delta t) = \int_{-\infty}^{\infty} a(t, t_o) \cdot w(t_o, \Delta t)\ e^{2\pi jut}\ dt \ . \tag{11}$$

From the theory of Fourier transforms we know that the Fourier transform of a product of two functions is equal to the convolution of the Fourier transforms of each function, so that we have

$$A(u, t_o, \Delta t) = A'(u, t_o) * W(u, t_o, \Delta t) , \tag{12}$$

where $A'(u, t_o)$ and $W(u, t_o, \Delta t)$ are the Fourier transforms of $a(t, t_o)$ and $w(t_o, \Delta t)$ respectively, and $*$ is the symbol for convolution.

We first look at the Fourier transform of $a(t, t_o)$. We rearrange (9) and get

$$a(t, t_o) = \sum_{i=1}^{N} \sqrt{\sigma_i} \cos [\Phi_i(t_o) + \Omega_i(t_o) . t] , \tag{13}$$

$$\Phi_i(t_o) = \frac{4\pi}{\lambda} [R + vt_o - z_i \cos\Theta - r_i \sin\Theta \cos(\varphi_{io} + \dot\varphi t_o)] +$$

$$+ \psi_i + \beta \tag{14}$$

$$\Omega_i(t_o) = \frac{4\pi}{\lambda} [v + r_i \sin\Theta \dot\varphi \sin(\varphi_{io} + \dot\varphi t_o)] . \tag{15}$$

The Fourier transform of (13) is

$$A'(u, t_o) = \int_{-\infty}^{\infty} \sum_{i=1}^{N} \sqrt{\sigma_i} \cos(\Phi_i(t_o) + \Omega_i(t_o)t) \, e^{2\pi j u t} \, dt \tag{16}$$

$$A'(u, t_o) = \frac{1}{2} \sum_{i=1}^{N} \sqrt{\sigma_i} \{e^{j\Phi_i(t_o)} \delta(u + \frac{\Omega_i(t_o)}{2\pi}) +$$

$$+ e^{-j\Phi_i(t_o)} . \delta(u - \frac{\Omega_i(t_o)}{2\pi})\} , \tag{17}$$

where $\delta(u - u_o)$ is the Dirac δ-function.

Thus the frequency spectrum of $a(t, t_o)$ consists of δ-functions at the frequencies $\pm \Omega_i(t_o)/2\pi$ multiplied by phase factors $\exp[\pm j\Phi(t_o)]$ and by amplitude factors $\sqrt{\sigma_i}$. This spectrum has to be convolved with the Fourier transform $W(u, t_o, \Delta t)$ of the window function $w(t_o, \Delta t)$. If we use the rectangular weighting function mentioned above, $W(u, t_o, \Delta t)$ is simply the well known $\sin x/x$-function multiplied by a phase factor and the width of the window $2\Delta t$:

$$W(u, t_o, \Delta t) = 2\Delta t . e^{2\pi j u t_o} . \frac{\sin 2\pi u \Delta t}{2\pi u \Delta t} . \tag{18}$$

The convolution of (17) and (18) gives (19).

$$A(u, t_o, \Delta t) = \frac{1}{2} \sum_{i=1}^{N} \sqrt{\sigma_i} \{e^{j\Phi_i(t_o)} 2\Delta t \, e^{2\pi j (u + \frac{\Omega_i(t_o)}{2\pi}) t_o} .$$

$$\cdot \frac{\sin 2\pi (u + \frac{\Omega_i(t_o)}{2\pi}) \Delta t_o}{2\pi (u + \frac{\Omega_i(t_o)}{2\pi}) \Delta t_o} + e^{-j\Phi_i(t_o)} 2\Delta t .$$

$$\cdot e^{2\pi j(u - \frac{\Omega_i(t_o)}{2\pi}) t_o} \frac{\sin 2\pi (u - \frac{\Omega_i(t_o)}{2\pi}) \Delta t_o}{2\pi (u - \frac{\Omega_i(t_o)}{2\pi}) \Delta t_o}\} .$$

$$(19)$$

This means that each δ-function of (17) is replaced by the $\sin x/x$-function of (18).

Thus each scattering center on the target produces two Doppler lines with a $\sin x/x$ shape in the spectrum of $a(t, t_o)$ at the center frequencies $\pm \Omega_i(t_o)/2\pi$. Let us concentrate on the lines at $+ \Omega_i(t_o)/2\pi$

$$\frac{\Omega_i(t_o)}{2\pi} = \frac{2}{\lambda} [v + r_i \sin\Theta\dot\varphi \sin(\varphi_{io} + \dot\varphi t_o)] . \qquad (20)$$

$\dot\varphi$ we have put equal 1. For convenience in the following we also shall put the constant factor $\sin\Theta$ equal to 1 as we shall see that it is only a scaling factor.

$(2/\lambda) r_i \sin(\varphi_{io} + \dot\varphi t_o)$ in (20) is the distance, measured in halves of a wavelength, of the i-th scattering center from the plane made up by the line of sight and the axis of rotation. The Doppler spectrum thus with a constant scale factor $1/(\lambda/2)$ gives the distribution of the scattering centers on a target perpendicular to the plane made up by the line of sight and the axis of rotation. This distribution on the frequency axis is shifted by an amount $v/(\lambda/2)$ from zero. If v is greater than the greatest value of r_i all $+ \Omega_i(t_o)$ are positive and all $- \Omega_i(t_o)$ are negative so that there is no overlapping of the spectrum of the $+ \Omega_i(t_o)/2\pi$-lines and those at $- \Omega_i(t_o)/2\pi$.

Fig. 3 illustrates the relation between Doppler frequency spectrum and the distribution of the scattering centers on a target for $\dot\varphi\sin\Theta = 1$. The Doppler lines can be looked at as projections of the scattering centers onto the frequency axis perpendicular to the line of sight, the intensity of the lines being proportional to the σ_i of the respective scattering center.

Our result thus has a very simple interpretation: As the target is rotating, the various scattering centers have speeds towards the radar that are proportional to their distance from the plane made up by the line of sight and the axis of rotation. In Fig. 3 scattering centers to the right of the axis of rotation are moving

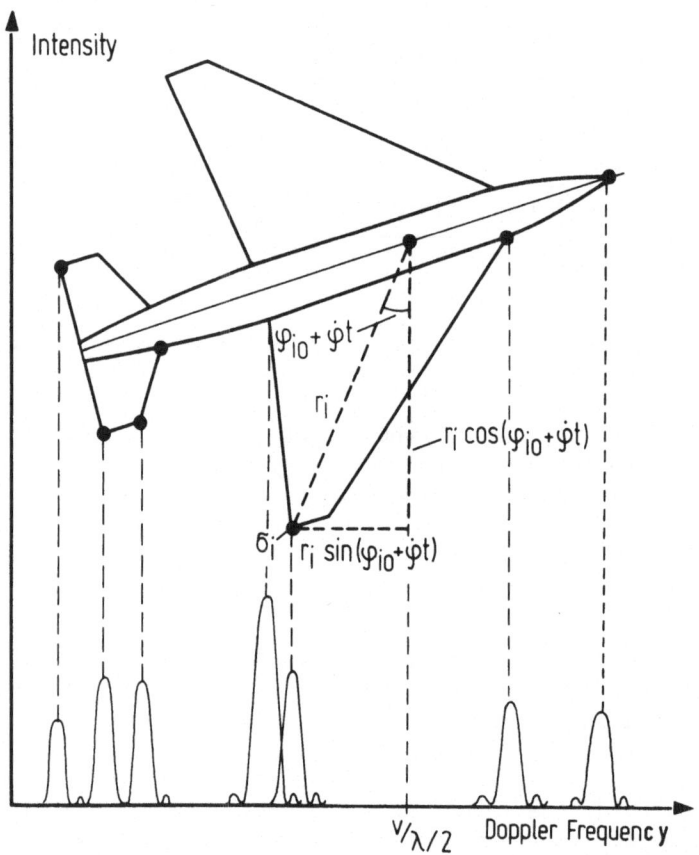

Fig. 3. Relation between Doppler spectrum and location of the
scattering centers on a rotating target.

towards the radar and thus have positive Doppler shift which is
the higher, the greater the distance from the axis, scattering
centers to the left are moving from the radar and thus have negative Doppler shift. Scattering centers exactly before or behind the
axis of rotation are moving tangentially and thus have zero Doppler shift relative to the Doppler shift of the axis $v/(\lambda/2)$.

As an example Fig. 4 shows a series of Doppler spectra of the radar
return from three scattering centers, made from different successive
aspect angles. The intensity is in a linear scale with 10 db dynamic range. The scattering centers were cylinders the axis of which
was parallel to the axis of rotation. The distances of the cylinders from the axis of rotation were 0.7, 2.1 and 6.4 wave lengths
(Fig. 5), wavelength was 3.2 cm. The spectra give no resolution
along the axis of rotation but give very good resolution in the
cross range direction perpendicular to the axis of rotation.

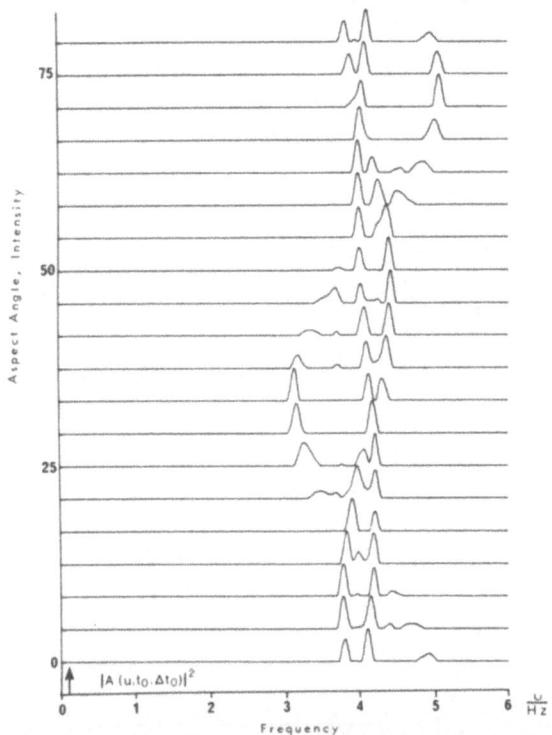

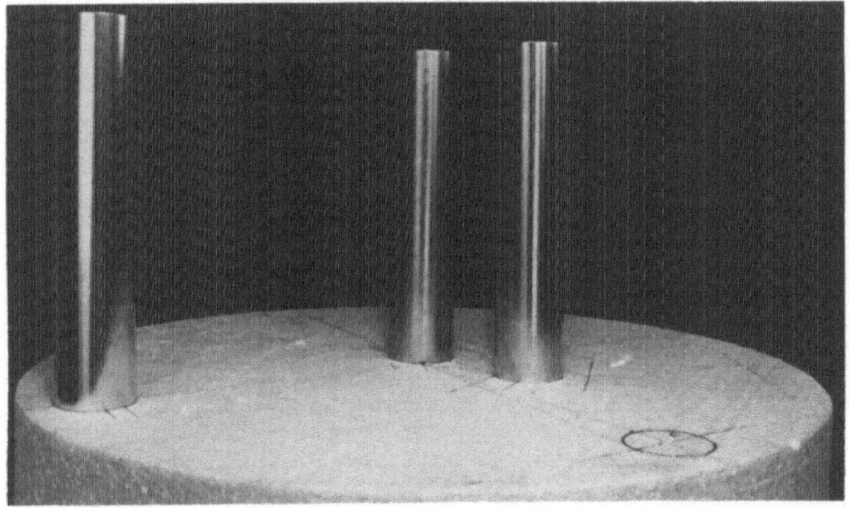

Fig. 4, 5. "Instantaneous" Doppler spectra of three scattering centers (cylinders) and image of these cylinders.

Fig. 6 shows spectra of a more realistic radar target, namely of a model of a Fokker Friendship aircraft, the total length of which was 14 wavelengths.

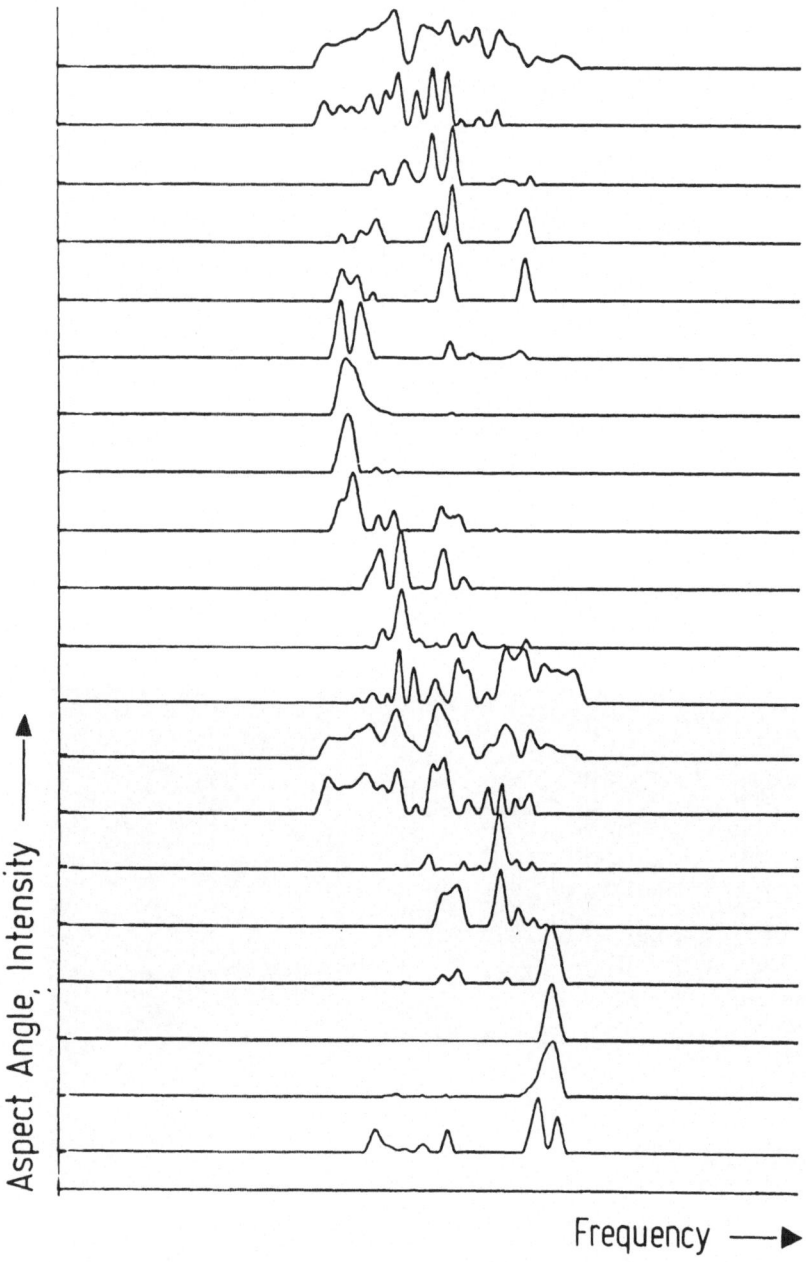

Fig. 6. "Instantaneous" Doppler spectra of a Model of a Fokker Friendship (rotating).

This figure shows, that Doppler frequency analysis is not restricted
to targets with fixed scattering centers and constant $\sigma_.$. For example
when a wing is perpendicular to the line of sight, all the illuminated edge of the wing is one long scattering center. In the Doppler
spectrum out of an aspect angle window that contains the aspect
angle in which the wing is perpendicular to the line of sight,
correspondingly we find a very broad Doppler line, the width of
which is proportional to the length of the wing and the intensity
of which is nearly constant over its width. For example in the 7th
spectrum from the bottom of Fig. 6 the left wing can be seen and
in the 9th spectrum you see the right wing.

Fig. 7 shows the lowest of the spectra of Fig. 6 together with a
photograph of the aircraft made in the center of the aspect angle
window for the Doppler spectrum. The Doppler lines are easily related to scattering centers on the aircraft. The major Doppler lines
in the spectrum correspond to scattering centers on the cockpit,
the tip of the engine, the corner between engine and wing and the
two nearly vertical edges of the tail.

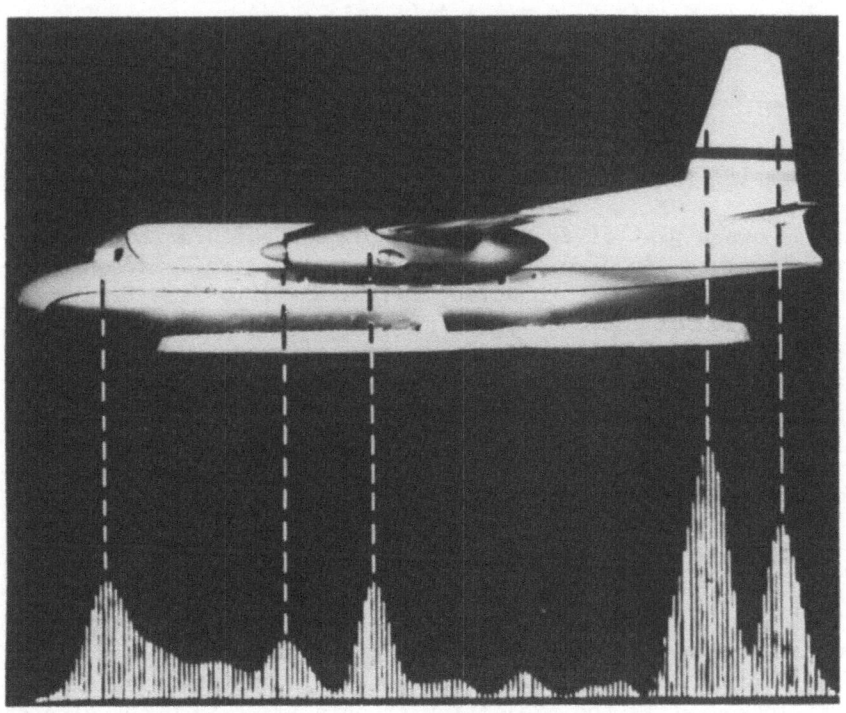

Fig. 7. The lowest of the spectra of Fig. 6 together with an
        image of the Fokker Friendship.

A very important point in the production of Doppler spectra of radar
returns is the choice of the widths of the aspect angle windows,
which is responsible for the width of the Doppler lines. Narrow
Doppler lines are important to get high resolution of scattering
centers which are close to each other. If we use a narrow window,
we have broad Doppler lines as the Fourier transform of the window
function becomes very wide, if we use a wide window, we have narrow
Fourier transforms of the window function, now however our Taylor
series of the signal function is no longer a good approximation for
$a(t, t_o)$, so that instead of the δ-function at $\pm \Omega_i(t_o)/2\pi$ we get
complicated broad spectra, and the convolution again gives broad
and even structured "Doppler lines". By an approximative method
it is possible to find the optimum width of the aspect angle windows
[3, 4]. The result is, that the optimum aspect angle window $\Delta\varphi_{opt}$
depends on the location on the target of the scattering center to
be resolved, so that a sort of focusing is necessary to obtain optimum resolution in the Doppler spectrum. For $-\pi \leq \varphi(t_o) \leq \pi$ we obtain

$$\Delta\varphi_{opt} = 3\sqrt{\frac{4 \cdot k_w}{(r/\lambda) \cdot (1 + \rho)}} \; ; \qquad \rho = \frac{2[\pi/2 - \varphi(t_o)]}{\Delta\varphi} \qquad (20)$$

if $\pi/2 - \Delta\varphi/2 \leq |\varphi(t_o)| \leq \pi/2 + \Delta\varphi/2$ and

$$\Delta_{opt} = \sqrt{\frac{k_w}{(r/\lambda) \cdot \cos\varphi(t_o)}} \qquad (21)$$

elsewhere; k_w is a constant factor that depends on the shape of
the weighting function. For a rectangular weighting function $k_w = 1$.

Fig. 8 shows a plot of $\Delta\varphi_{opt}$ in a plane perpendicular to the axis
of rotation and centered on it. If a target is projected onto this
plane along the axis of rotation, the parameter of the line crossing
the projection of a scattering center indicates the optimum width
of the aspect angle window for best resolution of this scattering
center. Fig. 9 in a similar way shows optimum cross range resolution Δx in the Doppler spectra, when the optimum windows are used.

3. TARGET IMAGING FROM DOPPLER SPECTRA

Let us project all scattering centers of a target parallel to the
axis of rotation onto the r-φ-plane at z = 0 of our coordinate system
In this plane we then get a two dimensional image of the target
(similar to Fig. 1) as seen from the direction of the axis of rotation. From the preceding section we know, that our Doppler spectra
can be looked at as the projection of this image into the direction
of the radar (Fig. 3) so that in the Doppler spectrum each Doppler
line is the projection of a scattering center. We shall investigate
now how to go the reverse way, that is how to find the image of the
target from the spectra. For this we use the image plane mentioned

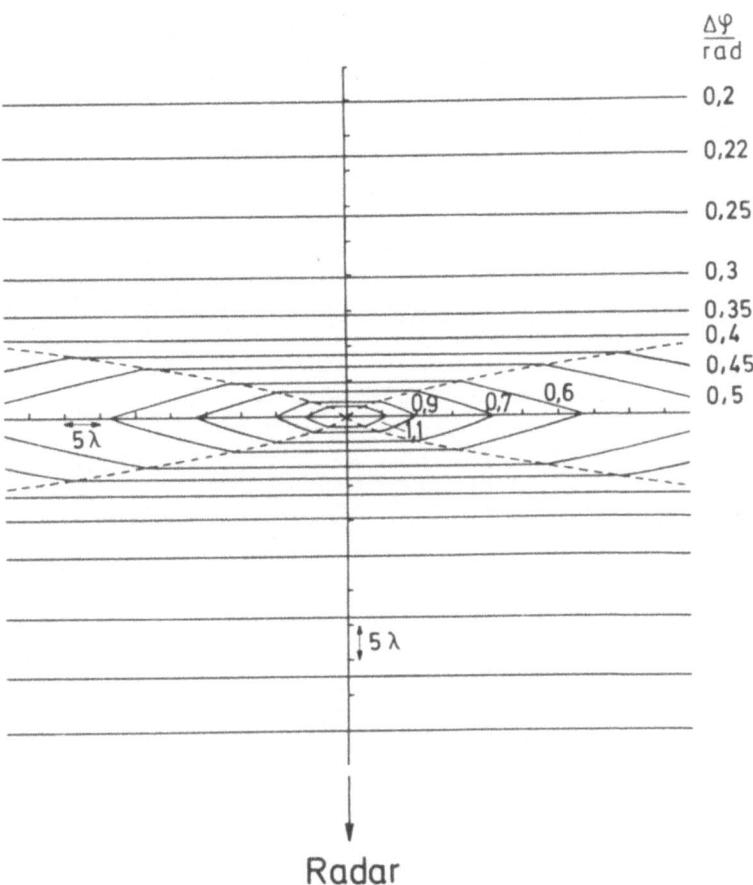

Fig. 8. Lines of constant optimum aspect angle window in a r-φ-
        plane of Fig. 2.

above and round this image plane, corresponding to their respective
aspect angles we arrange the Doppler spectra (Fig. 10). If we now
draw lines from the maxima of a Doppler spectrum towards the plane
(direction radar-target) each of these lines crosses the scattering
center of which it is the projection.

If we want to go from the spectra to the image however we only have
spectra and no image. We therefore do not know where the scattering
centers are located exactly, we only know that they are located
somewhere on the projection lines. If however we draw lines from a
second spectrum, the scattering centers must lie on the crossings
of these lines. In general however this in not yet unequivocal as
there are several crossings. If however we draw these lines for all
spectra, all lines corresponding to the same scattering center are

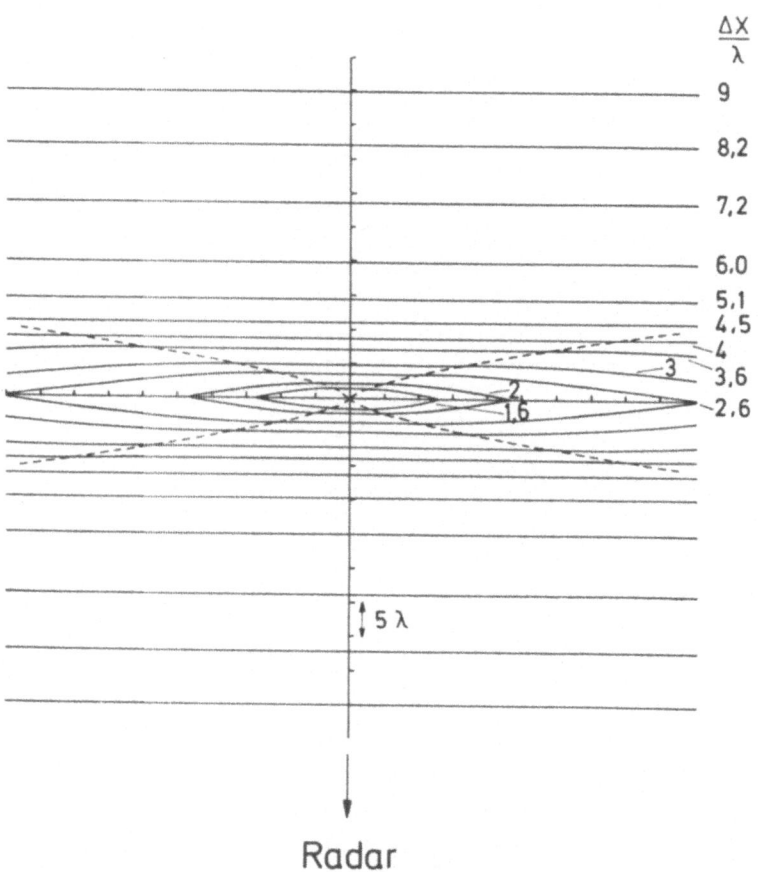

Fig. 9. Lines of constant optimum cross range resolution in a r-φ-
       plane of Fig. 2.

crossing at the same point, which then is identified as a scattering
center. Fig. 11 shows such a plot made from the Doppler spectra of
the three cylinders (Figs. 4, 5).

The reverse projection process by drawing lines of course is only
good for this demonstration, in general it is better to put the
spectra into a computer and to make a projection of the whole Doppler spectra and to add all these values in the image plane. If we
do this for the above mentioned spectra we obtain Fig. 12, it shows
the location of the three cylinders as seen from the direction of
the axis of rotation. Resolution is about one wavelength.

As now however we did not use the phase in the complex Doppler
spectra. If we integrate phase into our reconstruction process we

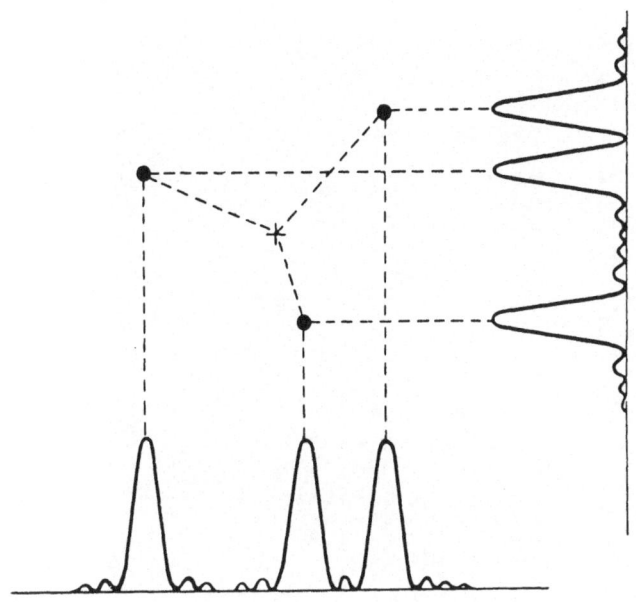

Fig. 10. Two Doppler spectra arranged according to their respective
         aspect angles and projection of their maxima into an image
         plane.

are able to get much better resolution.

From (19) we see that the phase in the Doppler lines at $u = -$
$- \Omega_i(t_o)/2\pi$ is $\Phi_i(t_o)$ and at $u = + \Omega_i(t_o)/2\pi$ is $- \Phi_i(t_o)$. We again
only look at the positive spectrum at $+ \Omega_i(t_o)/2\pi$. In the phase $\Phi_i(t_o)$
according to (14) only the terms $(4\pi/\lambda)$ vt_o and $(4\pi/\lambda)$ r_i $\sin\Theta$
$\cos(\varphi_{io} + \dot\varphi t_o)$ depend on t_o, all other terms are constant. While
$(4\pi/\lambda) vt_o$ is known if v is known or can be measured, the term
$(4\pi/\lambda)$ r_i $\sin\Theta$ $\cos(\varphi_{io} + \dot\varphi t_o)$ where we have put $\sin\Theta = 1$ again has
a simple meaning. Except of the factor 2π it represents the distan-
ce of the scattering center from the axis of rotation in a direction
parallel to the direction radar-target (Fig. 3), the distance again
is measured in halves of a wavelength. Thus after elimination of
$(4\pi/\lambda) vt_o$ the phase of each Doppler line except for constants is
proportional to the distance in the direction radar-target of the
scattering center from the axis of rotation.

If in our projection process (Fig. 10) we now shift the phases of
the spectra along the path of projection by an amount of 2π each
half wavelength the phases in all spectra are shifted by a constant
phase angle α, when the projection is off the axis of rotation. After an additional distance of $r_i \cos(\varphi_{io} + \dot\varphi t_o)$ the Doppler line at
$\Omega_i(t_o)/2\pi$, which corresponds to the i-th scattering center, has the
phase

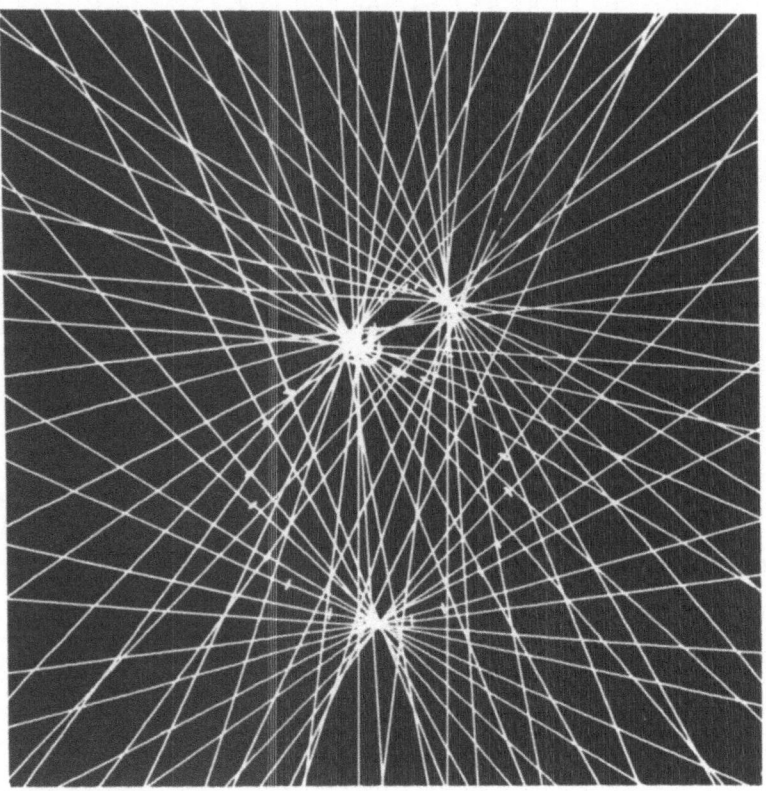

Fig. 11. Lines drawn from the maxima of the spectra in Fig. 4 into
          an image plane, resulting in an "image" of the three
          cylinders seen from above.

$$\Phi_i(t_o) + \alpha + \frac{4\pi}{\lambda} r_i \cos(\varphi_{io} + \dot\varphi t_o) = \frac{4\pi}{\lambda} (R - z_i \cos\Theta)$$

$$+ \psi_i + \beta + \alpha \tag{23}$$

which for a given i is the same constant for all t_o, that is for
all spectra. At the distance $r_i \cos(\varphi_{io} + \dot\varphi t_o)$ from the axis of rotation however is the location of the i-th scattering center in
the image plane.

Thus at the location where the i-th scattering center is expected
in the image plane, the Doppler lines corresponding to that scattering center for all spectra have the same phase, while this is not
the case in the environment. At the location of the scattering
center thus the amplitudes of all spectra add with the same phase,

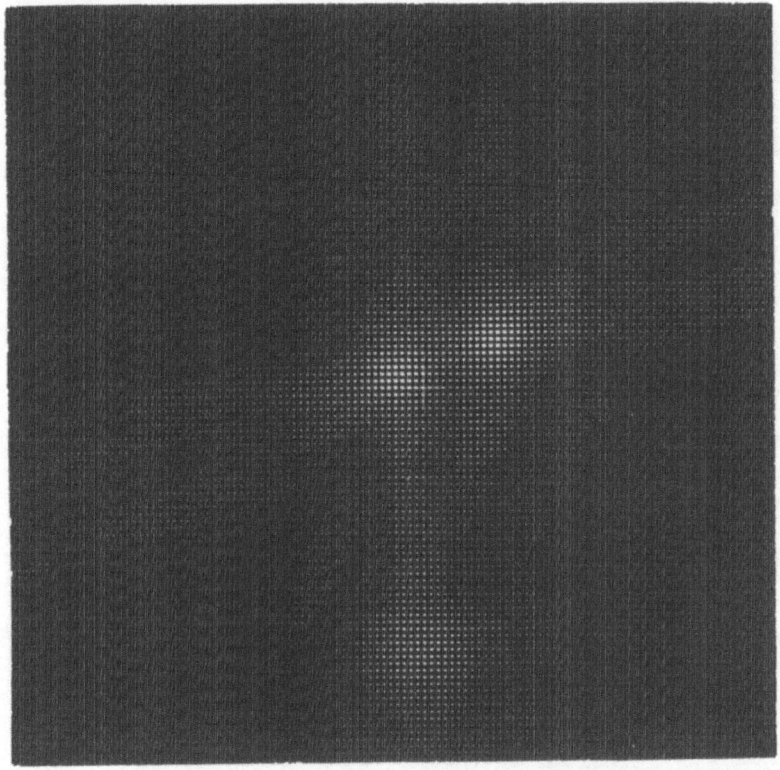

Fig. 12. Microwave image of the three cylinders of Fig. 5, seen
from above, produced by projecting the Doppler spectra
of Fig. 4 into an image plane (intensity only).

while in the environment we only have an addition of intensities.

The result of such a reconstruction process is seen in Fig. 13,
showing an image of the three cylinders mentioned above, which was
made from the spectra of Fig. 4. Resolution now is 0.19 wavelengths,
which is better than in Fig. 12 where the phase of the spectra was
not used. We again have no resolution in the direction of the axis
of rotation. The image does not show the surface of the cylinders
but only their axes which in our setup represent the origin of the
reflected waves.

It can be shown, that the intensity distribution at the location of
a scattering center is given by a Bessel function, if a range of
aspect angles of 360 degrees and an infinite number ob Doppler
spectra out of small aspect angle windows is used [4]. The distance
of the first zero of that Bessel function from the maximum gives

Fig. 13. Microwave image of the three cylinders of Fig. 5, seen
from above, produced by projecting the Doppler spectra
of Fig. 4 into an image plane and by using the phase of
the spectra. Aspect angle range 360 deg.

the resolution of 0.19 λ mentioned above.

For smaller ranges of aspect angles resolution decreases and becomes
dependent on the direction in the image.

Fig. 14 shows the images of the three cylinders with an aspect
angle range of 90 degrees. The images of the scattering centers
now are ellipses with their long axes in the direction of the center of the aspect angle range used for the Doppler spectra. Resolution as a function of the range of aspect angles α is given in
Fig. 15, where the two curves give the resolution along the long
and the short axes of the image ellipses (Sommerfeld-approximation
[4, 5]).

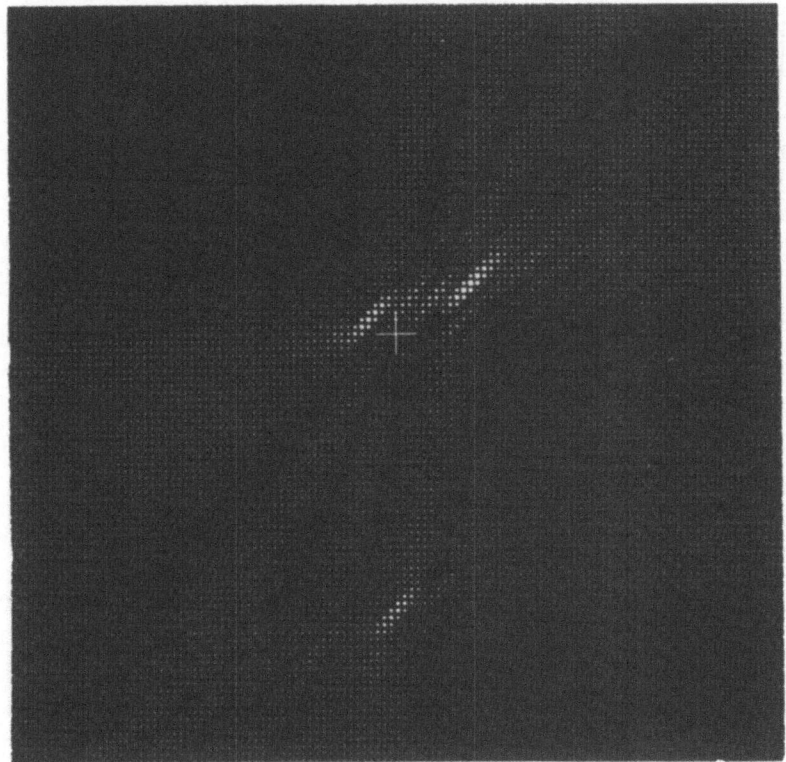

Fig. 14. Microwave image of three cylinders as in Fig. 13, aspect
         angle range 90 deg.

The aspect angle windows used for the individual Doppler spectra
for imaging of the scattering centers, in contrast to the optimum
$\Delta\varphi$ described above, when we analized the spectra directly, now ought
to be as small as possible. This however increases the number of
Doppler spectra used and thus increases computer time so that a
compromise has to be made [4]. Images in Figs. (13) and (14) were
made from spectra each of which was made from an aspect angle window of 12 degrees.

We saw, that from a relatively simple analysis of the Doppler return
from a radar target we can get a lot of information on the location
of the scattering centers on the target and thus on the shape of the
target and that with more effort it is possible to make highly resolved microwave images from these spectra.

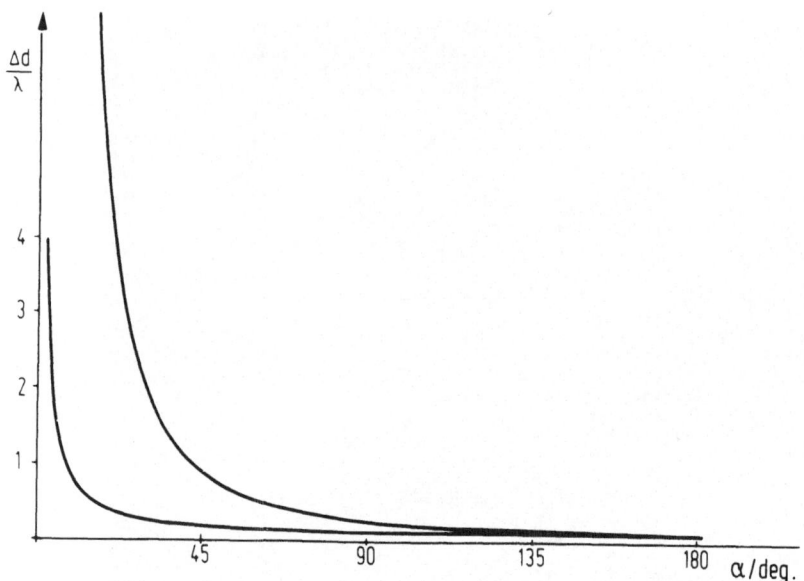

Fig. 15. Resolution Δd (in wavelengths) of a microwave image made
from complex Doppler spectra. Upper curve is resolution in
the direction of the center of the aspect angle range α,
lower curve is resolution in a direction perpendicular to
the center of the aspect angle range (Sommerfeld approximation).

REFERENCES

1. F.E. Nathanson, Radar Design Principles, McGraw-Hill Book
    Company, New York, 1969.
2. A.W. Rihaczek, Principles of High-Resolution Radar, McGraw-
    Hill Book Company, New York, 1969.
3. G. Graf, On the Optimization of Aspect Angle Windows for the
    Doppler Analysis of the Radar Return of Rotating Targets,
    to be published in IEEE Transactions of AP-S, May 1976.
4. G. Graf, Analyse des Dopplerfrequenzspektrums rotierender Kör-
    per, Deutsche Luft- und Raumfahrt FB 72-46, 1972.
5. A. Sommerfeld, Vorlesungen über theoretische Physik, Bd. IV,
    Optik, Dieterichsche Verlagsbuchhandlung Wiesbaden, 1959.

ON SOME PARALLELISMS BETWEEN RADAR AND SONAR TRANSMISSION CHANNELS

G. Tacconi *

Istituto di Elettrotecnica
Università di Genova and CNR
Italy

ABSTRACT. Problems of telecommunications, independently from the physical phenomena on which they are based, can be approached by means of the same mathematical model. Acoustic and electromagnetic wave propagation, with some restrictions, can be treated with nearly identical formal expressions. Common and analogous aspects of radar and sonar are presented, together with some concepts, models and methodologies elaborated for an underwater transmission channel, suitable for application, after appropriate changes, to an electromagnetic channel.

1. GENERAL

Science has many highly specialized sectional branches which often use the same basic disciplines for their data acquisition, reduction, analysis and interpretation methods, i.e. communication theory, statistics, signal processing, information theory and computer facilities [1-5]. These disciplines enable scientists to elaborate simulations and models in order to approach experiments with a high level of expected results [6]. It often happens that in one branch a new system of a common discipline is developed up to a very advanced stage and the knowledge of such methodologies could be of great interest to scientists in other fields.
A macroscopic example of such a transfer from one field to another is represented by sonar, which has drawn its fundamentals from those of radar [7].

---

* Professor of Electro-Acoustics at Electronic Engineering
  Faculty, Communications Section

Jeske (ed.), Atmospheric Effects on Radar Target Identification and Imaging, 239–253.

The two physical aspects of wave propagation (electromagnetic and acoustic) can be related to each other by means of the mechanical-electromagnetic-acoustic analogy with similar formal mathematical expressions, even with certain restrictions limiting the validity of such analogy [8].

FORCE - VOLTAGE - PRESSURE

VELOCITY - CURRENT - ACOUSTIC FLUX

MASS - INDUCTANCE - MASS

COMPLIANCE - CAPACITY - ACOUSTIC - CAPACITY

FRICTIONAL - ELECTRIC - ACOUSTIC PASSIVE
RESISTANCE   RESISTANCE   RESISTANCE

The well known analogy is based on the following laws:

(i) $F = m\dfrac{dv}{dt}$: First Newton's equation of force, valid without reservations.

(ii) $F = \int v\dfrac{dt}{C_m}$: Hook's law, valid only if the force is proportional to displacement.

(iii) $F = R_m v$: Linear approximation, better fulfilled the smaller the velocity.

2. THE TRANSMISSION CHANNEL

A very general communication system, as proposed by Shannon [9,10], is shown in Fig. 1.

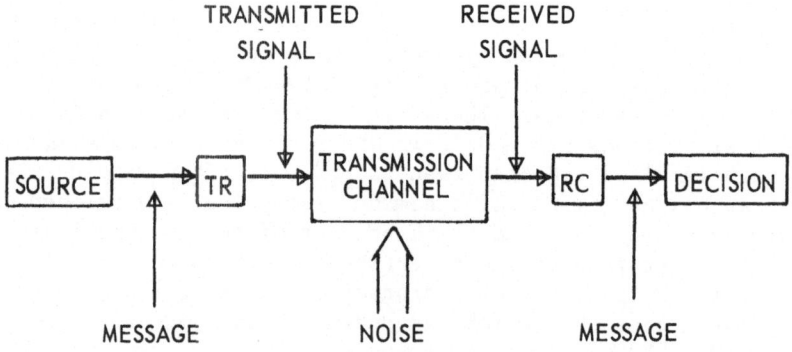

Fig. 1. The general transmission channel

The sonar system takes advantage of such modelling, which enables
to separate the various elements (blocks) represented as linear
filters [Fig. 2].

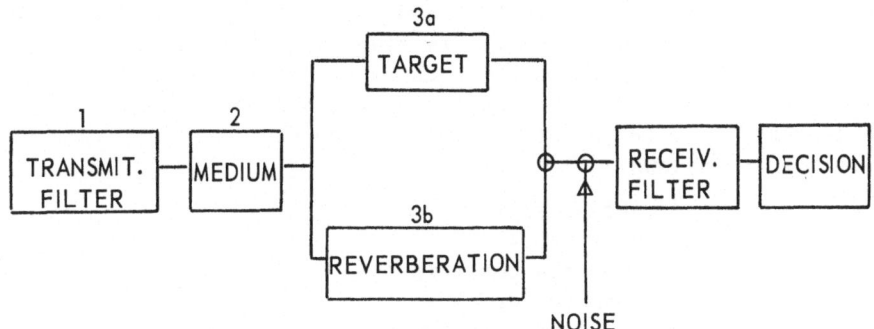

Fig. 2. The transmission channel in terms of filters

Obviously transmitting and receiving technologies in radar and
sonar are different and will not be considered here. In both of
them the advances in engineering, through the influence of such
disciplines as information theory and digital signal processing,
have posed many problems to the designing people. Transducers,
however, continue to be essentially analogue devices. As said
before, many concepts of information theory found useful in radar
had a great influence on sonar, i.e. the significance of the time-
bandwidth product for processing complex signals, plays an
important role in both. Consequently, the behaviour of the
general transmission channel is influenced by the time-bandwidth
product and in this respect can be outlined as follows [11]:

Transmitter	Medium	Receiver
Wave-form coding	Ambiguity [12]	Correlation
Energy transfer	and	Processing Gain
Matching	Scattering functions	Classification

The problem of optimizing the performance of a transmission channel
is, however, common and in order to approach it the characteristics
of the following must be known:

(i)   The transmission source.

(ii)   The medium.

(iii) The receiving system.

3. THE MEDIUM

The medium where the propagation takes place (atmosphere and ocean) is the component of the transmission channel model which normally escapes even the most rigorous scientific control. The physical nature and the behaviour in time and space of the medium are the origin of a large number of deformations produced on the transmitted signal; this is valid for both radar and sonar [13,14]. The main physical phenomena in the medium which produce unwanted "accidents" on a transmitted signal can be divided, with some overlapping, as follows:

> Attenuation
> Scattering
> Boundary conditions
> Propagation
> Non-linear phenomena

Parallelisms between radar and sonar are shown in Tables 1-5. The comment at the foot of each Table emphasizes the statistical nature of space and time dependence. It should also be pointed out that the items mentioned under radar and sonar must be considered globally, and that there is not necessarily any correspondence between items occurring on the same line.

TABLE 1

Attenuation

Radar	Sonar
Atmospheric : Wind shear [13]	Spreading : Geometrical effect [14]
Vertical falls	Absorption : Range (acoustic path)
Turbulence	Energy transformation
Stratification	
Precipitation : Rain fall	Leakage out of sound channel
Drop size	Volume disuniformities
Humidity	
Clouds	
Dependence from frequency	Dependence from frequency (viscosity).
For Both : Space non-uniformity and time stochastic processes	

TABLE 2(a)

Scattering [15]

Radar	Sonar	
Rain (drop size)	Volume Reverberation	
Snow		Pressure
Clouds	Density	Salinity
Wind shear		Temperature
Beam broadening	Turbulence [16]	
Birds	Plankton and fish	
Fall velocity distribution	Microstructure	
Turbulence [16]	Internal waves	

For Both : Space non-uniformity and
time stochastic processes

Scattering phenomenology implies problems of target detection and recognition. What is a target? Anything can be a target if the observer attributes it with certain a priori expected characteristics, i.e. for a meteorologist studying atmospheric phenomena, a target could be clouds, rain, etc., for airport radar operators it is airplanes, for a naturalist it is birds or fish, and so on. From the transmission point of view, a target is simply one of several causes which alterate the transmitted signal. Anyhow target detection and classification represents a very important problem in any type of communication. Consequently, Table 2(a) can be completed with Table 2(b), relating only to target problems.

TABLE 2(b)

Targets [15,18,19]

Radar	Sonar
Target glint	Target strength
Target cross section	Target cross section
Echo structure	Echo structure

For Both : Time deterministic or quasi-
deterministic processes

TABLE 3

Propagation

Radar	Sonar
Varying refraction index of medium (air) [13]	Varying refraction index or propagation velocity in sea water
Ray tracing approach	Ray tracing approach
Bending rays	Normal model approach (shallow waters) [17]
Horizontal stratification [20]	Horizontal stratification [21]
For Both : Space non-uniformity and time stochastic processes	

TABLE 4

Boundary Conditions

Sea backscatter (clutter) :	Surface reverberation :
Sea state Wave height Wind direction Grazing angle	Sea state Wave height Wind direction Grazing angle
For Both : Space non-uniformity and time stochastic processes	
Terrain backscatter [22] :	Bottom reverberation :
Shadowing hills Soil moisture or snow Land culture Grazing angle	Configuration Bottom type Grazing angle
For Both : Space non-uniformity	

The influences of non-linear effects on sonar and on radar are not, in general relevant. Theoretical and technical investigations on finite amplitude acoustics have recently given promising results [23,24] and also non-linear effects have been studied in the electromagnetic propagation phenomenologies [25]. Table 5 shows the main non-linear pehnomena in both fields.

<u>TABLE 5</u>

<u>Non-linear Phenomena in Propagation</u>

<u>Electromagnetics [25]</u>	<u>Acoustics [24]</u>
Non-linear effects in EM wave propagation	Finite amplitude in fluids
Parametric instabilities	Non-linear generation of secondary waves
Jomosphere layer formation	

4. THE TRANSMISSION CHANNEL INFLUENCED BY MEDIUM

In order to overcome, or to evaluate, the influence of the medium on the transmitted signal, several models have been studied by many scientists [6,26,27,28], and a large number of experiments have been carried out in order to check the worth of the models for different practical applications [29,30].

The medium in the broad sense is as shown in blocks 2, 3a and 3b of Fig. 2 and for the case of an underwater acoustic channel Fig. 3a can be considered like a composite filter [26,27,28,29]. This approach with due caution can also be considered for radar problems which are very similar [Fig. 3b].

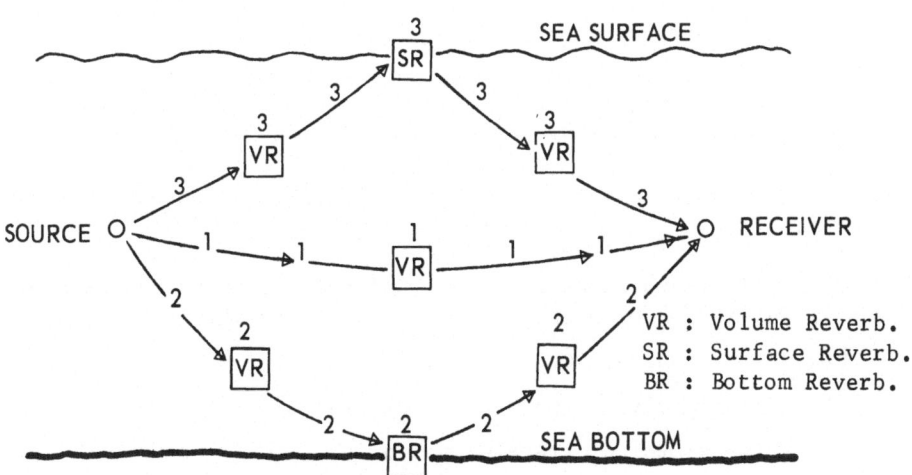

Fig. 3a. The underwater transmission channel.

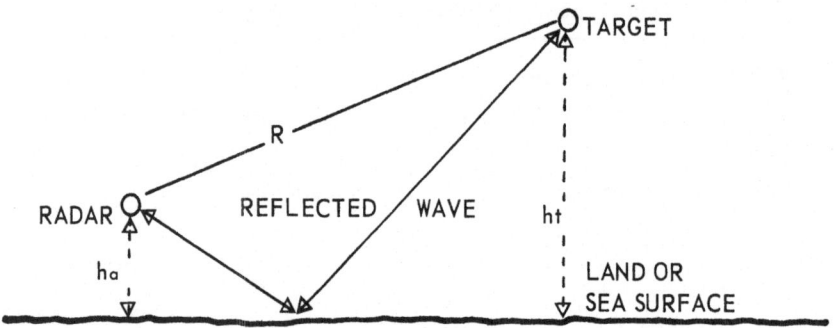

Fig. 3b.   Geometry for radar propagation above a surface.

Several categories of transmission channel must be distinguished as follows:

TABLE 6

Transmission Channel Categories

1.   Time Invariant & Space Invariant
2.   Time Variant (General) & Space Invariant
3.   Time Variant (Stochastic) & Space Invariant
4.   Time Invariant & Space Variant (General)
5.   Time Variant (Stochastic) & Space Variant (Stochastic)

The filter theory applied to each case [26,27] can be outlined as
follows:-

(i)   Time and Space Invariant Channel. This is represented
as a linear filter by means of the following relations:

$$\begin{cases} H(f) & : \text{ Transfer function} \\ h(\tau) & : \text{ Impulse response } = \mathscr{F}\left[H(f)\right] \end{cases} \qquad (4.1)$$

   $\mathscr{F}$ is the symbol of transformation.

(ii)   Time Variant and Space Invariant Channel. This is
defined completely by one of the four functions of the following
set:

$H(f,t)$: Transfer function

$h(\tau,t)$: Impulse response

$B(f,\varphi)$: Bifrequential function $= \mathscr{F}[H(f,t)]$ (4.2)

$S(\tau,\varphi)$: Diffusion function $= \mathscr{F}[h(\tau,t)]$

(iii) Time Stochastic Variant and Space Invariant Channel. The global transformation of the signal represented by $H(f,t)$ can be considered the combination of a large number of elementary transformations, which can be either of deterministic or stochastic nature. Since the medium behaviour is mainly a combination of stochastic processes, only the group of stochastic nature transformations is considered. Such a channel can be represented by means of the four variable functions obtained as autocorrelation functions (second order moments):

$$R_H(f,f',t,t') = \overline{H(f,t)\,H*(f',t')}$$
$$R_h(\tau,\tau',t,t') = \overline{H(\tau,t)\,h*(\tau',t')}$$
$$R_B(f,f',\varphi,\varphi') = \overline{B(f,\varphi)\,B*(f',\varphi')}$$
$$R_S(\tau,\tau',\varphi,\varphi') = \overline{S(\tau,\varphi)\,S*(\tau',\varphi')}$$

(4.3)

The analytical treatment of the above functions is quite complex. In order to simplify the approach some assumptions of a statistical nature [2] must be made. The main assumption is that the channel should be "wide sense stationary uncorrelated scattering" (WSSUS) [27]. With this assumption functions (4.3) become:

$$R_H(\Delta f,\Delta t)$$
$$R_h(\tau,\Delta t)$$
$$R_B(\Delta f,\varphi)$$
$$R_S(\tau,\varphi)$$

(4.4)

Functions (4.4) are fourier transforms of each other [28]. Only one of them may be considered as representative of the channel and the most convenient results to be the scattering function

$$R_S(\tau,\varphi)$$ (4.5)

which depends from the time spreading (τ) and the frequency smear (φ) of the received signal. This function for radar and sonar application is particularly useful because it describes the resolution limitations of the medium in the range-doppler space. Its convolution with the ambiguity function of the signal $\mathscr{A}x(t)$ [4,26,29] defines, provided the WSSUS condition exists, a combined ambiguity function of the signal and the medium.

In conclusion, a point target moving at constant speed in the range-doppler space is not seen as a point, but as a "spot", and the scattering function is an estimation of the energy density of this "spot".

(iv) Time Invariant and Space Invariant Channel. This type can be represented by the following set, analogue to Eq. (4.2).

$H(f,\sigma)$: Transfer function

$h(\tau,\sigma)$: Impulse response

$B(f,\rho)$: Bifrequential function $= \mathcal{F}\left[H(f,\sigma)\right]$

$S(\tau,\rho)$: Diffusion function $= \mathcal{F}\left[h(\tau,\sigma)\right]$

$\hfill (4.6)$

σ represents a spatial parameter which takes into account the transmitter and receiver position vectors.

(v) Time Variant and Space Variant Channel. This type can be represented by the equations

$H(f,t,\sigma)$

$h(\tau,t,\sigma)$

$\hfill (4.7)$

With the analogue procedure of the preceding cases, including the time and space WSSUS assumptions, and by reducing the space parameter σ into two coordinates, it is possible to define a scattering function of four variables in the four-dimension range-doppler-angle-space:

$$R_S(\tau,\varphi,u,v) \quad \begin{cases} u = \dfrac{f}{c}\,\sin\,\theta_H \\ v = \dfrac{f}{c}\,\sin\,\theta_v \end{cases} [27] \qquad\qquad (4.8)$$

Taking into account the above theories, fully developed by R. Laval and others [27,29,30], many experiments of underwater acoustic communications have been carried out, one of the most complex and examinant is that of the Azores Fixed Acoustic Range (AFAR), which has given the facility to verify experimentally some of the above outlined theoretical speculations [29]. From the analysis of an acoustic signal, received after its passage through an underwater channel by means of fixed submerged transmitter and receiver, the scattering function has been defined.

Figure 4 shows a real underwater transmission channel defined experimentally by means of the ray-tracing technique, provided a certain knowledge of the layered structure of the ocean is known [29]. In comparing the ray tracing of the channel with the computed scattering function shown in Fig. 5 [29], it is easy to recognize the contributions of different groups of "rays" to the time spreading (τ) and to the frequency smear φ, which are the variables of such function. The three dimensional diagram (scattering function) in Fig. 5 is the representation of Eq.(4.5).

The following considerations can be deduced:

(i)   Direct Path. The contribution of the direct path [Figs. 3 and 4] to the scattering function [Fig. 5] consists of the first arrival of the signal with a sensitive time-spreading and a relatively small frequency smear, both due to volume time-space variations only.

(ii) Bottom Reflected Paths. This contribution [Figs. 3 and 4] is characterized by a broader time spreading [Fig. 5], because of the bottom configuration (time invariant), which generates multipaths at close transit times. The frequency smear remains relatively small.

(iii) Surface Reflected Paths. The presence of surface reflected rays [Figs. 3 and 4], apart from the longer transit time, contributes to the scatter function [Fig. 5] as an evident frequency smear, due mainly to surface roughness which is time dependent.

CONCLUSIONS

The validity of such an estimation of the scattering function depends on some statistical assumptions (WSSUS) which, for an objective evaluation, heve to be carefully verified. An important interdisciplinary aspect of such investigations is the study of the oceanographic parameters encountered along a direct path during an acoustic propagation experiment. In this case, the correlations between the fluctuations of oceanographic parameters and those of the acoustic signal parameters will offer precious diagnostic tools of investigation in the respective fields. For reaching satisfactory practical results, energetic evaluations of the received acoustic signal are no longer accurate enough, and it is therefore necessary to consider other signal parameters, such as phase [31], frequency, bandwidth and time-bandwidth product, which are more significant.

ACKNOWLEDGEMENTS

The author is grateful to the US Naval Underwater Systems Center of New London, in the persons of Dr W. Von Winkle and Mr A.W. Ellinthorpe, who kindly permitted the use of Figs. 4 and 5 regarding the Azores Range. At the beginning of 1975 the Italian oceanographic ship BANNOCK of the National Research Council (CNR) participated in a joint oceanographic-acoustic experiment with ships of other nations, the concept of this experiment contributed to the content of this paper.

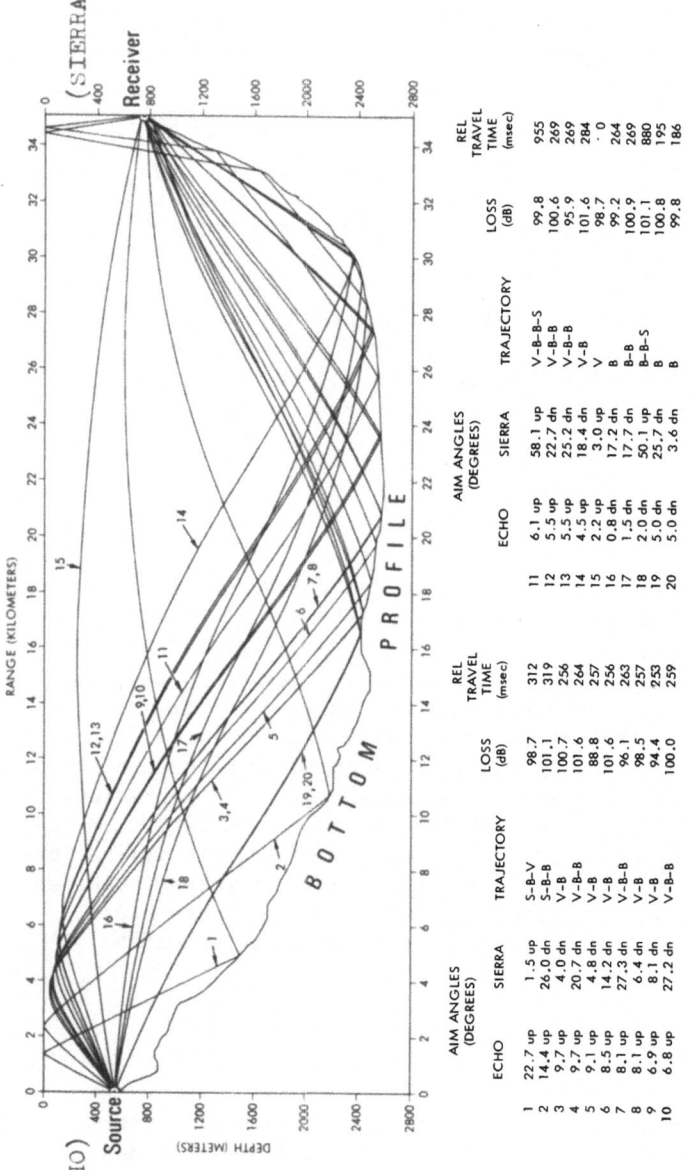

	AIM ANGLES (DEGREES)		TRAJECTORY	LOSS (dB)	REL TRAVEL TIME (msec)
	ECHO	SIERRA			
1	22.7 up	1.5 up	S-B-V	98.7	312
2	14.4 up	26.0 dn	S-B-B	101.1	319
3	9.7 up	4.0 dn	V-B	100.7	256
4	9.7 dn	20.7 dn	V-B-B	101.6	264
5	9.1 up	4.8 dn	V-B	88.8	257
6	8.5 up	14.2 dn	V-B	101.6	256
7	8.1 up	27.3 dn	V-B-B	96.1	263
8	8.1 up	6.4 dn	V-B	98.5	257
9	6.9 up	8.1 dn	V-B	94.4	253
10	6.8 up	27.2 dn	V-B-B	100.0	259

	AIM ANGLES (DEGREES)		TRAJECTORY	LOSS (dB)	REL TRAVEL TIME (msec)
	ECHO	SIERRA			
11	6.1 up	58.1 up	V-B-B-S	99.8	955
12	5.5 up	22.7 dn	V-B-B	100.6	269
13	5.5 up	25.2 dn	V-B-B	95.9	269
14	4.5 dn	18.4 dn	V-B	101.6	284
15	2.2 up	3.0 up	V	98.7	0
16	0.8 dn	17.2 dn	B	99.2	264
17	1.5 dn	17.7 dn	B-B	100.9	269
18	2.0 dn	50.1 up	B-B-S	101.1	880
19	5.0 dn	25.7 dn	B	100.8	195
20	5.0 dn	3.6 dn	B	99.8	186

NOTES:

1. "AIM ANGLES" ARE THE ANGLES AT WHICH THE ANTENNAS WOULD HAVE TO BE AIMED TO PUT THE RAY ON THE BORESIGHT AXIS. THEY ARE GIVEN WITH RESPECT TO THE HORIZONTAL.

2. THE TRAJECTORY IS EXPRESSED IN REVERSALS PROCEEDING FROM THE SOURCE TO THE RECEIVER: S IS A SURFACE REFLECTION, V A VERTEX, B A BOTTOM REFLECTION.

Fig. 4. Ray tracing of an acoustic underwater transmission channel [29]

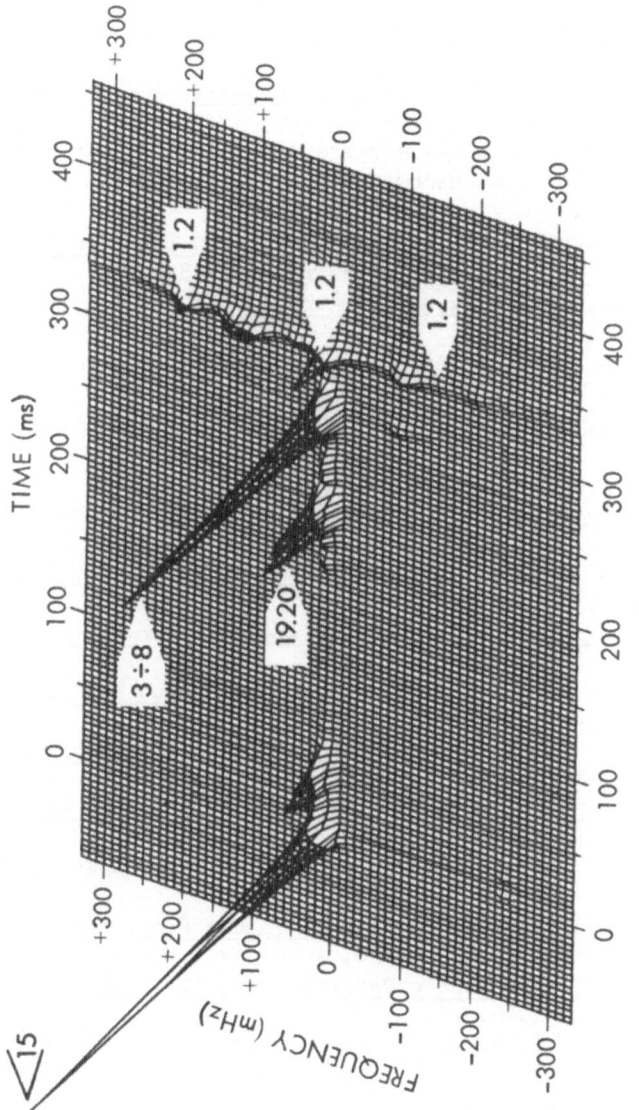

Fig. 5. Scattering function of the channel shown in Fig. 4 [29]

REFERENCES

1. Y.W. Lee, Statistical Theory of Communication, J. Wiley, 1963.
2. G.P. Patil, S. Kotz, J.K. Ord, Statistical Distributions in
    Scientific Work, NATO ASI Proceedings (Canada), Reidel Publ.
    Co., 1974.
3. B. Cook-Bernfeld, Radar Signals, Academic Press, 1967.
4. NATO ASI on Signal Processing, Loughborough, Academic Press,
    1972.
5. A.D. Whalen, Detection of Signals in Noise, Academic Press,
    1971.
6. SACLANTCEN Conference Proceedings on Oceanic Acoustic Model-
    ling, (Organizers: W. Bachman and B. Williams), La Spezia,
    1975.
7. P.M. Woodward, Probability and Information Theory with
    Applications to Radar, Pergamon Press, 1953.
8. F.A. Fischer, Grundzuge der Elektro-Kustik, Fachnerlag Schiele
    & Schon, Berlin, 1950.
9. C.E. Shannon, A Mathematical Theory of Communications, BSTJ,
    1948.
10. D. Middleton, Topics in Communication Theory, McGraw Hill,
    1965.
11. T.F. Hueter, JASA, p. 1025, Vol.51, No.3, 1972.
12. J. Max, Methodes et Techniques de Traitement du Signal,
    Masson & Cie, 1972.
13. A Sommerfeld, Partial Differential Equations in Physics,
    Academic Press, 1949.
14. R.J. Urick, Principles of Underwater Sound for Engineers,
    McGraw Hill, 1967.
15. AGARD Lecture Series No. 59, Determination and Use of Radar
    Scattering Characteristics, 1973.
16. V.I. Tatarskii, The Effects of the Turbulent Atmosphere on
    Wave Propagation, Israel Prog. for Scient. Transl., 1971.
17. I. Tolstoy, Resonant Frequencies and High Modes in Layered
    Wave Guides, JASA, Vol.28, 1956.
18. L. Meyer and H. Meyer, Radar Target Detection, Academic Press,
    1973.
19. H. Goldstein, Frequency Dependance of the Properties of Sea
    Echo, Phys. Rev., Vol.70, 1956.
20. J.R. Wait, Electromagnetic Waves in Stratified Media,
    Pergamon Press, 1970.
21. L.M. Brekhovskikh, Waves in Layered Media, Academic Press,
    1960.
22. P. Beckman and A. Spizzichino, The Scattering of Electro-
    magnetic Waves from Rough Surfaces, The McMillan Co., New
    York, 1963.
23. E. Fubini-Ghiron, Anomalie nella Propagazione di Onde
    Acustiche di Grande Ampiezza, Alta Frequenza, 1935.

24. H.L. Biørnø, Finite Amplitude Effect in Fluids, Conf.Proc.
     Copenhagen, 1973.
25. AGARD Conference Proceedings No. 138, Non-linear Effects in
     Electromagnetic Wave Propagation, 1974.
26. P.A. Bello, Characterization of Randomly Time-Variant Linear
     Channels, IEEE Trans. on Communication Systems, 1963.
27. R. Laval, Sound Propagation Effects on Signal Processing,
     NATO ASI on Signal Processing, Loughborough, 1972.
28. K.A.Søstrand, Mathematics of the Time-Varying Channel, NATO
     ASI on Signal Processing with Emphasis on Underwater Acoustics,
     Ensthede, 1968.
29. A.W. Ellinthorpe, The Azores Range, USNUSC, New London, 1973
30. B. Grandmaux, Le Canal de Transmission Acoustique Sous-Marin
     Considere comme un Filtre Lineaire Non Homogene, Proc. GRETSI,
     Nice, 1971.
31. G. Tacconi, Measurements of Phase Dispersion of an Underwater
     Acoustic Channel, Third Congress AIA, Perugia, 1975.

P A R T I I

ATMOSPHERIC INFLUENCES

ATMOSPHERIC EFFECTS: SOME THEORETICAL RELATIONS AND SAMPLE MEASUREMENTS

Alan T. Waterman, Jr.

Radioscience Laboratory, Stanford University,
Stanford, CA 94305

GENERAL OVERVIEW

The atmosphere interacts with radar through the dielectric and conducting properties of its constituents. In this introductory paper, we will discuss some of these interactions. The discussion will begin in broad, comprehensive terms and rapidly narrow down to a more detailed treatment of certain interactions --those involving refractive effects of clear air--and, more particularly, the statistical effects of turbulent scattering. Other interactions will be mentioned only briefly, leaving detailed treatment to other papers to follow.

If we first make a distinction on the basis of the atmospheric constituents, on the one hand we have the gaseous constituents --which will be given the emphasis here--and on the other the liquid and solid matter. The second category consists primarily of hydrometeors: rain and clouds. Hydrometeors are of importance in three respects. They attenuate electromagnetic waves, both through direct absorption and through scattering; they distort a wave propagated through them by their scattering effects; they add to the undesirable clutter by back scatter of the transmitted wave. These matters are discussed in the papers by Hodge and by Vogel. Aerosols are included among the solid constituents of the atmosphere. Their importance is limited primarily to scattering of optical waves--and therefore to lidars rather than radars in the centimeter or millimeter wave region--owing to their usually small size. The presence of birds, and their possible confusion with man-made radar targets, are discussed by Flock.

Jeske (ed.), Atmospheric Effects on Radar Target Identification and Imaging, 257–273.
All Rights Reserved. Copyright © 1976 by D. Reidel Publishing Company, Dordrecht-Holland.

Reducing the extent of our discussion to gaseous constituents, the major distinction to be drawn is that between dry air and water vapor. The well-known resonance absorption lines occur as shown in Table 1.

Gaseous Constituent	Resonant Frequencies	
H_2O	22 GHz	180 GHz
O_2	60 GHz	118 GHz

Table 1. Resonant absorption lines

At radio frequencies higher than those shown here, absorption is prohibitive. Except in specialized applications, it is desirous to avoid absorption, and so we will assume that is the case--i.e., that radar frequencies are chosen well away from these lines. Thus we further narrow our discussion down to effects arising from the real part of the refractive index. It is given by a formula expressing its dependence on temperature T, pressure P, and vapor pressure e:

$$n = 1 + \frac{77.6}{T} \left(P + 4800 \frac{e}{T}\right) \times 10^{-6}$$

For typical surface values, this is roughly

$$n \simeq 1.000300$$

with variations occurring in the fifth and sixth decimal places. At this point, we may distinguish between systematic variations of refractive index and statistical variations. The former are of greatest importance in air which has become horizontally stratified under stable conditions; while the latter occur as a consequence of turbulent mixing. We first make a few brief remarks about the systematic stratifications, before turning to a lengthier discussion of turbulent scattering.

SYSTEMATIC REFRACTION

Under conditions in which the refractive index varies sufficiently smoothly that its fractional change in a wave length is small, geometrical optics are applicable, and much can be learned by tracing the trajectory of rays through the atmosphere.

Fermat's principle leads to the relation

$$n\ r\ \cos\ \beta = C$$

which is a modified form of Snell's law. Here n is the refractive index (taken to be a function of height--or distance from the earth's center--only), r is the radial distance from the earth's center, β is the angle of the ray trajectory with respect to the local horizontal, and C is a constant. When applied to a spherical geometry with effective earth-radius R, the distance D along the surface of the earth may be expressed as an integral over the height range covered by the ray trajectory:

$$D = RC \int \frac{dr}{r\sqrt{n^2 r^2 - C^2}}$$

Comparable expressions may be written for the length of the ray trajectory through the atmosphere--both the physical length and the electric length. In the particular circumstance in which the refractive index dependence on height can be expressed in terms of a power of the radial distance from the earth's center,

$$n = n_0 \left(\frac{r}{R}\right)^b$$

$$= n_0 \left(1 + \frac{h}{R}\right)^b$$

$$\simeq n_0 \left(1 + \frac{b}{R}\ h\right)$$

which, as can be seen, is very nearly a linear variation with height h, the above integral can be evaluated in closed form:

$$D = \frac{R}{b+1}\ (\beta - \beta_0)$$

in which β_0 is the angle of the ray trajectory at the earth's surface. Thus given the variation of refractive index with height, it is possible to trace the trajectories of families of rays through the atmosphere, either by breaking the trajectory up into segments in each of which the simplified relationships given above apply, or by going directly to a numerical integration. A family of rays traced out in this fashion is useful in pointing out graphically regions of non-standard behavior: (a) ducting, with its associated extended ranges of coverage, (b) radio holes, regions inaccessible to radar coverage, and

(c) atmospheric multipath, with its associated distortions and wild signal fluctuations.

Systematic refractivity structures in the atmosphere can also be given a more rigorous, wave treatment in a limited number of special--usually oversimplified--cases. An example is given in the paper by Langenberg.

RANDOM SCATTERING

We now turn to our topic of primary emphasis: scattering from clear air. To clarify the emphasis, a three-fold distinction may be made:

(a) weak forward scattering
(b) strong forward (saturation) scattering
(c) backscatter

The third of these categories can act as a source of clutter in radars of sufficient sensitivity, though in most instances returns from clear air are too weak to be of serious concern. They are more useful for their role in atmospheric probes. Strong scattering from the clear air is of importance only at large ranges or under conditions of extreme atmospheric inhomogeneity. It is generally less well understood than weak scattering. Our concern is thus with the first category, weak (or single) scattering. This mechanism is now well understood, largely through the work of Tatarskii. Our treatment will draw on the somewhat equivalent analysis of Lee and Harp, largely because of its greater propensity for physical visualization.

WEAK SCATTERING CONCEPTS

A wave reflected from a point source--a single, not-too-large target--would propagate as a spherical wave in the absence of atmospheric perturbations. Since the atmosphere is not entirely homogeneous, but contains small variations from point to point of temperature and humidity, and thus of refractive index, the wave front departs from true sphericity. It has small perturbations in both phase and amplitude. The nature of these small perturbations can be seen in Fig. 1. Consider the effect of a thin slab of atmosphere at some point between target and receiver, and normal to the propagation path. Let the variations of refractive index within the slab be decomposed into spatial Fourier components, and concentrate on just one such component, of wave number κ. The refractive index, for this component, varies sinusoidally across the path with period $2\pi/\kappa$. Rays from the transmitter which pass through the crests

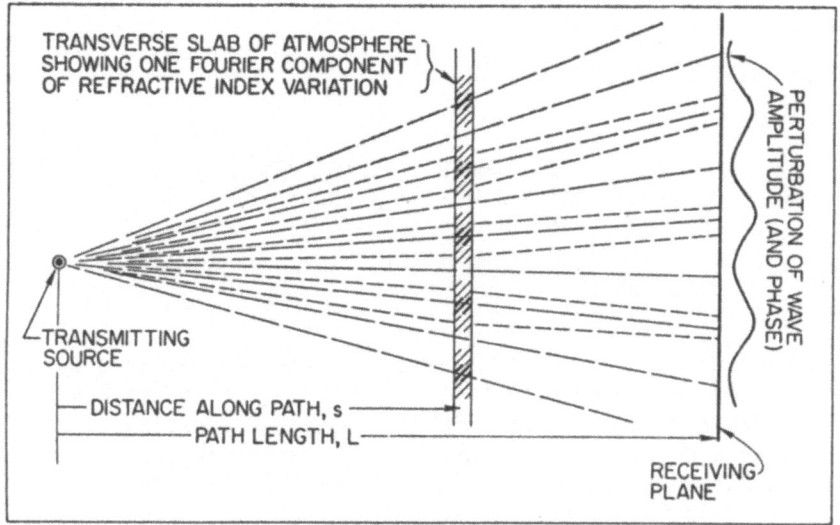

Fig. 1.   Geometrical interpretation of weak scattering.

of this sinusoid and through the troughs of the sinusoid are
undeflected in this trajectory toward the plane of the receiver.
Rays which pass through other portions of the sinusoid are bent
ever so slightly toward the rays passing through the crests.
Thus at the plane of the receiver the rays are periodically concentrated and spread out, so that the received field is periodically intensified and weakened. The spatial periodicity of
this variation in field strength is the projection of the
periodicity of refractivity in the perturbing slab. (In "weak"
scattering, the deflection is never so great as to allow the
rays to cross.)

Similarly, as regards phase variations, rays passing through
crests of refractivity in the slab are retarded upon reaching the
receiving plane, while those passing through troughs are (in a
relative sense) advanced. Thus the spatial periodicity of phase
variations at the receiving plane is also a projection of the
refractivity periodicity in the slab.

The total variation will be a sum--or integral--of all such
variations taken over all wave numbers in the slab and over all
slabs in the path. It is weighted by the spectral distribution
of refractive index variations and the relative contribution of
each wave number at each location to the total variation. For
example, the variance of small amplitude fluctuations is

proportional to

$$\int d\kappa \int ds \left[\kappa \Phi(\kappa)\right]\left[\sin^2\left(\frac{\kappa^2 s}{2k}\left(1 - \frac{s}{L}\right)\right)\right]$$

Here, κ = wave number of atmospheric refractivity
     $\Phi(\kappa)$ = 3-dimensional refractivity spectrum
       k = wave number of propagating wave
       s = distance from transmitter
       L = path length

For phase fluctuations, the sine-squared factor is replaced by a cosine-squared factor (of the same argument). The covariance between signals received at two laterally spaced locations, separated by a spacing d, includes another factor in the integrand, $J_0(\kappa sd/L)$. If there is a lateral drift of the atmosphere with velocity $v_\perp$ transverse to the path, the time lagged covariance includes this same Bessel-function factor with a modified argument, $J_0(\kappa sd/L + v_\perp \tau)$, in which the velocity may also be a function of position along the path. To find the covariance between two frequencies, of wave number k_1 and k_2, the sine squared factor is replaced by the product of two sines, one containing k_1 in its argument, and the other k_2. Taking the Fourier transform of the time lagged covariance yields the corresponding spectrum. Additional factors must be included if the filtering effects of using large antenna apertures are to be taken into account.

MEASURED VALUES OF SIGNAL FLUCTUATIONS

Now let us turn to measurements of these quantities. The examples below will be drawn mostly from a 28 km line-of-sight path operating at 35 GHz. Fig. 2 shows the covariance (at zero time lag) of amplitude fluctuations received at two laterally spaced locations, from a single transmitter, as a function of receiver spacing. Measured values are indicated by circles (with error bars shown), and the curves are computed theoretically using five different spectral slopes in the assumed refractivity distribution--the exponent -11/3 (i.e., h=11) corresponding to a Kolmogorov distribution. If geometrical configurations other than a single-source-two-receiver arrangement are used--i.e., a parallel-path arrangement or a crossed-path arrangement--then the space covariance of amplitude fluctuations is expected to behave as shown in Fig. 3. Here two curves are drawn for each geometrical configuration, corresponding to different values of the exponent of the refractivity spectrum. It can be seen that the extent of this region over which the received signal fluctuations remain correlated is strongly dependent on the nature of the source.

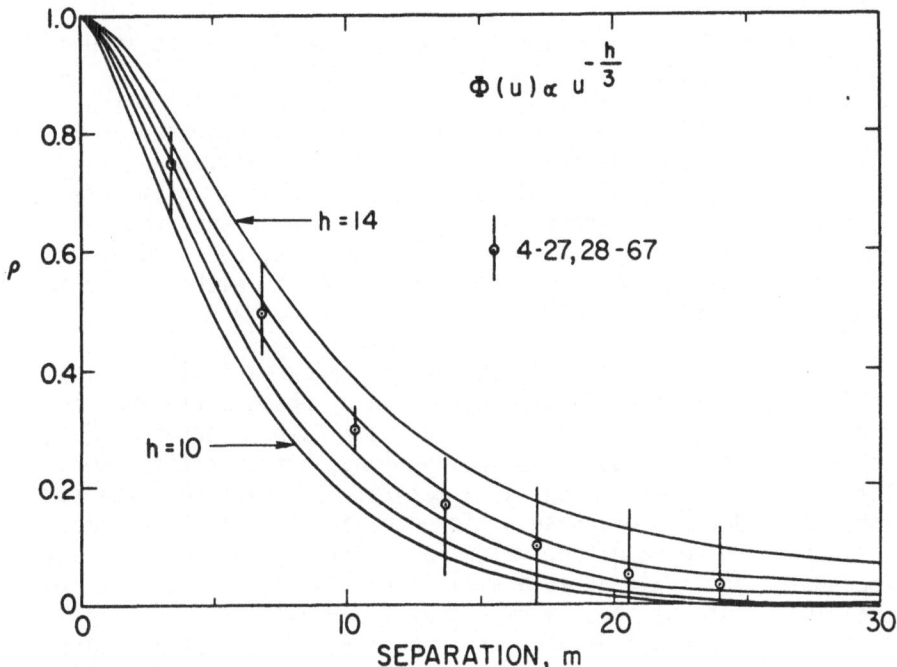

Fig. 2.   Amplitude Covariance Functions for Different Spectra
          $\Phi(u)$, with Experimental Data.

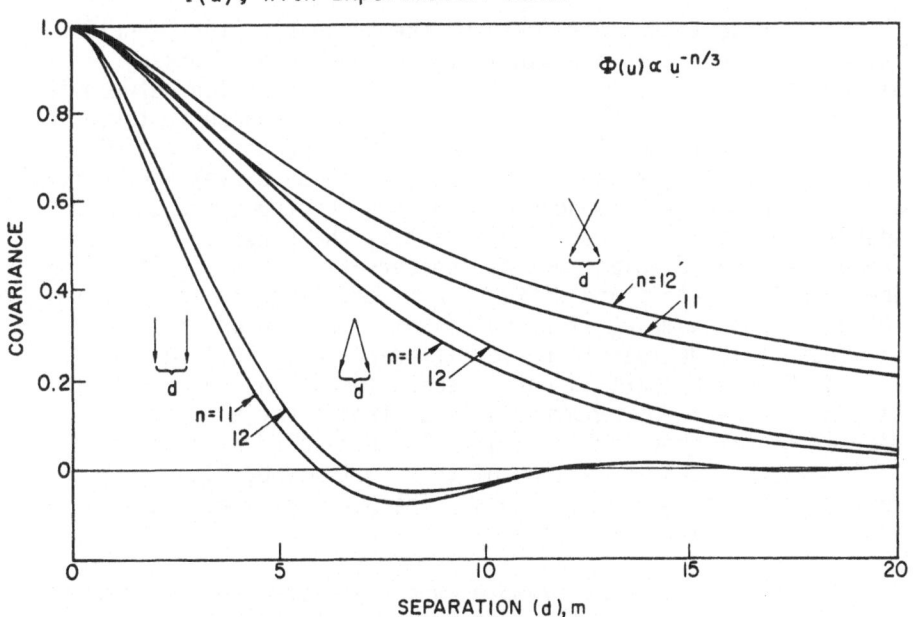

Fig. 3.   Amplitude Covariance as a Function of Path Separation
          d, for Three Path Geometries.

Next we consider measured values of the difference in phase
of the signal received at two laterally spaced points. Fig. 4
shows the variation, throughout a 40-hour period, of the 160-
second mean difference in phase between points separated by 24
meters. What is significant here is the variability, not the
absolute value (which in this case merely reflects an imperfect
alignment in initial phase settings). It can be seen that on
only a few occasions does the phase difference, averaged over a
160-second period, depart by as much as 20°, in either direction,
from the long-term average.

Another measure of this phase-difference variation is shown
in Fig. 5, where the standard deviation of phase difference is
plotted against time of day over the same 40-hour period. Each
sampling period is again 160 seconds long, but this time the
lateral antenna separation is 12 meters. The left ordinate is
labeled in degrees of phase difference, and the right ordinate
is labeled in terms of equivalent angle-of-arrival, expressed in
seconds of arc. Fig. 6 is a similar record taken on a different
occasion. Both of these show angle-of-arrival variations to be
less than 10 seconds of arc except on rare occasions--a result
typical for the path in question. Thus, although the signal
fluctuations are only weakly correlated at these receiver separations, the phase-front distortions are relatively small.

To pursue this point further, the phase structure-function
may be examined. It is essentially the variance of phase
differences expressed as a function of the separation between the
two points at which the difference in received phase is taken.
Fig. 7 shows some sample measurements, plotted to logarithmic
scales. Weak scattering theory predicts these curves should have
a 5/3 slope, for a Kolmogorov turbulence spectrum, and that is
seen to be the case here, independent of the strength of turbulence (vertical displacement of the curves). This result allows
one to extrapolate phase-difference fluctuations to larger separations. For example, if we note from Figures 6 and 7 that a
variance of 10 degrees in phase difference between receiving
points 12 meters apart is not exceeded too often, their extrapolation via the 5/3 dependence tells us that a variance of 180
degrees (total phase ambiguity) would be reached at a separation
of 68 meters. Considerations of this sort may be significant
for synthetic-apertures antennas. The paper by Gjessing will
analyze this point more fully.

We now turn to some examples which include the time dependence
of signal fluctuations--not just their magnitude. Fig. 8 shows
covariances between signal amplitudes as received at two laterally
spaced antennas, as a function of time lag. Eight curves are
shown, corresponding to the covariances between antenna #1 and

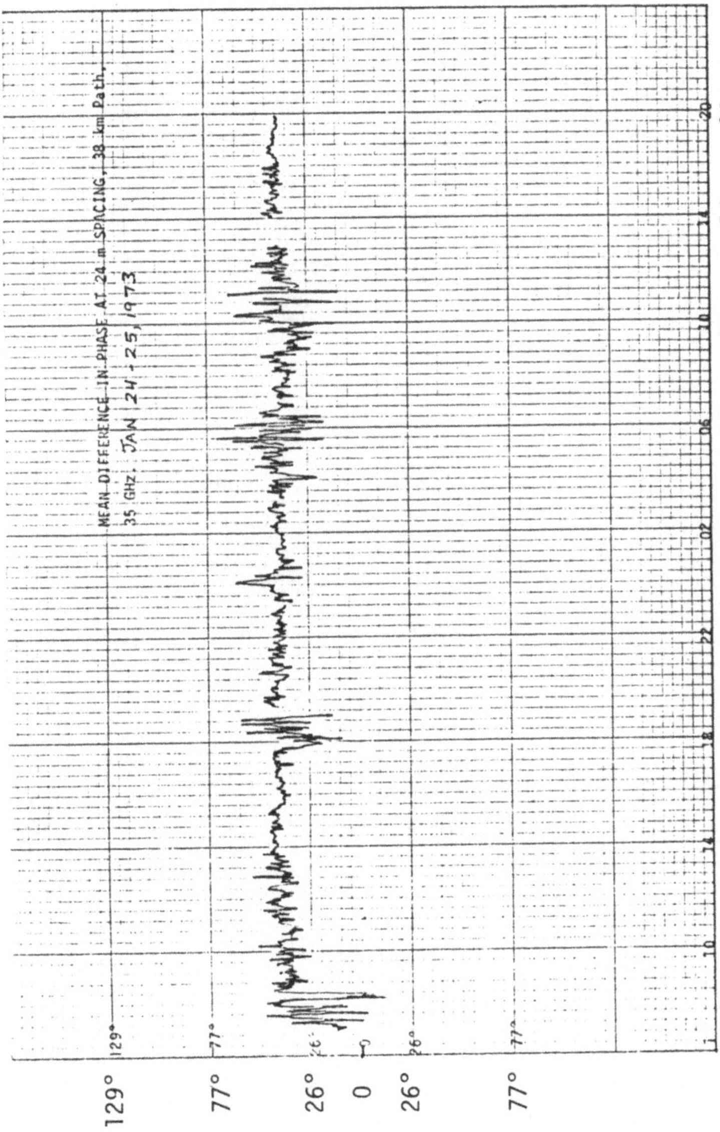

Fig. 4. Mean Difference in Phase at 24 m Spacing, 38 km Path, 35 GHz.
         Jan. 24-25, 1973.

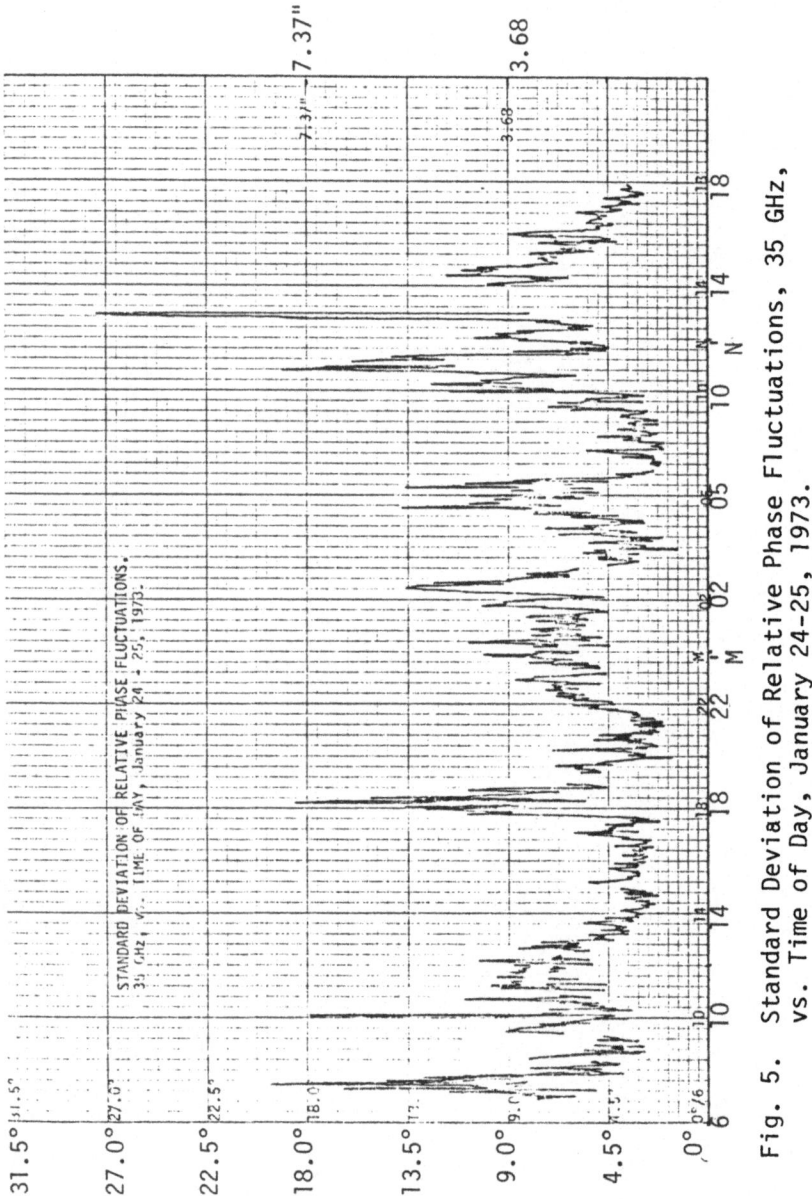

Fig. 5.   Standard Deviation of Relative Phase Fluctuations, 35 GHz,
vs. Time of Day, January 24–25, 1973.

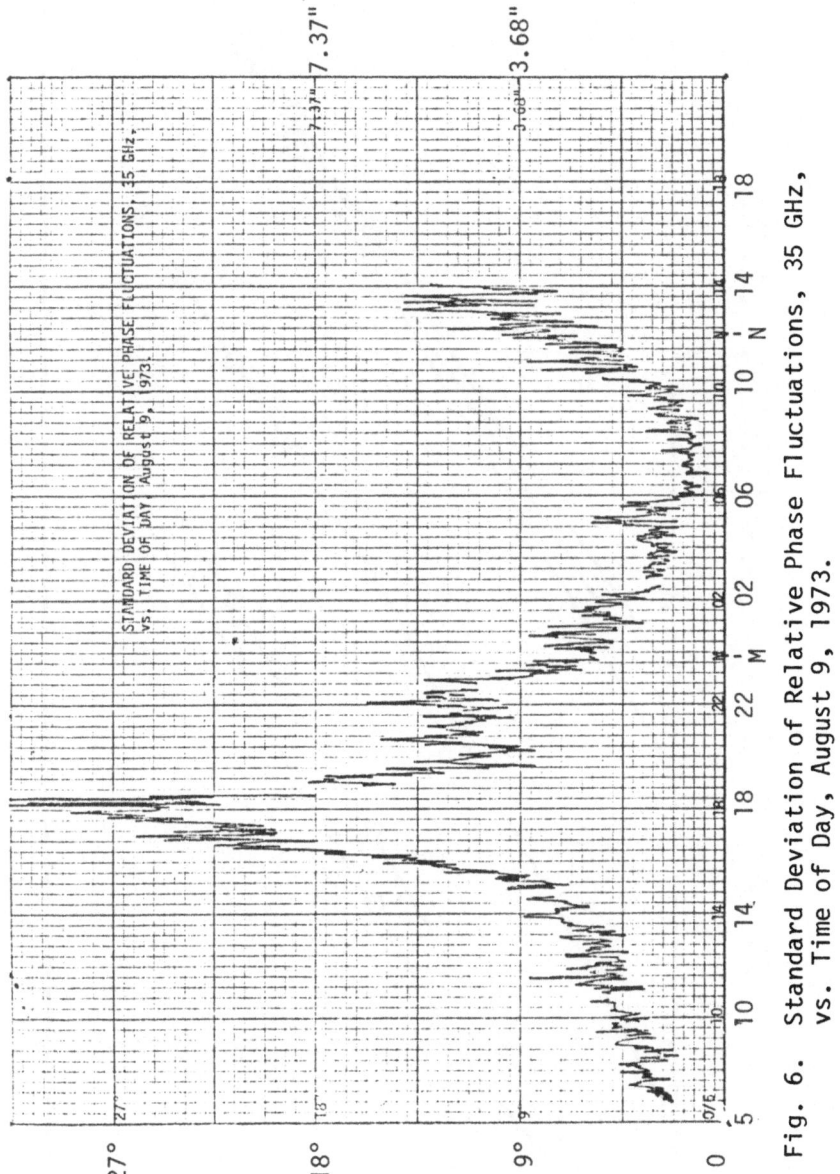

Fig. 6. Standard Deviation of Relative Phase Fluctuations, 35 GHz, vs. Time of Day, August 9, 1973.

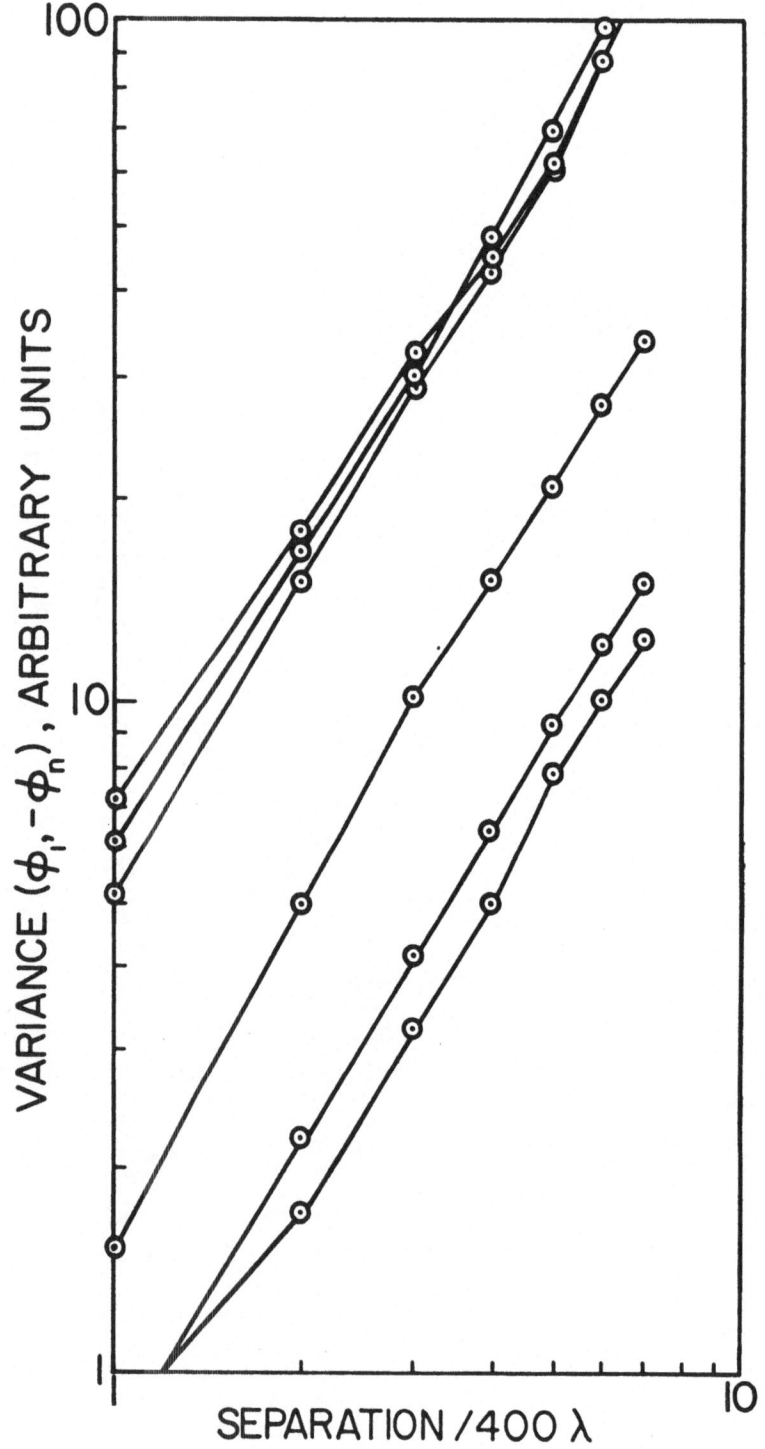

Fig. 7. Measured phase structure-functions.

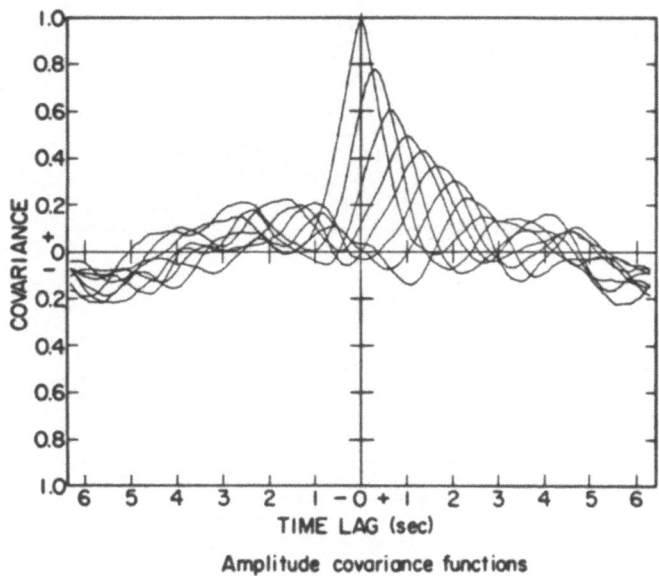

Amplitude covariance functions

Fig. 8. Measured Covariance Functions.

each of the other antennas (including the autocovariance of #1
with itself). Notable points are the following: (a) the time
lag of the peak value of the covariance increases as the separation between receiving antennas increases, (b) the peak value
of the covariance decreases gradually with increasing antenna
separation, and (c) the width of the autocovariance curve (with
peak at unity value and zero lag) is roughly consistent with
the rate of drift of peak covariance. Thus the overall picture
is one of a medium whose structure is changing relatively slowly
as it drifts across the path; in short, the data support Taylor's
frozen-turbulence hypothesis. Confidence in this picture is
affirmed by Fig. 9 which shows the theoretical curves computed
for an atmospheric model having a Kolmogorov spectrum and uniform
wind and turbulence distribution along the path. The similarity
to the measured curves of Fig. 8 is evident.

The spectrum of signal fluctuations may be obtained by taking
the Fourier transform of the time-lagged covariance function;
alternatively, it may be obtained by passing the fluctuation
signal through a bank of filters. Fig. 10 gives some examples.
Here normalized spectral density is plotted against a normalized
frequency. The quantities whose spectra are shown are differences
in received phase as measured at three antenna separations:
3.44, 10.32, and 24.00 m (corresponding to the normalized

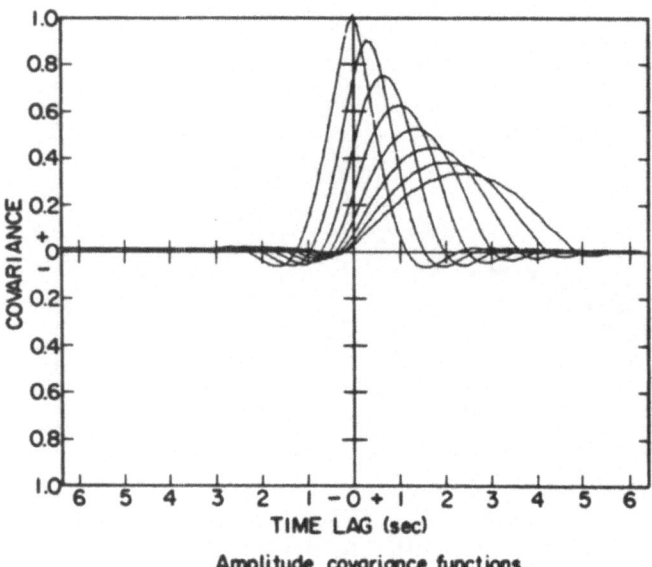

Amplitude covariance functions

Fig. 9. Covariance vs. time lag for uniform wind field.

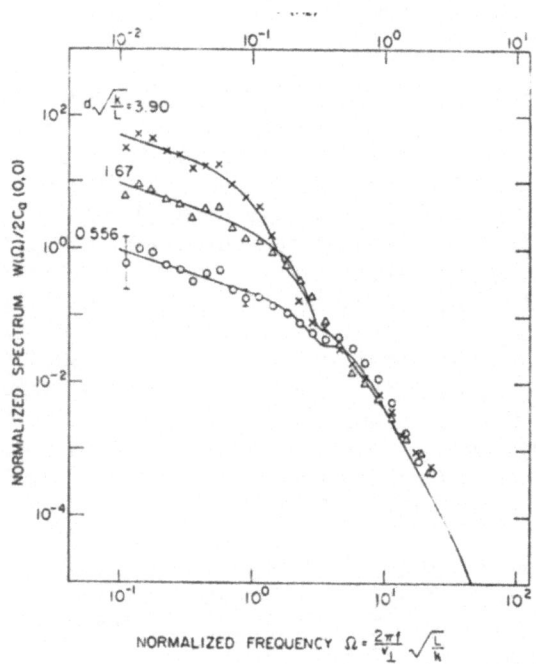

Fig. 10. Experimental and theoretical phase-difference
          fluctuation power spectra at mm wavelengths. Measured
          $v_\perp$ = 2.4 m sec^{-1}, calculated $v_\perp$= 3.8 m sec^{-1},
          calculated C_n = 3.1 x 10$^-$ m $^{1/3}$ on July 13, 1967, at
          2100.

separation markings of 0.556, 1.67 and 3.90 respectively). In the abscissa, a normalized frequency of $\Omega = 10$ corresponds to an actual frequency of about 0.97 Hz in the measured data. To make the theoretical model account for this frequency scale, a transverse wind drift across the path at 3.8 m/s is required. On this occasion the wind measured at mid-path at ground level (a few hundred meters below the line-of-sight ray trajectory) was 2.4 m/s--a reasonably consistent figure. Thus our confidence in the theory is further reinforced, and the spectrum gives us a feel for the time scale on which the signal is fluctuating.

Another characteristic of weak line-of-sight scattering is aperture smoothing--the reduction of higher frequency fluctuations where large antennas are used. Fig. 11 shows an example. These curves are spectra of amplitude fluctuations and amplitude-difference fluctuations--as labeled. The dashed curves marked $\sigma = 0$ are theoretical spectra computed for "point" receiving and transmitting antennas, whereas the other curves take into account the actual finite extent of the paraboloid antennas used. It can be seen that the measured points follow these latter curves up to a higher fluctuation frequency--before system noise and other complications set in.

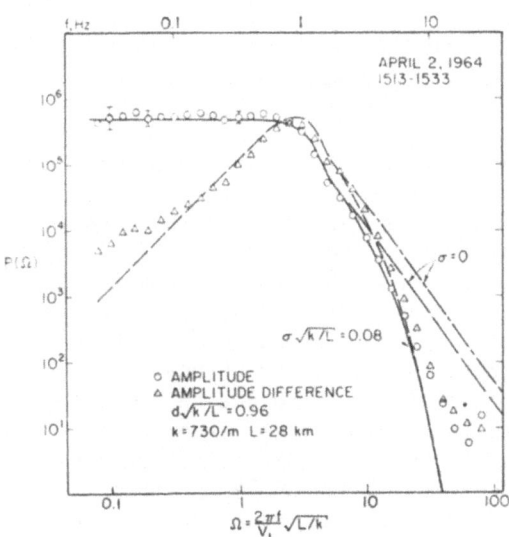

Fig. 11.   Amplitude scintillation power spectra, as a function of normalized fluctuation frequency Ω.

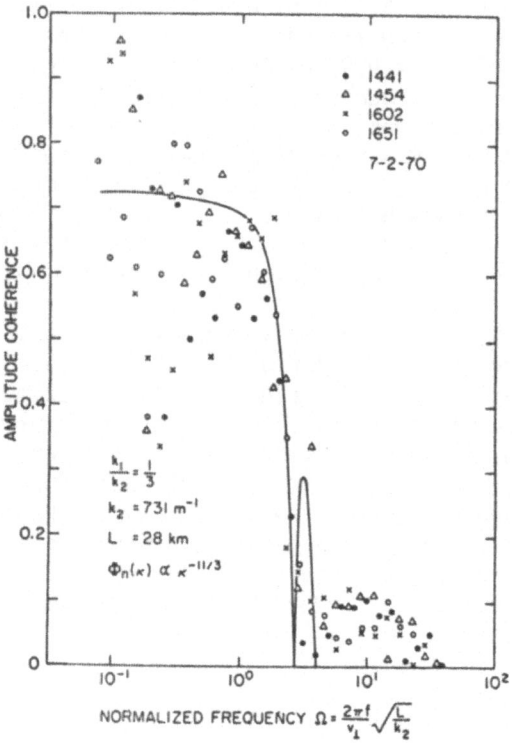

Fig. 12.   Experimental and theoretical coherence of amplitude
            fluctuations at radio wavelengths.

        Finally, the coherence between signal fluctuations at two
widely separated carrier frequencies shows interesting characteristics. Fig. 12 shows the magnitude of the coherence as a
function of normalized fluctuation frequency; here the two
carrier frequencies are 11.63 and 34.89 GHz--having a ratio of
1:3. The variously designated data points entered are measured
values, while the solid curve is computed from theory. The
cut-off at a normalized fluctuation frequency of Ω = 2.6 is
evident both in the theoretical curve and in the data, in spite
of the wide spread in the data at lower frequencies. Scattering
by small scale sizes (higher fluctuation frequencies) is
uncorrelated between the 11.63 and 34.89 GHz transmissions.
Scattering by larger scale sizes is significantly better--
though imperfectly correlated; the theoretical coherence
approaches a limiting value of 0.72 for this 1:3 carrier-
frequency ratio. The normalized cut-off frequency of Ω = 2.6
translates into a cut-off scale size (of atmospheric scatterers)
of about 15 m for the circumstances of this experiment.

CONCLUDING REMARKS

The above examples indicate some of the principal characteristics of the effects of weak scattering on a microwave or
millimeter wavelength signal propagated through the atmosphere
under normal conditions. For the most part, these scattering
effects are small and would not sifnificantly affect the
performance of an ordinary radar--in contrast with the more
drastic effects of refractive bending, as with radio holes, for
example. However, for systems which require coherence over
large apertures or for appreciable time intervals, these effects
can be of critical importance. The paper by M. C. Thompson
provides additional examples. Gjessing's paper explores some
of these matters in greater extent and specificity. The paper
by D. W. Thomson indicates how meteorological information can be
used to advantage, while that by J. Richter describes a specific
system for this purpose.

Acknowledgement: The examples used here were taken from work
supported largely by the U. S. Air Force Cambridge Research
Laboratories under contracts F19628-68-C-0055 and
F19628-71-C-0050.

NOTE:

Figures 2, 3, 7, 8 and 9 are taken from Lee and Harp, Proc. IEEE
57, 4, 375; Apr. 1970.

Figure 10 is taken from Mandics, Lee and Waterman, Radio Science 8,
3, 185; Mar. 1973.

Figure 11 is taken from Lee, Radio Science 6, 12, 1059; Dec. 1971.

Figure 12 is taken from Mandics, et al., Radio Science 9, 8 & 9,
723; Aug.-Sept. 1974.

DISTORTION OF RADAR PULSES BY ATMOSPHERIC LAYERS

K. J. Langenberg

Fachrichtung Elektrotechnik, Universität
des Saarlandes, D-66 Saarbrücken,
Fed. Rep. Germany

ABSTRACT. The paper presents a theory which allows to
calculate the atmospheric distortion of radar pulses
exactly provided the influence of the atmosphere consists in ducting the transmitted signal through a
ground layer. The duct model used is that of Kahan
and Eckart consisting of a discontinuous drop of the
otherwise constant relative permittivity at the upper
duct boundary. The earth is assumed to be an ideal
conductor and ideally plane. The source of the electromagnetic field is taken to be a vertical magnetic
dipole within the surface layer. Assuming the transmitted signal to be the building up of a carrier
frequency the transient behaviour of the electric
field strength at any distance in the duct layer can
be given exactly. The numerical analysis leads to
numerous curves being thoroughly discussed especially
regarding the results which would be given by the mode
theory if it is applied to the same problem.

Jeske (ed.), Atmospheric Effects on Radar Target Identification and Imaging, 275–287.
All Rights Reserved. Copyright © 1976 by D. Reidel Publishing Company, Dordrecht-Holland.

1. INTRODUCTION

The correlation between the propagation of electromagnetic waves in the atmosphere and meteorological
conditions has been investigated by radiometeorolo-
gists for the last thirty years. Today the interest
is mainly on the effect of the atmosphere on target
identification and imaging. Because of the variability of the atmospheric refractive index the signal velocity can be highly variable resulting in distortion
of the signal pulse which propagates to the radar target. This paper treats the special problem of propagation of a modulated carrier frequency impulse in an
abnormal stratification of the lower troposphere, a
so-called surface duct. Such inversions occur particularly in the boundary layer along the sea.
In considering our problem we will restrict ourselves
to one of the simpler duct models. It was developed
by Kahan and Eckart [1] and assumes a discontinuous
drop in the otherwise constant refraction index at
the upper duct boundary. The earth is assumed to be
an ideal conductor and ideally plane, which corresponds well to the situation at sea in the microwave
frequency range.
We wish to extend the steady state duct propagation
theory of Kahan and Eckart to transient excitation
when no restrictions on the distance between receiving
and transmitting end are made. We therefore use a
method originally due to Cagniard [2] which has been
simplified by de Hoop and Frankena [3,4]. Recently
an extension of the method to the case of total reflection phenomena occurring in duct propagation has
been published by the author [5].

2. FORMULATION OF THE PROBLEM

Fig. 1 shows the duct model of Kahan and Eckart. A

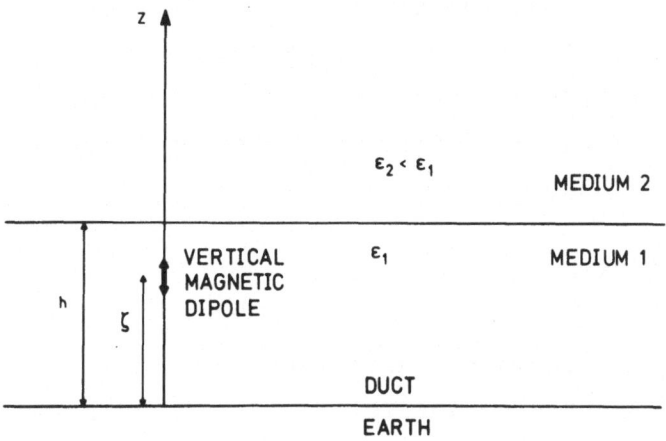

Fig. 1. Geometry of the problem

dielectric layer is assumed of relative permittivity ε_1 overlying an infinitely conducting plane earth which is confined by the plane $z = 0$ of a cartesian coordinate system (x,y,z). At the height h this permittivity decreases discontinuously to the value ε_2 . The relative permeability μ is assumed to be constant throughout the half-space $z > 0$. We refer to the layer as medium 1 and to the half-space $z > h$ as medium 2. The fields which belong to the two media are marked by correspnding indices. The source of the field is assumed to be a vertical magnetic dipole in medium 1 at the point $x = y = 0$, $z = \zeta > 0$ whose moment is given by $\vec{p}_m = \{0; 0; F(t)\delta(x,y,z-\zeta)\}$, t being the time variable. Regarding F(t) we make the causality assumption F(t) = 0 for $t < 0$. This guarantees us the uniqueness of our solution.

Physically, F(t) describes the time variation of the current which flows in an equivalent small loop an-

tenna, whose transient radiation field is not very
different from the one radiated by an ideal magnetic
dipole [6].

Our aim is to determine the electrical field strength
$\vec{E}^{(1)}(x,y,z;t)$ exactly at some fixed point (x,y,z)
within the duct layer as a function of time, having
chosen $F(t)$ as a modulated carrier frequency pulse.

3. METHOD OF SOLUTION

The method will be explained using the flow chart of
Fig. 2; for the corresponding formulas the reader is
referred to [7].

The method is essentially based on the application of
two functional transforms with respect to time and to
the horizontal spatial coordinates. The starting point
is the wave equation for the electrical field strength
in the original space $(x,y,z;t)$.

The first step is the application of a Laplace transform with respect to time, which yields a three-dimensional Helmholtz equation in the spatial coordinates. This step uses the initial conditions: electric
field strength being zero for t equal to zero. The
next step is the similar use of a two-dimensional
Fourier transform with respect to x and y, yielding
a one-dimensional Helmholtz equation in the height
coordinate z. This is of course an ordinary differential equation, whose solution can be written down
immediately in terms of unknown integration constants.
These constants can be determined as solution of an
algebraic system of equations resulting from the
boundary conditions for the electromagnetic field on
the earth's surface and at the upper duct boundary.

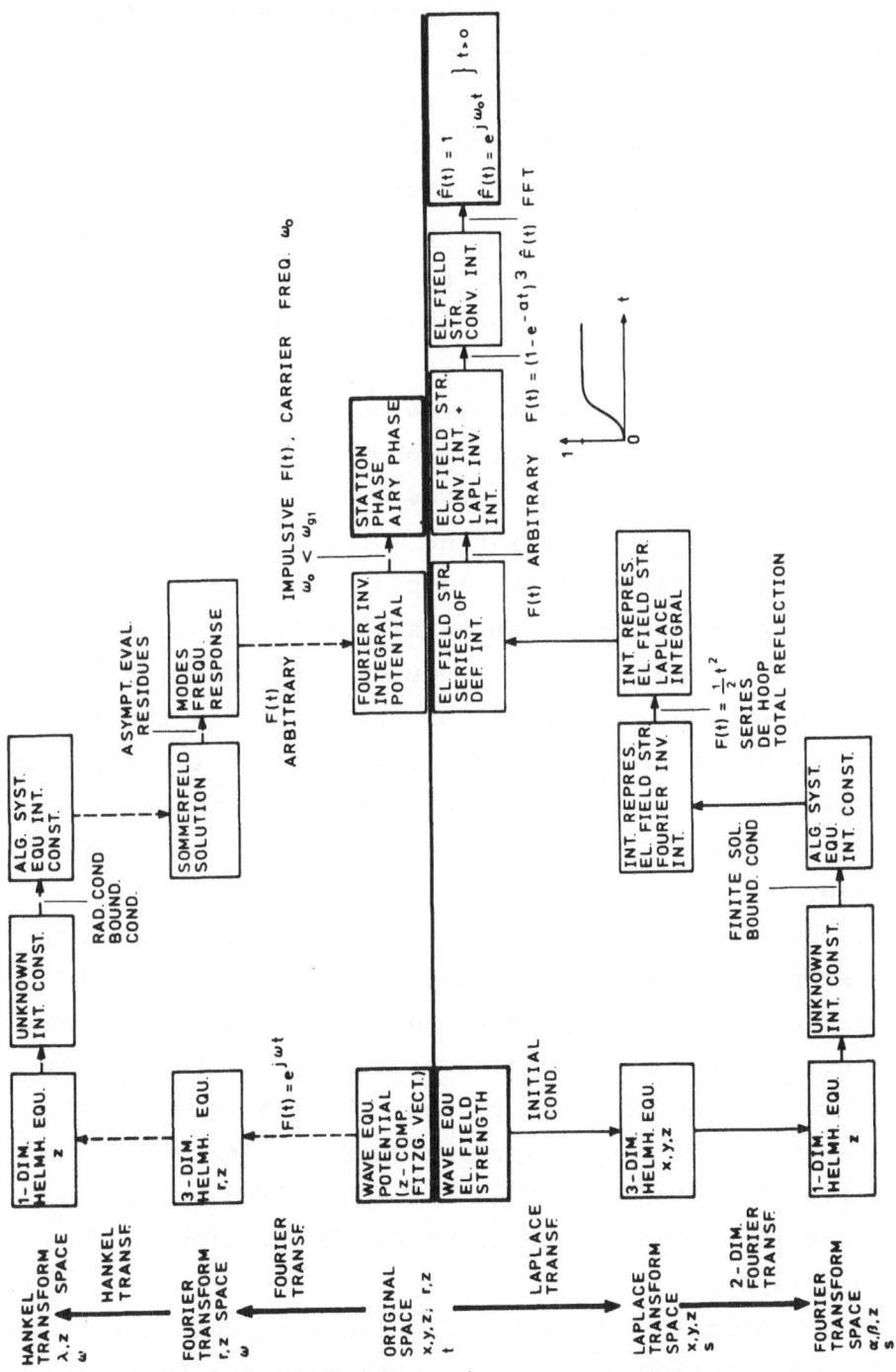

Fig. 2. Method of solution (---: mode theory;
_____ : exact theory)

A further condition used is the condition of finiteness of the field at infinity.

Thus results an integral representation of the Laplace transform of the electric field strength in terms of two-dimensional inverse Fourier integrals. Now we do not use the inverse Laplace integral to come back to the original space; we use instead a method of de Hoop [3], which can be extended to our present problem [5].

This method works if we first choose $F(t) = \frac{1}{2}t^2$ and if we secondly use a series expansion of the integrands of our integral representation. Physically, this series expansion yields a representation of the solution in terms of the images of the primary source which result from the successive reflections at the earth's surface and at the upper duct boundary. The result of the application of the method is the representation of the Laplace transform of the electric field strength as an explicit Laplace integral. So one is able to read off the solution in the time domain.

We can describe our present result as follows: it is a representation of the electric field strength in the original space in terms of an infinite series of definite integrals with finite boundaries. Physically spoken we have got the response of the communication channel "duct layer over sea surface" to the excitation function $F(t) = \frac{1}{2}t^2$ at some fixed but arbitrary situation of the point of observation in the duct layer.

An arbitrarily time varying antenna current now yields the convolution of this response with the third derivative of $F(t)$ plus certain inverse Laplace transforms [7] which vanish if we choose $F(t) = (1-e^{-at})^3\hat{F}(t)$

$\hat{F}(t)$ being another arbitrary time function. The convolution integral has been numerically evaluated by
means of the fast Fourier transform for the cases
$\hat{F}(t) = 1$ and $\hat{F}(t) = e^{j\omega_0 t}$; that's to say, the second
case corresponds to a cw antenna current, which is
switched on according to a rise time being proportional to a^{-1}.
The numerical results for $\hat{F}(t) = 1$ are described in
[7].
Before presenting the results of our exact theory for
the case $\hat{F}(t) = e^{j\omega_0 t}$, an alternative approximate
procedure, the well-known mode theory will be brief-
ly reviewed. This will be done for two reasons; first-
ly, to point out, where the mode theory is an approximate theory and secondly, to be able to discuss the
exact theory's results in terms of the cut-off frequencies which are given by the mode theory.

4. REVIEW OF THE MODE THEORY

Again we refer to the flow chart of Fig. 2.
The mode theory uses a Fourier transform with respect
to time leading to a frequency space with the variable
ω , and a Hankel transform with respect to the horizontal distance of the receiving and transmitting
end.
Until the box "Sommerfeld solution" the procedure is
mainly the same as the one discussed in chapter 3.
The Sommerfeld solution itself is an inverse Hankel
transform in the Fourier transform space, i.e. the
monochromatic solution of the problem.
To come back to the time domain, there are two approximations involved; the first one is the asymptotic

evaluation of the Sommerfeld solution for large distances in terms of residues [1], the so-called modes.
The single modes can be described by their frequency
response, which is shown in Fig. 3.

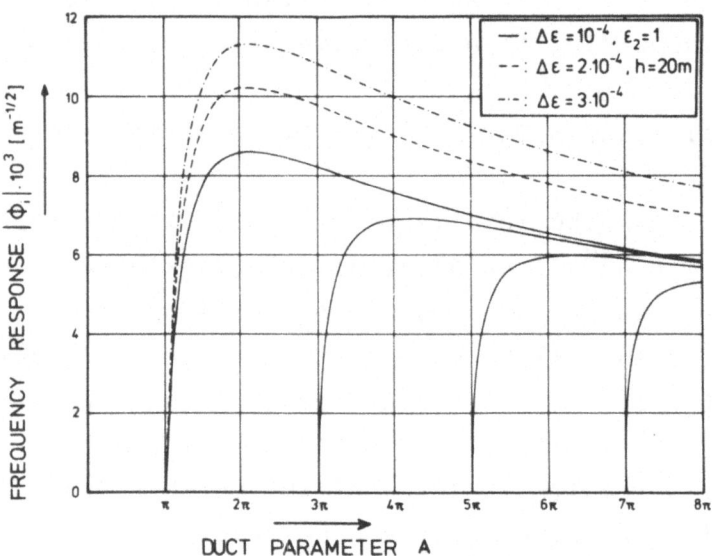

Fig. 3. Frequency response of the first four modes

We see the sudden occurrence of single modes at discrete values of the duct parameter A, which is pro-
portinal to ω ; the corresponding frequency values
f_{gi} (i = 1,2,3,...) are called cut-off frequencies.
For $\Delta\varepsilon = \varepsilon_1 - \varepsilon_2 = 4.10^{-4}$ and h = 20 m, the first
four of them are given by f_{g1} = 188 MHz, f_{g2} = 563 MHz,
f_{g3} = 940 MHz, f_{g4} = 1316 MHz [7].
Now, from the monochromatic behaviour to the time
variation, we have to Fourier transform the product
of the frequency response and the spectrum of F(t),
the arbitrary excitation function. Now the second
approximation is involved in the evaluation of this
inverse Fourier integral; the method used is the
method of stationary phase and it only works, if the

carrier frequency ω_0 of the transmitted signal is
less than the cut-off frequency ω_{g1} of the first mode.
The well-known result is the typical waveform of the
space wave, the rider wave and the Airy phase [8].
This restriction yields the fact, that only the high
frequencies of the spectrum of F(t) can propagate in
the layer, the carrier frequency itself is absolutely cut off. The comparison of the exact theory's results gives us better knowledge of the mechanism of
duct propagation.

5. DISCUSSION OF THE RESULTS

Fig. 4 shows the absolute value of the electric field
strength as a function of some normalized time τ ,
which is normalized to the arrival time of the primary wave front originating from the dipole source
itself; that's to say $\tau = 0$ refers to the arrival
time of this wave front. One can see, that the received signal has nonzero values for τ being negative; there is something before the primary wave
front. These so-called Mach waves have a conical
wave front, traveling at the upper duct boundary of
the layer with the phase velocity of the half-space
above the layer, which is greater than the one inside the layer, resulting in those early arrival
times.
Throughout Fig. 4 the transmitter receiver distance
is r = 15000 m; the antenna current F(t) has been
chosen to be the building up of a carrier frequency,
varying from 25 MHz to 2 GHz. Beginning with the discussion of the results for 25 MHz, which is small
compared to the cut-off frequency of the first mode,

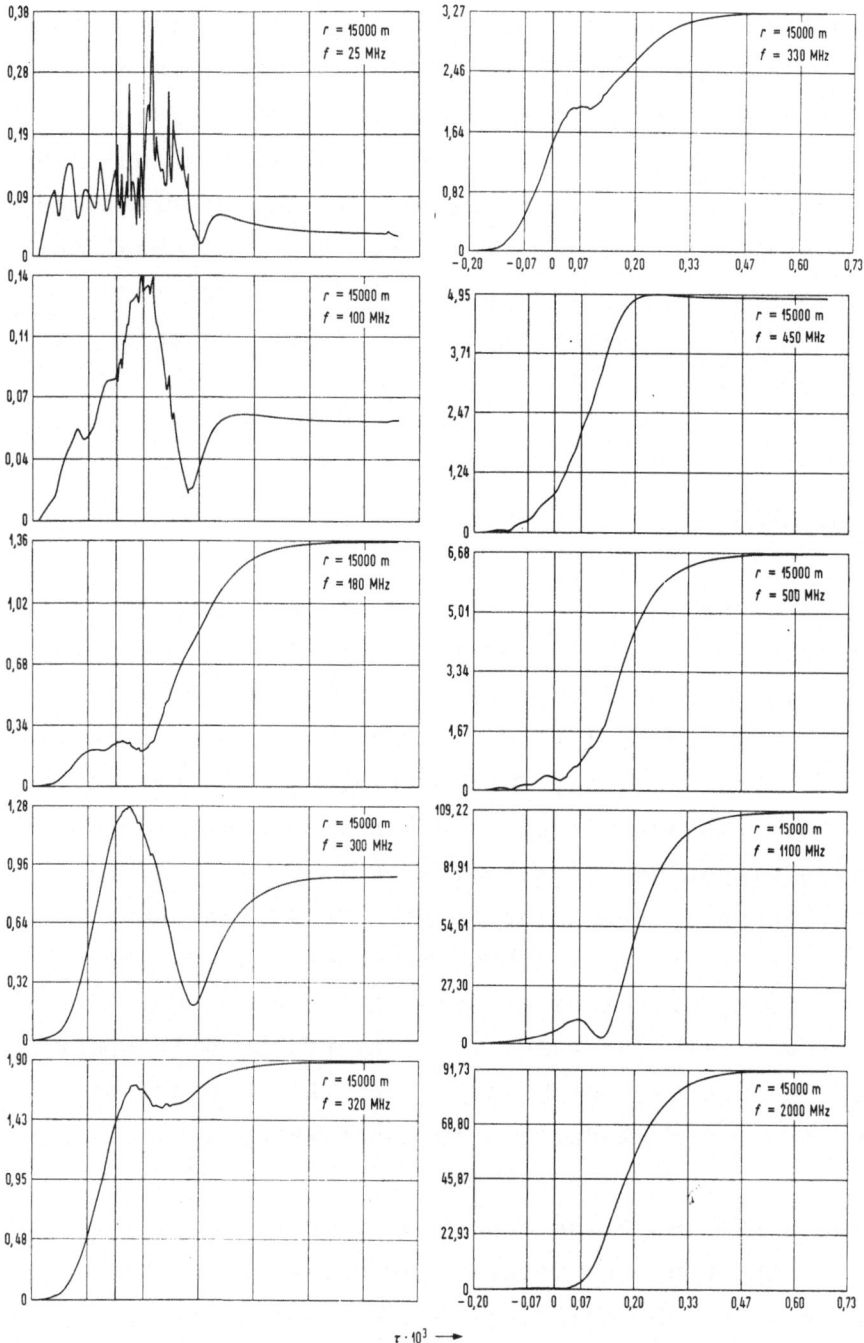

Fig. 4. Time history of the electric field strength.

we see, that the carrier frequency is not absolutely
cut off, because the received signal becomes constant
for large times as well as the envelop of the input
signal. Nevertheless the transmitted signal suffers
great distortion for such low carrier frequencies;
for example, the peaks due to the different image
sources can still be distinguished. The possibility
of distinguishing these sources vanishes with increasing carrier frequency (100 MHz) and the amplitude
of the carrier frequency itself becomes more pronounced. The value of 180 MHz is just below the cut-
off frequency of the first mode, resulting in a received signal which already is very similar to the
input signal. But the distortion increases once again
if the carrier frequency is again increased; if it
equals 300 MHz, the maximum of the spectrum of the
emitted signal is in the neighbourhood of the first
mode's frequency response maximum, so that the steep
increase of this frequency response at the cut-off
frequency affects the signal considerably. A small
increase of the carrier frequency (320MHz, 330 MHz)
shifts the signal's spectrum into that part of the
frequency response, where it is slowly varying, so
that the distortion of the received signal is reduced.
At 450 MHz, the distortion of the signal is nearly
negligible; the only indication that it propagated
through the duct is the early arrival time due to
the Mach waves.
A further increase of the carrier frequency yields
the influence of the higher order modes. The value
of 500 MHz is just below the cut-off frequency of
the second mode, which affects the leading tail of
the received signal once again. At 1100 MHz we find
ourselves in the third mode's maximum; the result is

similar to the case of 300 MHz. Finally, at 2 GHz, there is absolutely no influence of the duct on the received signal; even the arrival time coincides with the free space arrival time.

6. CONCLUDING REMARKS

The examples, which have just been discussed are based on some distance r, which can be considered as great, where this "great" can be defined clearly by the exact theory [5]. And for those distances the mode theory should not yield bad results; that's what we learned from the examples justshown: for great distances, the dispersion characteristics of a ducting layer can be explained principally assuming the existence of certain discrete modes. Only one prediction of the mode theory is absolutely wrong: carrier frequencies which are less than the cut-off frequency of the first mode are not fully cut-off. But if we want to investigate the distortion of the received signal for distances which are not great, only the exact theory yields good results [7]. For example, the distance r = 5000 m shows principally the same behaviour as r = 15000 m if we increase the carrier frequency from zero to a certain upper limit. Yet the characteristic distortion of the signal which can be explained by single modes does not occur at the frequencies predicted by the mode theory: the cut-off frequencies are approximately halved.
If we choose r = 1500 m, there is no indication of characteristic cut-off frequencies . Carrier fre-quencies above 100 MHz do not show any distortion.

REFERENCES

1. T. Kahan, G. Eckart, Théorie de la propagation
    des ondes dans le guide d'ondes atmosphérique.
    Annales de Physique 5 (1950),641-705

2. L. Cagniard, Réfléxion et Réfraction des Ondes
    Séismiques Progressives. Gauthier-Villard,
    Paris 1939.

3. A..T. de Hoop, A modification of Cagniard's
    method of solving seismic pulse problems.
    Appl. Sci. Res. B8 (1959), 349-356

4. A. T. de Hoop. H. J. Frankena, Radiation of
    pulses generated by a vertical electric dipole
    above a plane non-conducting earth.
    Appl. Sci. Res. B8 (1960), 369-377

5. K. J. Langenberg, The transient response of a
    dielectric layer.
    Appl. Phys. 3 (1974), 179-188

6. K. J. Langenberg, Transient fields of small loop
    antennas.
    IEEE Trans. Ant. Prop. AP-24 (1976)

7. H. Finkler, K. J. Langenberg, Das Einschwingver-
    halten der elektrischen Feldstärke in einem at-
    mosphärischen Bodenwellenleiter bei beliebiger
    Empfängerentfernung in Abhängigkeit von der
    Trägerfrequenz.
    AEÜ 29 (1975), 37-46

8. K. J. Langenberg, Propagation of an electro-
    magnetic pulse in a duct between ground ahd
    atmospheric layer.
    AGARD Symp. on Tropospheric Radio Wave Propa-
    gation, Conf, Publ. No. 70 (1970)

MEASUREMENTS OF WAVE-FRONT DISTORTION

M. C. Thompson, Jr.

U.S. Dept. of Commerce,
Institute for Telecommunication Sciences,
Boulder, Colorado

INTRODUCTION

In a typical microwave imaging system, the angular position of a point on the object is estimated from the differential time or phase delay of the signals from that point as received at two or more observing points. The delay along each of the paths involved is determined by the geometric path length and by the average signal speed along each path. The latter factor is determined by atmospheric conditions along the particular path. In the atmosphere, these conditions vary randomly in both space and time and, as a result, the differential delay measurements are subject to corresponding variability and consequent uncertainty.

In the following we will consider the effects to be expected from present theories and compare the results of direct measurements in controlled experiments.

MECHANISM

The transit time, τ, of a cw signal propagated over a path of length L through a medium of uniform refractive index, n, is given by:

$$\tau = \frac{L}{v} = \frac{nL}{c} \tag{1}$$

where v is the phase velocity and c the velocity of electromagnetic waves in vacuum.

Jeske (ed.), Atmospheric Effects on Radar Target Identification and Imaging, 289–299.

If n is known, (1) is easily inverted to provide L from a measurement of τ.

In practice, n is dependent on the state and composition of the atmosphere and can be expressed in terms of refractivity N, where,

$$N \equiv (n - 1) \times 10^6 = \frac{77.6P}{T} + \frac{3.73 \times 10^5 e}{T^2} \tag{2}$$

in which P is barometric pressure and e is water vapor pressure, both in millibars, and T is temperature in degrees Kelvin. The values of the three quantities, P, T, and e, vary randomly in both space and time resulting in a random distribution of index throughout the medium. Thus, (1) can be written as

$$\tau(t) = \frac{\overline{n}(t)L}{c} \tag{3}$$

where

$$\overline{n}(t) \equiv \frac{1}{L} \int_0^L n(s,t)\,ds \tag{4}$$

is the average refractive index along the ray path through the non-uniform medium.

INDEX STATISTICS

The variability in index is due to several physical processes and ranges from slowly-varying annual and seasonal changes which tend to occur nearly simultaneously over large spatial scales, e.g., hundreds of kilometers, to turbulence with spatial scales of centimeters or less and time scales of hundredths of a second. In the present case we are principally concerned with the latter type of effects.

For these purposes, it is convenient to describe the expected time-behavior of the refractive index and the resulting effects on imaging systems in terms of their variance or "power" spectra.

Figure 1 shows a sample spectrum of time fluctuations of index at a "point," (actually a sampling volume of a cubic meter or less), measured by a microwave refractometer. The spectral shape between about 10^{-5} Hz (1 cycle per day) and 30 Hz is usually found to be of the form

$$W_n(f) = Af^{-k}. \tag{5}$$

The parameter A describes the intensity of the index variations and may vary by a factor of 100 between day and night in a given

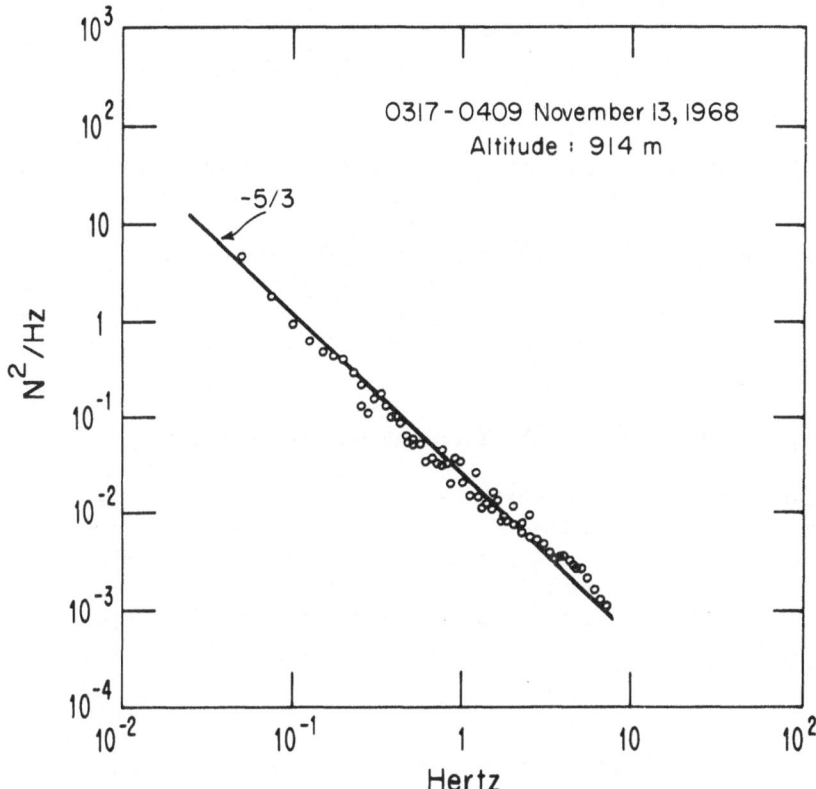

Figure 1. Example of refractive index spectrum computed from airborne refractometer data. Solid line shows slope of -5/3.

location or between different heights above sea level at a given time. The exponent k is influenced to some extent by wind speed and the atmospheric stability but often has a value close to the -5/3 value expected for the wave number spectrum of turbulent wind velocity fluctuations. Thus the form often used in engineering calculations is

$$W_n(f) = Af^{-5/3} \tag{6}$$

and a line is drawn on figure 1 with this slope to show the general agreement with the observations.

TRANSIT TIME STATISTICS

From this description of the random index fluctuations at each point along the propagation path, we can predict the form expected for the spectrum of n or τ.

If the variance of n at a point is σ^2_n, the variance of $\bar{n}$ is given by

$$\sigma^2_{\bar{n}} = \frac{1}{q} \sigma^2_n \ , \tag{7}$$

where q is the effective number of independent estimates available for σ^2_n.

By analogy the spectral densities are related by

$$W_{\bar{n}}(f) = \frac{W_n(f)}{q(f)} \tag{8}$$

where q is a function of the fluctuation frequency.

Structural features which are small compared with the path length may be characterized by the maximum distance, ℓ, over which their index variations are appreciably correlated. If the range of structural sizes moves into and out of the path vicinity with equal velocity V, each size can be associated with a component, f, in the spectrum of the path-average fluctuations, where

$$f = \frac{V}{\ell} \ . \tag{9}$$

The correlation distance, ℓ, is thus a function of f, and the corresponding form of q(f) in (8) is

$$q(f) = \frac{L}{\ell} = \frac{Lf}{V} \ . \tag{10}$$

Substituting (10) in (8) gives

$$W_{\bar{n}}(f) = \frac{V}{Lf} \cdot W_n(f). \tag{11}$$

For path lengths of tens of kilometers or less and the scales associated with turbulence in the lower atmosphere, we expect (6) to be applicable and (11) becomes

$$W_{\bar{n}}(f) = Bf^{-8/3}. \tag{12}$$

Figure 2 shows an example of $W_{\bar{n}}$ based on measured variations in phase of arrival. The solid line indicates a slope of -8/3.

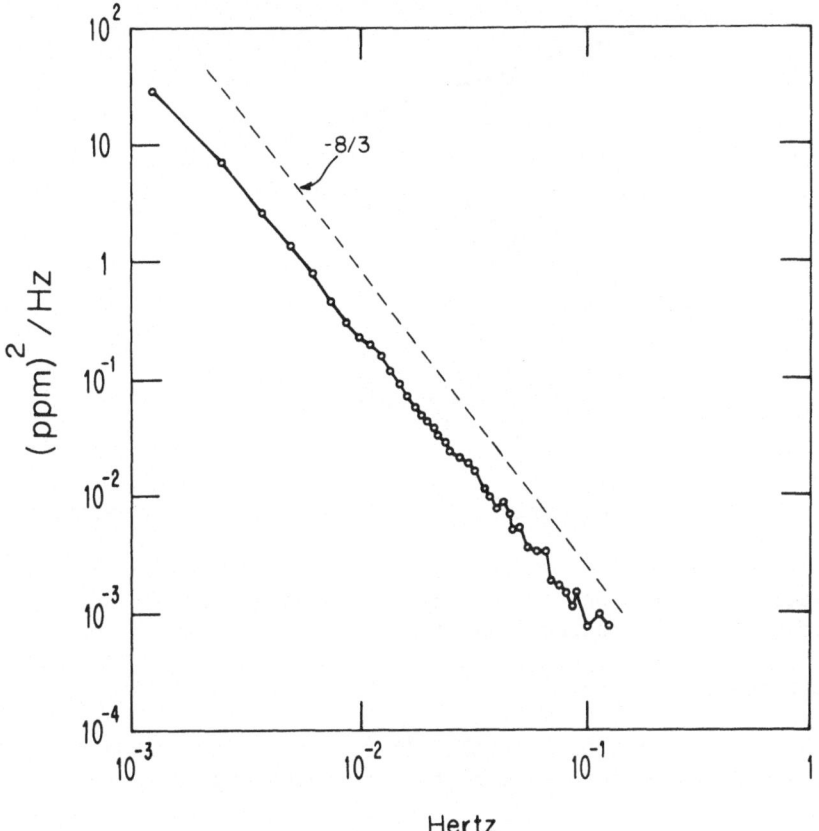

Figure 2.   Example of observed phase spectrum with dashed line indicating slope of −8/3.

DIFFERENTIAL TRANSIT TIME STATISTICS

The angular position of a target relative to a reference baseline is usually determined from measurement of the difference in transit times $\Delta\tau$ (or phases of arrival) of signals received at two points separated a distance μ along the baseline (see figure 3):

$$\Delta\tau = \frac{L}{c}(\bar{n}_2 - \bar{n}_1) = \frac{L}{c} \cdot \Delta\bar{n}. \tag{13}$$

In this case the error arises from the fact that the average index values along the two paths are not equal.

For the divergent paths involved in most imaging systems, the random index field will include spatial scales from hundreds of meters to centimeters. Each scale contributes to the difference

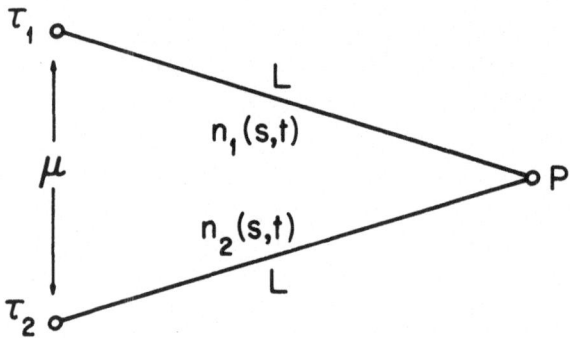

$$\Delta \tau = \tau_1 - \tau_2 = \frac{L}{C}(\bar{n}_1 - \bar{n}_2) = \frac{L}{C} \Delta \bar{n}$$

Figure 3. Geometry for differential transit time measurements.

$\bar{n}_2 - \bar{n}_1$ and thus to $\Delta \tau$. Those scales that are large compared with the baseline length tend to affect both paths similarly and the differential effect is less than that experienced by the individual paths. Conversely, scales that are small compared with μ produce effects on the individual paths which are uncorrelated.

The effects of this spatial filtering are illustrated in figure 4 and can be described by the following approximation:

$$W_{\Delta \bar{n}}(f) = \left(\frac{f}{f_\mu}\right)^2 W_n(f), \quad f < f_\mu \tag{14a}$$

$$= 2W_n(f), \quad f \geq f_\mu \tag{14b}$$

where

$$f_\mu \propto V/\mu$$

and V is the velocity of the index irregularities with respect to the two paths, in the direction of the baseline.

When the form of $W_n(f)$ is taken as $Bf^{-8/3}$, 14a and 14b become:

$$W_{\Delta \bar{n}}(f) = \frac{1}{f_\mu^2} \cdot Bf^{-2/3}, \quad f < f_\mu \tag{15a}$$

$$= 2 Bf^{-8/3}, \quad f \geq f_\mu. \tag{15b}$$

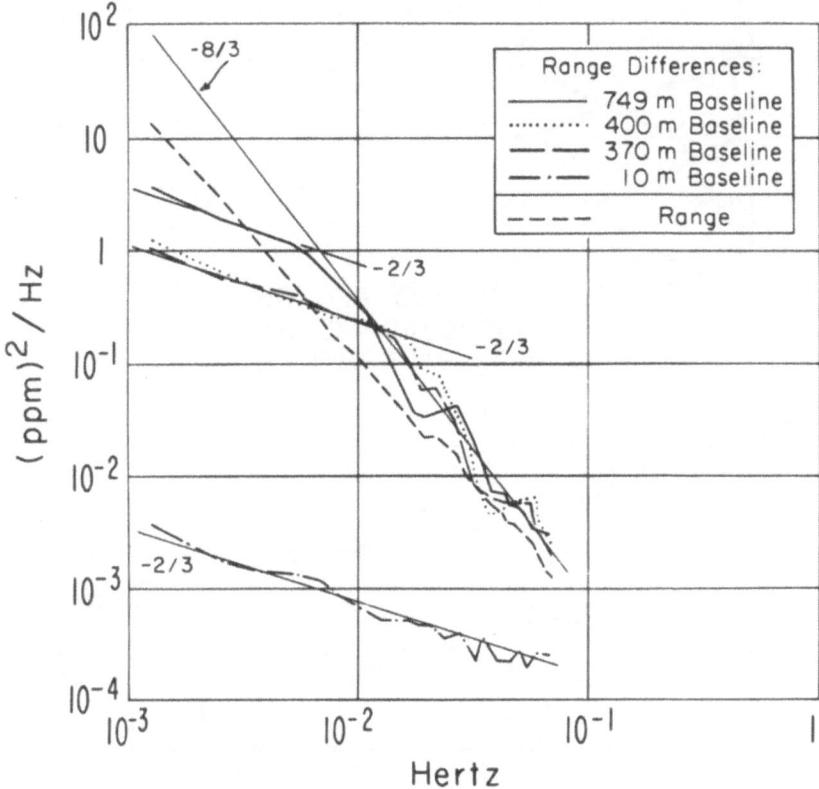

Figure 4. Example of observed spectra of range (phase) and range differences for different baseline lengths.

The solid line in figure 4 shows the form and the general agreement with the observations.

PREDICTION BASED ON C_n(h)

Although the spectral forms of the above three quantities (n, $\bar{n}$, Δn) seem to be reasonably predictable, their intensities under atmospheric conditions are subject to large spatial variations. As noted earlier, the parameter A in (5) can vary by a factor of 100 for a change in altitude of a few hundred meters. This phenomenon is shown in figure 5 which illustrates both the height variability and the changes in the height distribution with time. Thus for a path extending through even a kilometer thickness of the lower atmosphere, it is necessary to take account of this variation.

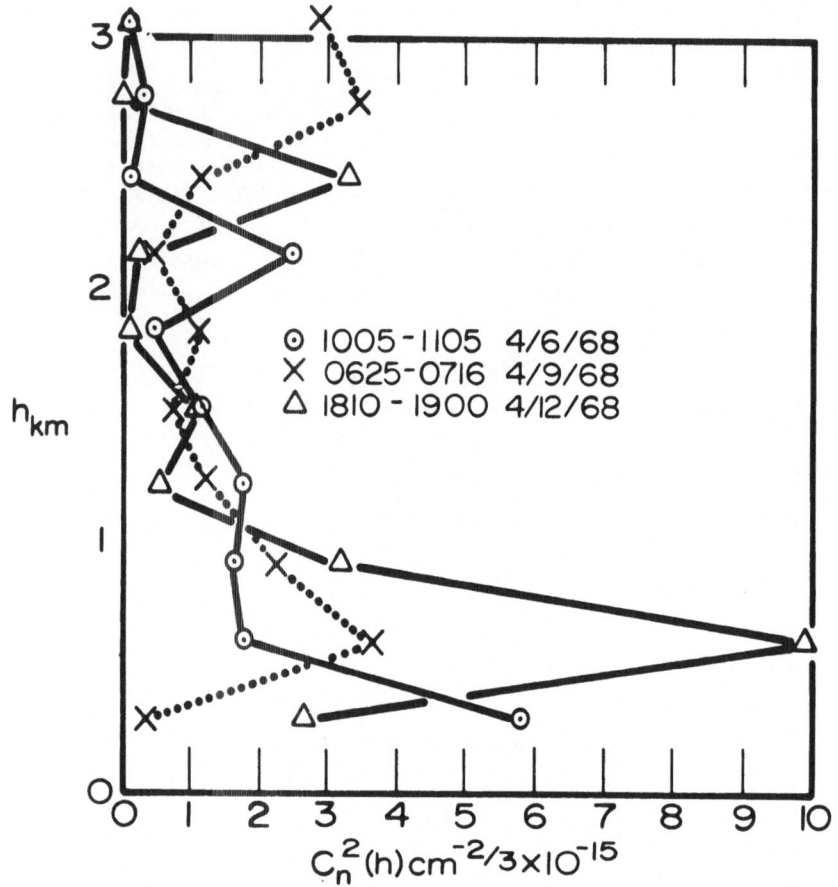

Figure 5. Examples of variations in C_n^2 with height above
surface obtained from airborne refractometer data.

A method for doing this has been proposed by Porcello [1],
in which he calculates the phase-front structure function

$$D_\phi(\mu) \equiv \overline{[\phi(x + \mu) - \phi(x)]^2} \tag{16}$$

where $\phi(x + \mu)$ and $\phi(x)$ are the phases of the signals at two
points separated by a distance μ on a line perpendicular to the
direction of propagation.

The expected value for $D_\phi(\mu)$ is obtained from

$$D_\phi(\mu) = 1.46 \left(\frac{2\pi}{\lambda}\right)^2 \left(\frac{\mu}{L}\right)^{5/3} (\csc \Psi)^{8/3} \int_0^{h_0} C_n^2(h) h^{5/3} dh, \tag{17}$$

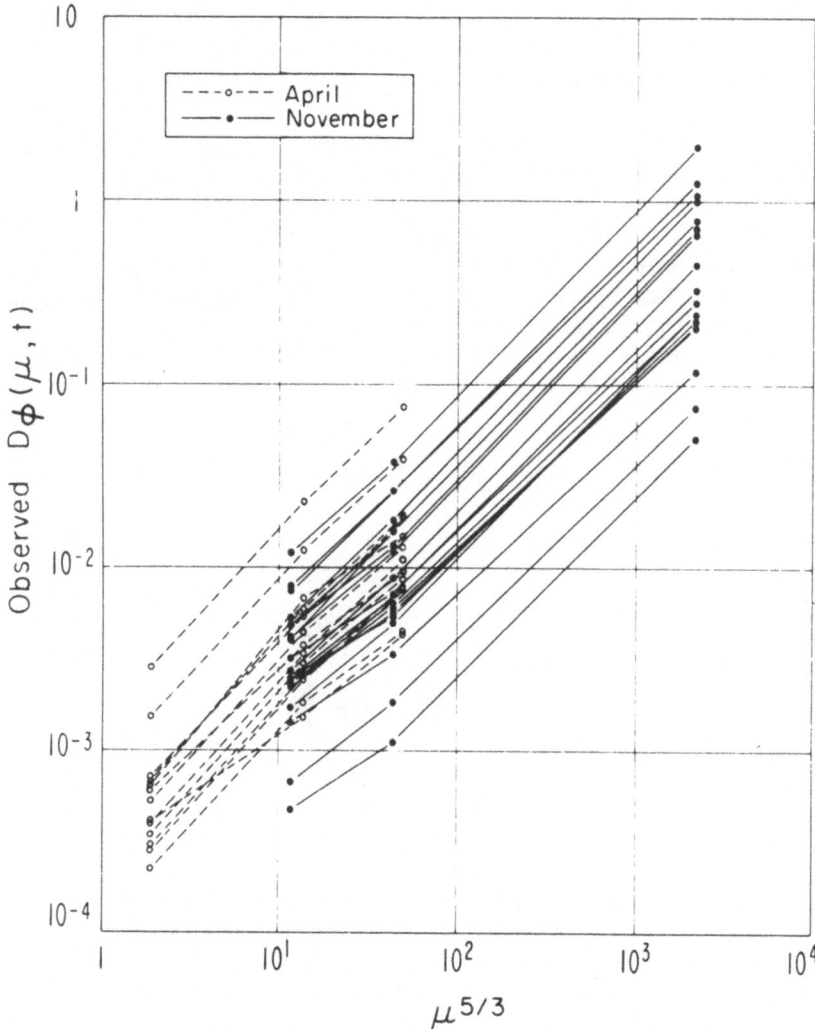

Figure 6.   Observed phase structure function.

where h_0 is height of the baseline above the lower path terminal,

$$\psi = \sin^{-1}\left(\frac{h_0}{L}\right)$$ (18)

and

$$C_n^2(h) \equiv \frac{D_n(r,h)}{r^{2/3}} \qquad\qquad (19)$$

is the refractive index structure function parameter.

This method was examined experimentally using a 65 km path in Hawaii extending from near sea-level (30 m.) at one end to about 3000 m. at the other [2]. Four antennas with separations from 1.5 to 100 m. were used to make direct measurements of $D_\phi(\mu)$. Simultaneously, $C_n^2(h)$ was determined from direct measurements of the index structure using an airborne microwave refractometer flown in the vicinity of the path. A check of the expected dependence of the observed $D_\phi(\mu)$ on $\mu^{5/3}$ is illustrated in figure 6 (from 20-minute data samples).

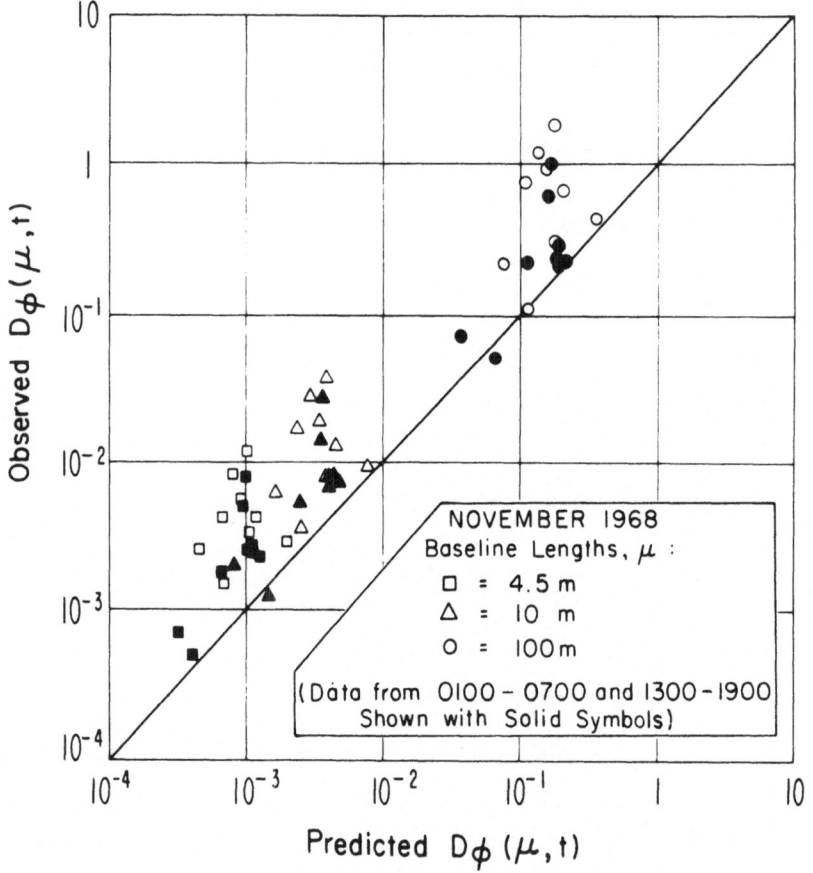

Figure 7.   Comparison of observed and predicted phase structure function.

Figure 7 shows the comparison of observed and predicted values of D_ϕ. In any one series of observations, the correlation between the observed and predicted values is quite high. However, the former values are, with few exceptions, too large. Between 0100-0700 and 1300-1900, it is expected that the statistics of the turbulent structure are most nearly stable. During other times of day, the diurnal warming or cooling may result in changing $C_n(h)$ distributions that are not accurately indicated by the aircraft data. To examine the possibility of this affecting the results, the data points are divided into two groups. The predicted results were based on C_n estimates at height intervals of 300 m. and the variability with height shown in figure 5 suggests that insufficient resolution of $C_n(h)$ may account for some of the discrepancies between the predicted and observed effects.

REFERENCES

[1] Porcello, L. J., Turbulence-induced phase errors in synthetic-aperture radars, IEEE Trans. Aerosp. Electron. Sys., AES-6, 636-644, (1970)

[2] Janes, H. B., and M. C. Thompson, Jr., Comparison of observed and predicted phase-front distortion in line-of-sight microwave signals, IEEE Trans. on Ant. and Prop., AP-21, No. 2, 263-266, (1973)

ON THE INFLUENCE OF ATMOSPHERIC REFRACTIVE INDEX IRREGULARITIES ON THE RESOLUTION PERFORMANCE OF A RADAR

Dag T. Gjessing

Norwegian Defence Research Establishment,
P O Box 25, N-2007 Kjeller, Norway

1. INTRODUCTION

The degree of sophistication in the field of radar target identification and imaging has now reached the state where in many cases the fundamental limitation to system performance is constituted by the propagation medium itself. In order to optimize the radar performance under the adverse conditions imposed by the atmosphere one has to understand the detailed nature of the atmospheric phenomena, and one needs information about the extent to which the atmospheric limitations prevail.

It is the object of this presentation to study the statistical properties of atmospheric factors which constitute a hindrance to optimum radar performance. Specifically, basing our calculations on atmospheric data which are readily accessible from routine observations, we shall seek information about the probability distributions of radar resolution and target distortion.

In the absence of ground based or elevated ducts capable of trapping the radio wave thus causing severe distortions of the radar coverage diagrams, the main characteristics of the radar performance are determined, as we shall see, by two factors:

- The rate at which the mean refractive index decreases with height. This is commonly described in terms of an effective earth radius a .

- The irregularity spectrum of refractive index in the spatial region through which the radio waves propagate.

Jeske (ed.), Atmospheric Effects on Radar Target Identification and Imaging, 301–318.

In this paper we shall focus our attention on the nonducting case, the ducting mode having previously been treated by several authors (1).

The current paper is based on two earlier contributions, and in this discussion we shall limit ourselves to giving essential results from these two papers. The first work (2) establishes the relationship between the characteristic properties of an electromagnetic wave having propagated through an irregular refractive index structure and parameters a and n characterizing the refractive index structure. The parameter a is related to the rate at which the average properties of refractive index change with height (effective earth radius) while n describes the spatial irregularity spectrum of refractivity ($\Phi(K) \sim K^{-n}$).

The second work (3) on which the current paper is based shows how the parameters a and n can be determined from conventional radiosonde measurements.

We shall show that, on the basis of probability distributions of a and n, obtained over a substantial period, we can calculate the probability distributions of the pertinent characteristic radar parameters resolution and distortion.

2. RELATION BETWEEN RADAR- AND RADIOMETEOROLOGI-CAL PARAMETERS

Knowing the detailed structure of refractive index in the spatial region between radar and target it is, in principle, possible to calculate the effect of the atmosphere on the radar performance. It should be noted, however, that such a detailed case by case study serves a limited practical purpose since such information about the propagation medium is not readily available. We shall have to rely on conventional routine meteorological data if long term world wide statistics are sought. Consequently, in this presentation we shall be discussing the gross properties of atmospheric influence only.

2.1 Influence of the atmosphere on the capability of a radar to measure height (distortion effects)

Knowing the refractive index profile we can calculate the ray bending from Snell's law. We are thus able to calculate the total bending to which a ray is subjected when propagating from the radar to the target (and back along the same path). This bending is dependent on the initial direction ϕ of the ray relative to the isosurface of refractive index (4). In this simple treatment we shall limit ourselves to near horizontal directions of the radar beam.

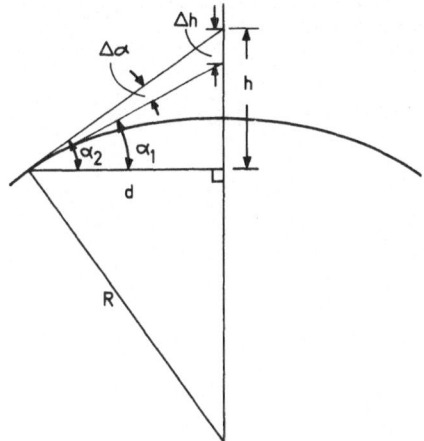

Fig. 1. Geometry related to the measurement of height

From this simple geometry shown in fig. 1, then, we have the following geometrical relationships:

$$\frac{1}{a} = \frac{1}{R} + \frac{dn}{dz} \cos \phi$$

$$a = k R$$

$$K = \frac{1}{1 + R\frac{dn}{dz} \cos\phi}$$

Direction to target for the case of no ray bending is given by

$$\alpha_o = \frac{d}{R}$$

R being the real earth's radius.
Similarly, with a bending corresponding to an effective earth's radius "a" the ray direction is

$$\alpha_1 = \frac{d}{a}$$

i.e. $$\frac{\Delta h}{h} = \frac{(\alpha_o - \alpha_1)d}{\alpha_1 d}$$

and $$\frac{\Delta h}{h} = \frac{\alpha_o}{\alpha_1} - 1 = (k-1) \qquad (1)$$

where k is the ratio of effective earth's radius to real radius.

There is thus a simple and direct relationship between the relative height error and the effective earth radius.

2.2 Influence of an irregular refractive index structure on radar resolution

For the purpose of illustrating the physics of the problem, we shall consider three simple antenna systems (see fig. 2).

APERTURE ANTENNA

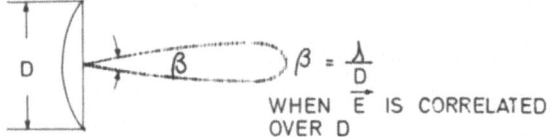

$$\beta = \frac{\lambda}{D}$$

WHEN $\overline{E}$ IS CORRELATED OVER D

ARRAY ANTENNA

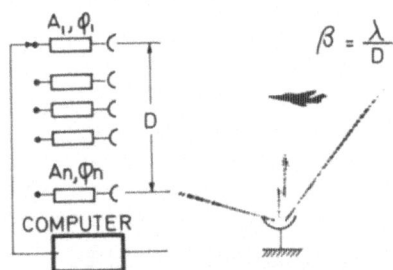

$$\beta = \frac{\lambda}{D}$$

SYNTHETIC APERTURE ANTENNA

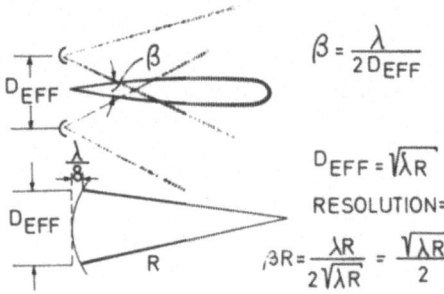

$$\beta = \frac{\lambda}{2 D_{EFF}}$$

$$D_{EFF} = \sqrt{\lambda R}$$

RESOLUTION=

$$\beta R = \frac{\lambda R}{2\sqrt{\lambda R}} = \frac{\sqrt{\lambda R}}{2}$$

Fig. 2. The antenna systems to be considered

An aperture antenna gives us automatically the Fourier transform of the field strength distribution over the aperture and hence the angular power spectrum of the wave illuminating the antenna. Thus if D be the diameter of the paraboloyd and λ the wavelength, then the width of the radiation pattern β is given by

$$\beta = \frac{\lambda}{D}$$

Note that this requires the electric field $\vec{E}$ to be coherent over the antenna aperture D.

If the correlation distance of the radio wave as a result of refractive index irregularities is less than D, we will suffer a degradation in the resolving power of the antenna. The effective antenna aperture D_{eff} is then reduced to the correlation distance of field strength.

The degree to which the resolving power is degraded depends on the severity of the phase and amplitude perturbations relative to the spherical wavefront. From antenna design considerations we know that $\lambda/8$ irregularities in the antenna surface will have measurable effects on its performance. The resolving power of an antenna is thus very sensitive to perturbations in the wavefront.

Similarly, the resolving power of an array antenna is given by the distance D over which the antenna elements are distributed. Measuring the amplitude and phase at each of the antenna elements, information about the radio wavefront is obtained and from this we achieve information about the target. As in the case of the aperture antenna above, the resolving power of the array antenna is λ/D provided the refractivity irregularities do not destroy the correlation properties of field strength over the set of array elements. The array antenna, however, has a marked advantage over the aperture antenna in that temporal averaging can be carried out at each of the antenna elements, thus minimizing under certain conditions the effects of the random medium. The degree to which this is possible depends on the temporal correlation properties of the field strength.

Finally, the synthetic aperture antenna samples the radio wavefront as the sampling antenna is shifted along a direction perpendicular to the direction of the propagating wave. This system has, contrary to the array antenna, no way of keeping track of the temporal variations of field strength at a point and is thus strongly dependent on a well behaved wavefront.

In the examples to follow in chapter 4, we shall focus our attention on the resolving power of a sidelooking radar under conditions of atmospheric limitations. There are several schemes by which a synthetic aperture antenna is made to give optimum resolution. Let us discuss the case where the phase curvature departs less than $\lambda/8$ from the plane. For this case the effective (synthetic) aperture is given by

$$D_{eff} = 2\sqrt{\lambda R}$$

such that the width of the synthetic "antenna diagram" is given by

$$\beta = \frac{\lambda}{2\sqrt{\lambda R}}$$

and the corresponding target resolution Δ at the distance R is given by

$$\Delta = R\beta = \frac{\sqrt{\lambda R}}{2} \tag{2}$$

As in the two cases above, if the field strength of the arriving wave is not correlated over the synthetic aperture $2\sqrt{\lambda R}$, a loss in resolving power is suffered.

Then, finally, let us discuss how the correlation distance of field strength is related to the spectral properties of the refractive index irregularities. Referring to Lee and Harp (2), we find that the covariance function $C(d)$ in a plane normal to the direction of wave propagation is given by

$$C_a(d) = 4\pi^2 k^2 \int_0^k dK \int_0^L dsK\Phi(K) J_0(\frac{dsK}{L})\sin^2(\frac{K^2 s(L-s)}{2 k L}) \tag{3}$$

for the amplitude covariance function; similarly, for the phase covariance we have

$$C_p(d) = 4\pi^2 k^2 \int_0^k dK \int_0^L dsK\Phi(K) J_0(\frac{dsK}{L})\cos^2(\frac{K^2 s(L-s)}{2 k L}) \tag{4}$$

where $k = 2\pi/\lambda$, $\Phi(K)$ is the spectrum of refractive irregularities, and L the length of transmission path.

Fig. 3 shows a plot of the amplitude covariance function for different spectra $\Phi(K)$. Here $\Phi(K)$ is written in the form

$$\Phi(K) \sim K^{-n/3}$$

and n is given values from 10 to 14.

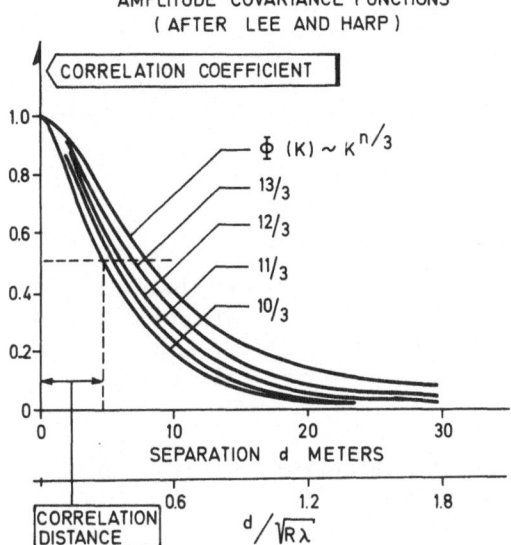

Fig. 3. Theoretical relationship between correlation distance
and spectrum slope n
(Path length is 28 km, frequency 35 GHz)

If we define correlation distance to be the separation d
corresponding to a 0.5 correlation coefficient, we see that this
correlation distance increases with increasing exponent n.

It then only remains, before we present the probability
distributions of radar resolution and distortion, to discuss the
radiometeorological parameters a and n and the parameters
obtainable from radiosondes.

3.   THE RELATION BETWEEN THE PARAMETERS a AND
      n CHARACTERIZING THE TRANSMISSION MEDIUM, AND
      RADIOSONDE PARAMETERS

This topic has been considered in some degree of detail in
two earlier publications (3), (5). We shall give a summary of
the results here. As we have already mentioned, and as will be
substantiated in the following section, there are two radiometeorological parameters which are of dominating importance with respect to the current application. One is the effective earth radius
a, the other is the spectrum slope n of the refractive index
irregularity spectrum (i. e. $\Phi(K) \sim K^{-n}$, where $\Phi(K)$ is the spectral
density function and K the wavenumber of the irregular refrac-
tive index structure). We shall now discuss the relationship between purely meteorological factors and the parameters a and n.

3. 1 Determination of effective earth radius   a   from radiosonde
      measurement

The refractivity N (where $N = (n-1)10^6$, n being the refractive index) is obtained from meteorological parameters by
the Debye relationship

$$N = 77.6 \frac{P}{T} + 3.73 \times 10^5 \frac{e}{T^2}$$

where P is the total pressure in millibars, T the absolute temperature, and e the water vapour pressure in millibars. Thus
from knowledge about the vertical profile of P, T, and e as obtained from a conventional radiosonde, we can calculate the
refractivity profile.

Knowing the N profile, we can calculate the ray bending
from Snell's law (4). For convenience we now introduce the
effective earth's radius   a   as follows:

$$\frac{1}{a} = \frac{1}{R} + \frac{dn}{dz} \cdot 10^{-6}$$

assuming a near-horizontal direction of the radio beam. We
observe, then, that on the basis of P, T and e data from a conventional radiosonde ascent, we can obtain the radiometeorological parameter a appropriate for a given path geometry. The
solid lines in fig. 4 show probability distributions of the ratio
a/R based on 230 radio soundings at Sola in south-western Norway
during 1966.

These were calculated from the average lapse in refractivity over the height interval from ground to the height indicated.
Note that the greater the extent of the height interval, the smaller
is the variation of a/R, and indeed also the mean value of a/R.
This reflects the exponential behaviour of the mean N profile,
since at high altitudes there is little or no bending and there are
only small variations in the N gradient.

3. 2 Determination of spectrum slope

Here the reader is referred to (3), where an empirical relationship was found between the atmospheric stability (a somewhat modified version of the well known Väisälä - Brunt frequency
ν^2) and the slope n of the refractive index irregularity spectrum.
The conventional Väisälä - Brunt frequency appears as the numerator in the Richardson's number and is normally written as

$$\nu^2 = \frac{g}{T} (\frac{dT}{dz} + \frac{g}{C_p})$$

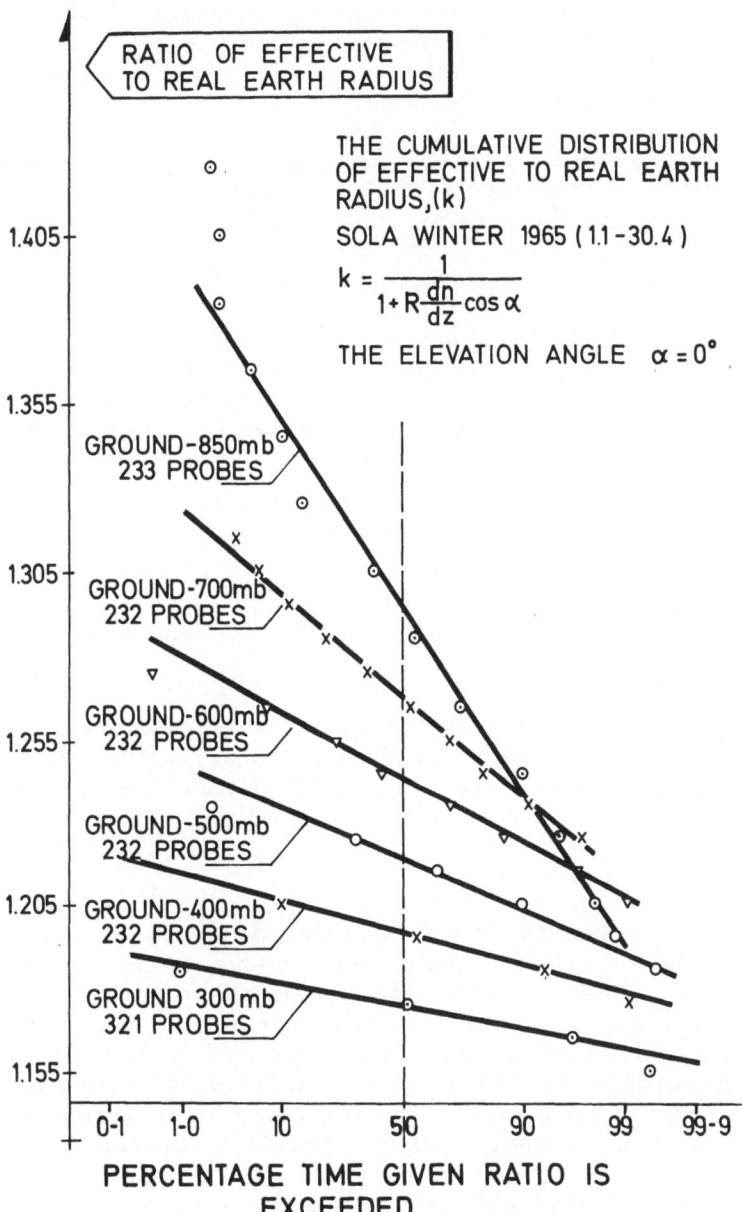

Fig. 4. Probability distribution of the ratio of effective to real
earth radius

Here g is the gravitational constant, T the temperature, dT/dz the vertical temperature lapse rate, and C_p the specific heat at constant pressure. We see that the expression within the brackets is a measure of the difference between the actual temperature lapse rate and the adiabatic lapse rate.

Comprehensive multiple regression analyses (3) have shown that the correlation of the spectrum slope n with atmospheric parameters describing the dynamic state of the atmosphere was significantly improved when a particular weighting function was placed on the temperature contribution to the stability.

Specifically, by adding a number which is determined by the temperature at the 850-mb surface (1500 m altitude) to the Väisälä - Brunt frequency, the n versus ν^2 correlation was improved. For our particular purpose, therefore, a modified version of ν^2 was used:

$$\nu^2 = \frac{g}{T}\left(\frac{dT}{dz} + 5.50 \times 10^{-3}\, T(850) + \frac{g}{C_p}\right)$$

Here T, measured in degrees K, and dT/dz, in degrees per 100 m, are average values obtained over the 850 - 400 mbar levels (1.5 to 7-km altitude).

Having obtained ν^2 from the results of a temperature profile determined by a conventional radiosonde observation, the spectrum slope n is found, using the expression for the n versus ν^2 regression line:

$$n = 40.6 + 708\,\nu^2$$

It should be noted that this relationship is the result of a multiple regression analysis, and as such is an empirical or statistical relation. On the basis of fluid dynamical considerations, such a relationship is to be expected (6). Furthermore, on the basis of a study on the scattering of radio waves from stratified layers in the troposphere (7) it is clear that large refractive index gradients, i.e. large dn/dz, are associated with large values of spectrum slope n. It could well be, therefore, that large values of n (and correspondingly large values of ν^2) are indicative of the existence of atmospheric layers with large refractive index gradients and are therefore phenomenologically not directly related to the total atmospheric stability.

Fig. 5 shows probability distributions of spectrum slope n. The two upper curves are for the radiosonde station Sola in southwestern Norway for summer and winter 1966. These curves correspond to the "a" distributions of fig. 4 and will form the basis of the calculations to be presented in the following sections. For

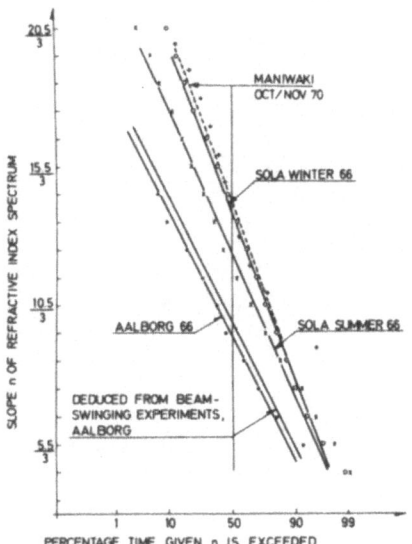

Fig. 5. Distributions of the slope of the spectrum of refractive index irregularities as deduced from radiosonde observations and beam-swinging experiments.

comparison, fig. 5 also shows n distributions for Maniwaki and for Aalborg, northern Denmark, as obtained on the basis of radiosondes. The "predicted" n distribution for Aalborg is compared with that deduced from radio beam-swinging experiments (2), (4).

Since the Aalborg beam-swinging results were used as the basis for establishing the n versus ν^2 correlation, the comparison between the predicted distribution and the distribution deduced from radio experiments is only partly justified. We do observe, however, that all the distributions are very close to being Gaussian, and that the curves are very close to being parallel, indicating the same standard deviation.

It may be of interest in this context to direct our attention to some radio experiments designed for the specific purpose of analyzing the refractive index irregularity structure of the atmosphere (8).

Fig. 6 shows a set of n distributions obtained from specific radio experiments and provides further justification for the choice of the radiometeorological parameters n. The figure shows that the variability of the spectrum slope n, the way this is defined and determined experimentally, is indeed a very variable quantity.

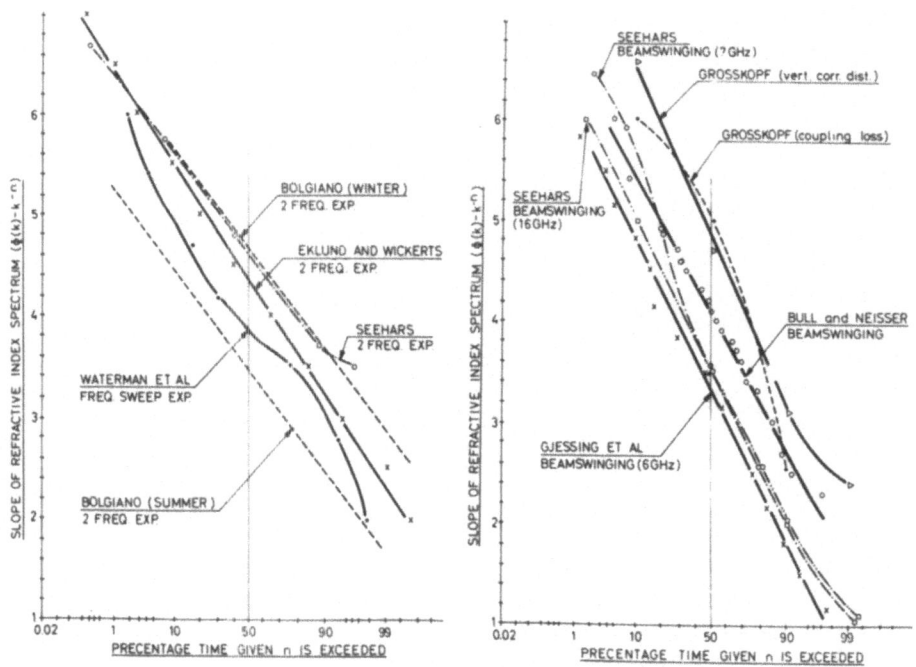

Fig. 6.   Probability distribution of spectrum slope n as determined from radio experiments

4. RADAR TARGET DISTORTION AND RESOLUTION AS DETERMINED FROM RADIOSONDE STATISTICS

It then only remains to present the implications of the n and a distributions on radar performance.

There are, as we have discussed previously, two radar parameters which we, on the basis of radiosondes, have information about. One is target distortion (height error), the other is target resolution. These results will now be given.

4.1 Distortion effects, height error

The simple and well known effects of ray bending will first be considered. On the basis of distribution of the ratio k of effective/real earth radius, we compute the distribution of fractional height error $\Delta h/h$ from equation 1, above.

The height error corresponding to zero elevation angle is shown in fig. 7. For the sake of illustrating the severity of height errors, numbers are given for the specific case of a target at a height of 1500 meters.

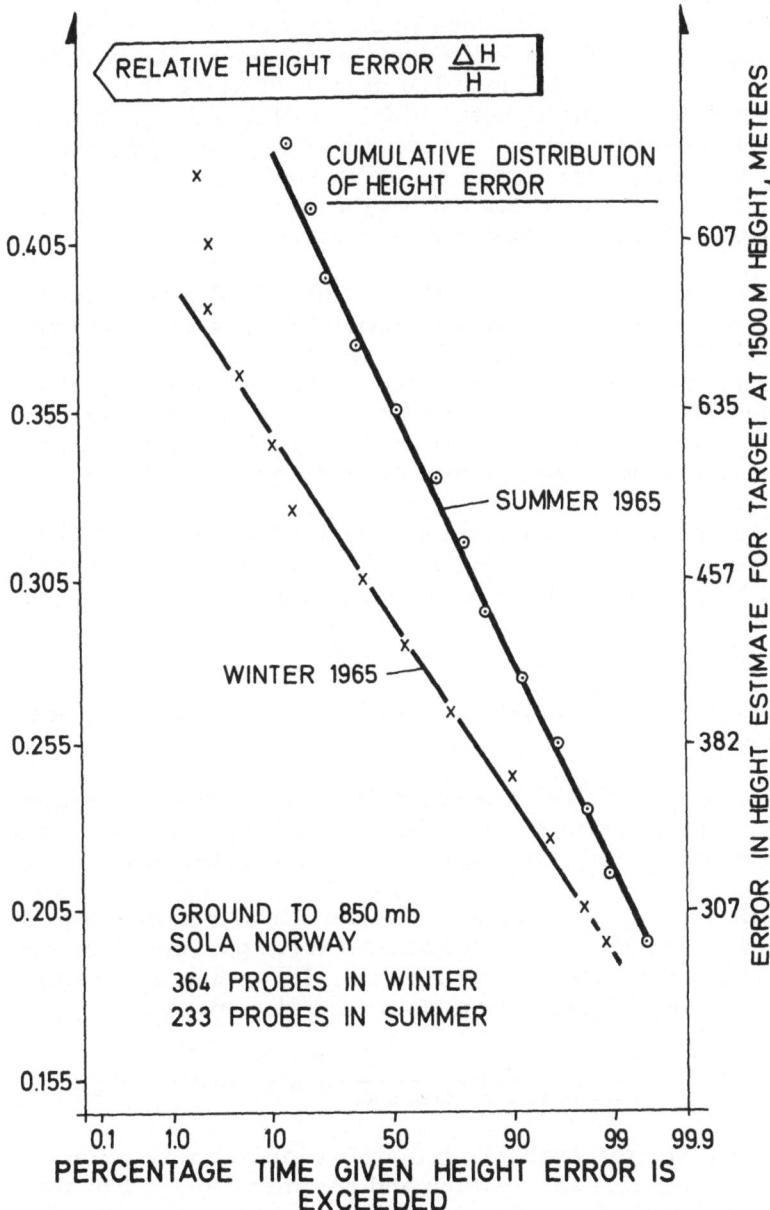

Fig. 7. Probability distribution of height error correlated on the basis of radiosonde information

We note that height errors are considerable and so is the
variation in height error. Specifically, if no information about
the atmosphere is available, the median winter error will be
440 meters and the 10 - 90% variation in height error will be
200 meters. Note that the median k value for summer is 0.355
and that for winter 0.295 (not 4/3). It is clear that these refraction effects, although serious in connection with measurement of target position, have little effect in connection with target identification and imaging. The degree to which the target
is distorted depends on the detailed form of the refractive index
profile. Such details are not revealed through the k-parameter.
However, the $\cos \phi$ dependence shown in the k-expression in
fig. 1 indicates that the fractional height error is dependent on
the direction at which the ray leaves the transmitter.

Obviously, if the angle ϕ between the initial direction of
the wave and the isosurface of refractive index is zero, the refraction effects are maximal. If the waves on the other hand are
launched in a direction perpendicular to the constant refractivity
surface then we experience no bending. If the vertical extent of
the target is large, it is clear from the expression for k that
refraction effects will lead to noticeable target distortion.

As an example, consider a 10° vertical angular distribution
of the target. For a typical value of 1.33 for k, we find that the
fractional height error $\Delta h/h$ will change approximately 2% over
the target.

4.2 Effects on radar resolution of an irregular refractive index structure

We have shown that it is possible to obtain the probability
distribution of spectrum slope n from radiosonde observations.
Such a distribution we have shown in fig. 5, chapter 3.2.

Furthermore, we have given a theoretical relationship
(illustrated in fig. 3) between the atmospheric parameter n and
the correlation distance of field strength d. It then only remains
to compute the radar resolution from the relationship:

$$\text{Resolution } \Delta = \frac{\lambda}{\text{correlation distance d}} \times \text{distance to target.}$$

Fig. 8 gives the probability distribution of correlation distance d for Sola, Norway, winter and summer 1966. The corresponding distribution for radar resolution is given in fig. 9. We
have here used a 35 GHz radar and a target range of 28 km.
This predicted probability distribution is compared with the resolution that one would achieve with a non-focused sidelooking radar

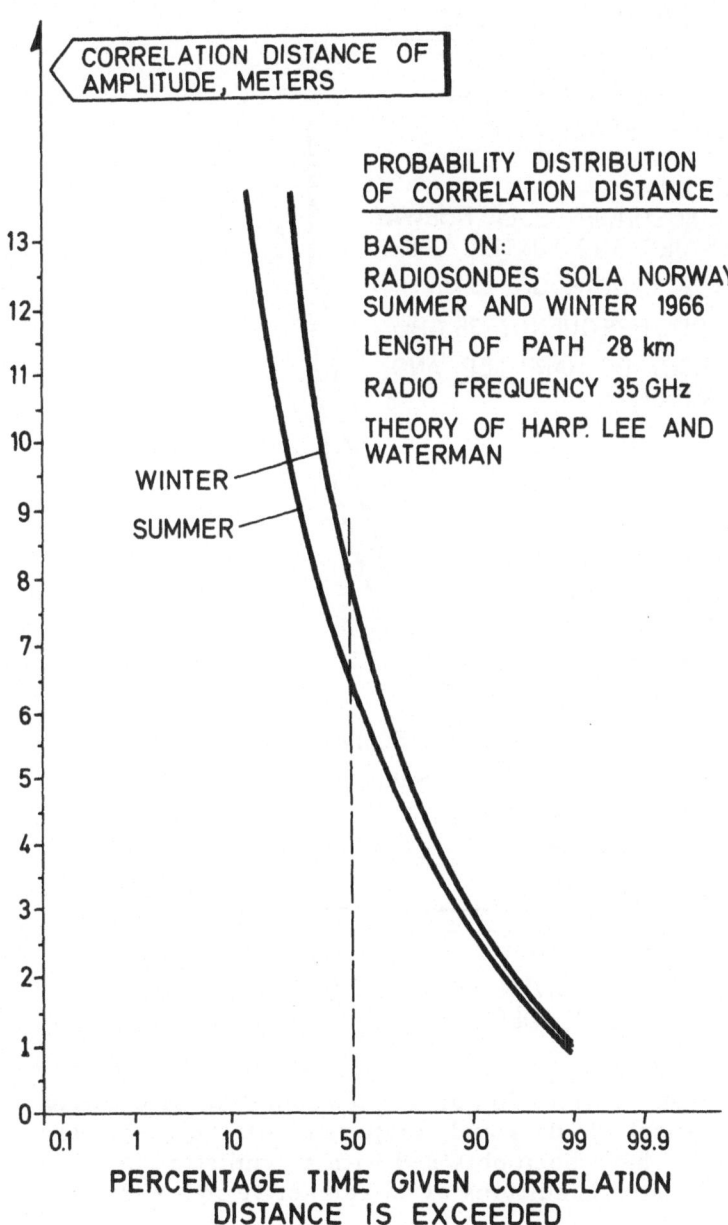

Fig. 8. Probability distribution of amplitude correlation distance for a 28 km path at 35 GHz calculated from radiosonde data

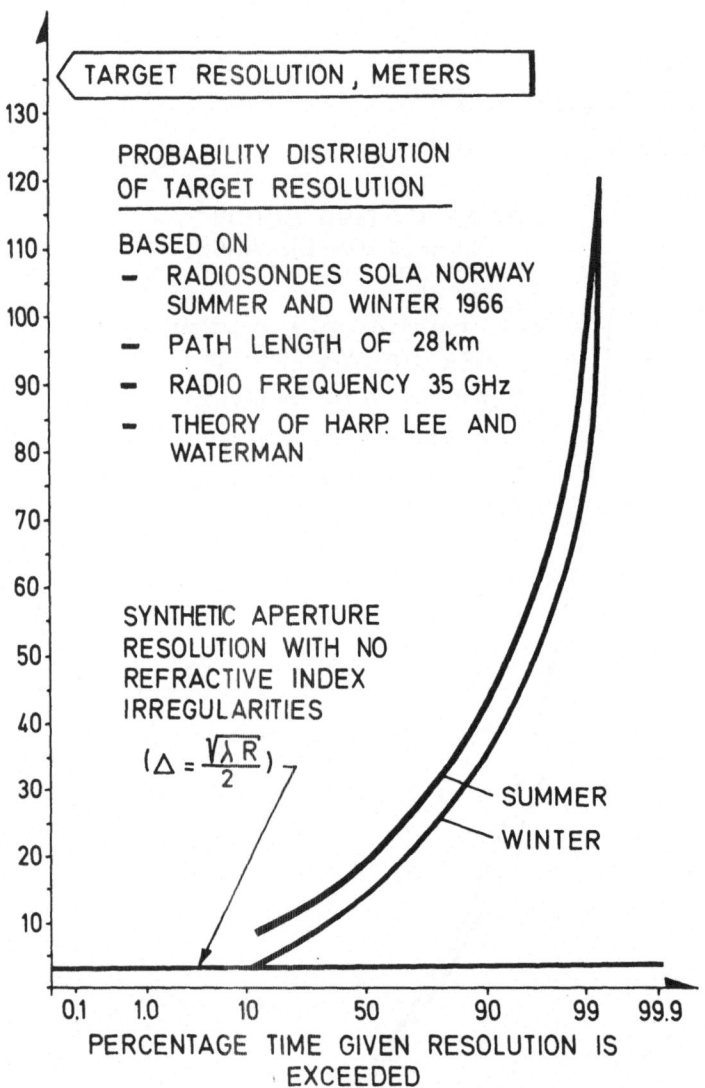

Fig. 9. Probability distribution of target resolution calculated
         from radiosonde data and compared with the resolution
         that would have been obtained with a synthetic aperture
         antenna under conditions of no refractive index irregu-
         larities.

under conditions with no refractivity irregularities. We see that in as much as 90% of the total time the atmosphere is a limiting factor. This suggests that, if optimum resolution is required from a sidelooking radar or any other high resolution system making use of electromagnetic waves, the atmospheric conditions under which the measurements are made should be carefully chosen. It should be emphasized again, however, that the predicted atmospheric influence should be taken as pessimistic worst case values. We have specified a covariance larger than 0.5 over the aperture for this to be effective and we have assumed that phase and amplitude perturbations are sufficiently large for the contributions to resolution of the antenna aperture outside the 0.5 covariance area to be insignificant.

5. CONCLUDING REMARKS

Modern high resolution systems making use of radio and optical waves will suffer a degradation resulting from systematic and random variations in the atmospheric refractivity. The variability of these effects is large. In only some 10% of the total time are the atmospheric effects expected to constitute no hindrance to optimum system performance. This 10% figure relates to a non-focused synthetic aperture system and a 28 km imaging range. If the range is larger, the probability of optimizing performance goes down.

If optimum performance is a requirement, the atmospheric conditions under which the measurements are performed should be closely watched.

REFERENCES

1.   D. T. Gjessing, Diffraction of radio waves by terrain obstacles in radio ducts, NATO Advanced Study Institute Series: Modern Topics in Microwave Propagation and Air-Sea Interaction, ed. by A. Zancla, D. Reidel Publishing Co., Holland/ USA, 1973.

2.   R. W. Lee and J. C. Harp, Weak scattering in random media, IEEE 57, p. 375, 1969.

3.   D. T. Gjessing, A. G. Kjelaas and J. Nordø, Spectral measurements and atmospheric stability, J. Atmosph. Sci. 26, 3, 462-8, 1969.

4.   B. R. Bean and E. J. Dutton, Radio Meteorology, US Dept of Commerce, NBS, Monograph 92, US Government Printing Office, Washington DC, 1966.

5.   D. T. Gjessing, H. Jeske and N. Klint Hansen, An investi-
      gation of the tropospheric fine scale properties using radio,
      radar and direct methods, J. Atmosph. Terr. Phys. 31,
      1157-82, 1969.

6.   H. A. Panofsky, Spectra of atmospheric variables in the
      boundary layer, Radio Sci. 4, 1101-9, 1969.

7.   D. T. Gjessing, On the use of forward scatter techniques
      in the study of turbulent stratified layers in the troposphere,
      Boundary Layer Meteorol. 4, 377-96, 1973.

8.   D. T. Gjessing, Atmospheric structure deduced from for-
      ward-scatter wave propagation experiments, Radio Sci. 4,
      1195-1210, 1969.

PROPAGATION EFFECTS DUE TO SPECIFIC ATMOSPHERIC STRUCTURES

Claus Fengler

Department of Electrical Engineering,
McGill University, Montreal, Que., Canada *)

ABSTRACT. Describing propagation effects differently idealized cases of the atmospheric structure can be distinguished.
In the atmosphere with embedded discontinuities the reflection processes at boundaries of different kinds of air masses
have to be taken into account. Using a model the variations
of phase and amplitude of the field strength of the travelling wave corresponding to the boundaries are treated. Experimental data evaluated with respect to various weather
situations are presented and interpreted as reflection
processes at atmospheric boundaries. The implication for
radar systems is considered.

1. INTRODUCTION

From the earliest applications of electromagnetic waves in
open air it was evident that there exists on their propagation considerable influence of the atmosphere and the
ground. There may be refraction round the earth's surface,
or reflection, refraction , scattering or ducting by the
ground or by atmospheric structures. In the frequency ranges
above VHF, i.e. above 30 MHz, the effects associated with
the structure of the troposphere are dominant. Furthermore,
specific effects of the upper atmosphere (the ionosphere)
also need to be considered.

---

*) on leave of SIGMA Association, Hamburg, Germany

Jeske (ed.), Atmospheric Effects on Radar Target Identification and Imaging, 319–333.
All Rights Reserved. Copyright © 1976 by D. Reidel Publishing Company, Dordrecht-Holland.

2. THE STRUCTURE OF THE ATMOSPHERE

The regions of the atmosphere which influence the propagation of electromagnetic waves are the ionosphere, from
80 km to above 500 km, as well as the troposphere below
12 km and in particular below the 3 km level. For reason
of completeness the influence of the earth's magnetic field
in the higher atmosphere, the so-called magnetosphere, must
also be mentioned, but will not be discussed here further.

The parameters of both the troposphere and ionosphere vary
with time and space. In addition to geographic there are
also temporal variations which may be diurnal, seasonal,
and non-periodic. The average structure is well known as
there are the corresponding parameters influencing the propagation of electromagnetic waves which in the ionosphere
include principally the concentration of free electrons and
the frequency of the incoming wave and in the troposphere
the temperature T [°K], humidity e [mb], and the pressure p
[mb]. The characteristic parameter for the electromagnetic
field is the index of refraction n which is determined by
the above mentioned atmospheric parameters (Figures 1 and
2).

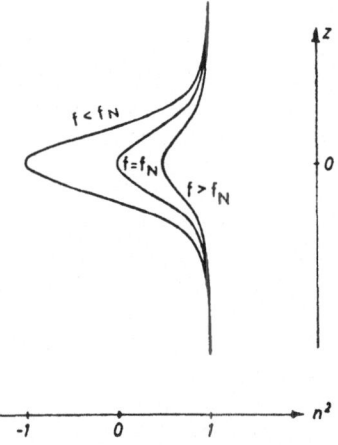

Figure 1   Average profile of index of refraction in the
            ionosphere; vertical incidence, no magnetic field.

$$f_N = \sqrt{\frac{N_e \cdot e^2}{4\pi \varepsilon_0 \, m}}$$

z : height above ground
$N_e = N_e(z)$: concentration of
                free electrons
e : charge of electron
m : mass of electron
ε_0 : permittivity of vacuum

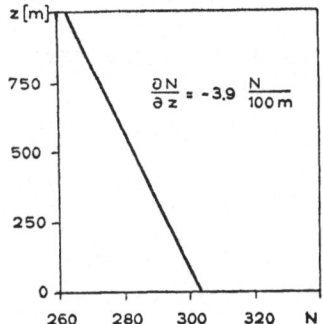

Figure 2   Average vertical profile of refractivity N in
            the troposphere. $N = (n - 1) \cdot 10^6$

In predicting the propagation properties of electromagnetic
waves in the atmosphere only the average properties can be
taken into account including when possible their periodic
and local variations (Bean and Dutton 1968).

Seasonal and diurnal variations correspond to the sun's
radiation which determines the number of free electrons in
the ionosphere and which causes the temperature in the lower
troposphere to fluctuate in relation to ground heat storage
and cloud activity. It is known that in consequence of the
sun's radiation the corresponding refractive index field
exhibits only a slow and smooth variation.

Discontinuities of various sizes may be embedded in the
slowly varying structure of the atmosphere. The microscale
discontinuities due to turbulent fluctuations which are
always present are assumed to be small in comparison with
wavelength in the VHF, UHF and even higher frequency ranges.
In addition to the microscale discontinuities there are larger scale ones of metric to decakilometric dimensions. They
are large as compared to the wavelengths involved and as a
consequence they give rise to specular reflection processes
at their boundaries.

Discontinuities in the ionosphere are regions of increased
free electron concentrations, which occur, for example,
round magnetic field lines.

The common property of this type of discontinuity in the
troposphere as well as in the ionosphere is the rapid change
of the index of refraction at the boundaries.

Two limiting types of atmospheric structures can be identified, as they affect the behaviour of electromagnetic waves:

A) The well mixed atmosphere, approximated by a normal stratified medium without embedded discontinuities.

B) The atmosphere with embedded wide-spread discontinuities
   such as ground or elevated inversion layers in the vertical profile and discontinuities due to weather fronts
   in the horizontal profile.

In the following only the type B atmospheric structure will
be considered, in particular concerning the troposphere.

The following types of refractive-index discontinuities can
be distinguished in the vertical profile:

a) Ground-based discontinuities, arising from a higher temperature and relative humidity near the ground than some
   hundred meters aloft. These discontinuities are largely
   due to the diurnal variation of radiation (Figure 3),
   see C. and G. Fengler 1975. This is true over land, while
   over sea there is a layer due to evaporation (Jeske 1965).

b) Elevated discontinuities, which are characterized by a
   stronger decrease of N with height than in the average
   vertical profile (Figure 4). They can arise by
   - subsidence of air in high pressure areas,
   - radiation from the upper surface of clouds and dust and
   - as residual part of diminishing ground inversions.

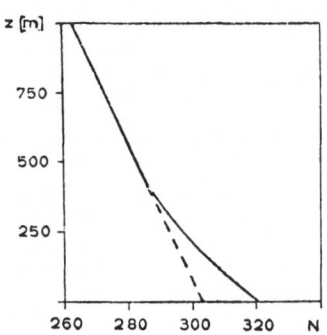

Figure 3   Dependence of refractivity N on height z in the
           presence of a ground-based inversion

Special cases of the above are subrefraction (a layer with
a smaller decrease of N with height than in the average)

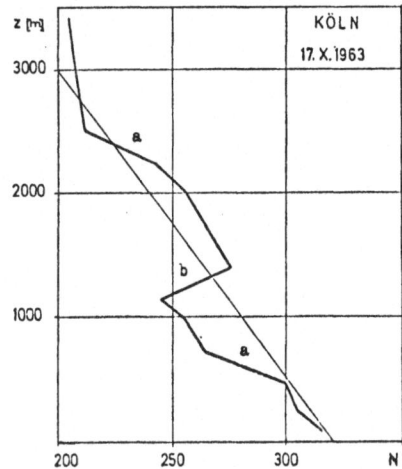

Figure 4   Dependence of refractivity N on height z
             characterizing elevated discontinuities.

a: region of superrefraction $\left|\dfrac{dN}{dz}\right| \geqslant 15.7 \dfrac{N}{100\ m}$

b: region of subrefraction $\left|\dfrac{dN}{dz}\right| \leqslant 3.9 \dfrac{N}{100\ m}$

and superrefraction (layers with a very strong decrease of
N with height). If several reflections at the same boundary
follow each other under a suitable angle of incidence, a
ducting process results.

In the horizontal profile there are refractive-index discontinuities which are due to boundaries of different air
masses, called fronts. Such weather fronts are defined by
a rapid change of pressure, temperature, humidity and wind
at ground, and are mostly associated with rain and cloudi-
ness. The principal types are

- warmfronts, moving towards an area of low temperature
  (Figure 5) and

- coldfronts, moving towards an area of high temperature
  (Figure 6).

The latter shows a more pronounced change in meteorological
parameters.

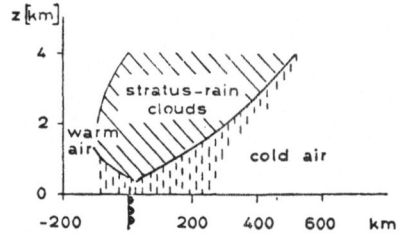

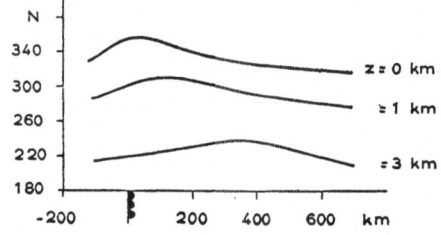

Figure 5   Cross section of idealized warm front in accor-
             dance to Pogosyan and horizontal N-profile of
             warm front (sketch) in accordance to Bean-Dutton.

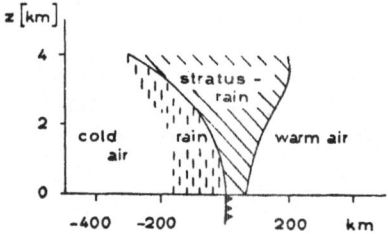

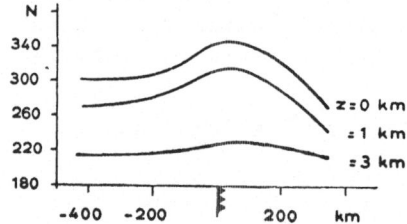

Figure 6   Cross section of idealized cold front, indicating
             stratus clouds and rain, in accordance to Pogosyan
             and horizontal N-profile of cold front (sketch) in
             accordance to Bean-Dutton.

3. REFLECTION AT BOUNDARIES

The interaction of electromagnetic waves with the structure
of the atmosphere results in

a) bending of the wave front if the variation of n is small
   within the distance of a wavelength (ray treatment),

b) diffraction round the earth's surface,

c) scattering at microscale inhomogeneities, an effect which
   is always present due to the everpresent turbulence, or

d) reflection and refraction at boundaries of embedded discontinuities whose effects are emphasized in this paper.

In order to obtain an overview on the processes at boundaries we introduce a plane boundary which separates the region with $n = n_1$ from the region with $n = n_2$. As a simplifica-
tion it is assumed that $n_1 = 1$ and $n_2 = 1 + \Delta n$. First the boundary is considered to be an abrupt transition (step function)
and later as a continuous transition in order to get a better representation of reality (Figure 7), see Fengler 1973.

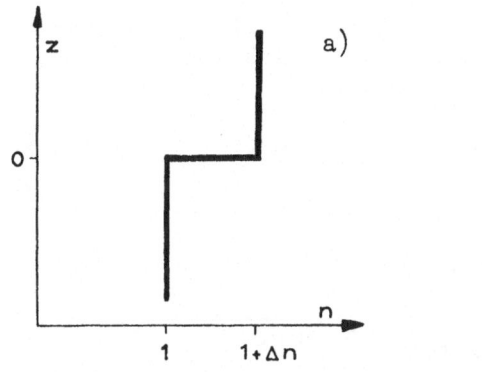 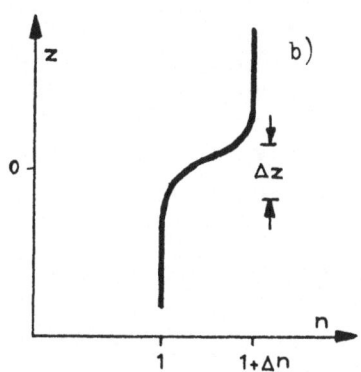

Figure 7   Models of refractive-index profiles. a) Abrupt
             transition, b) continuous transition (unsymmetric
             Epstein profile).

In treating the reflection process, there are introduced:
an incident wave E_i, a reflected wave $E_r = R \cdot E_i$ (R = reflection coefficient), and a transmitted wave $E_t = T \cdot E_i$
(T = transmission coefficient), see Figure 8.

Of interest are only those cases where T modifies the inci-

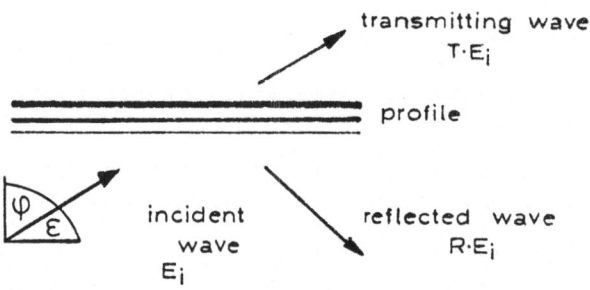

Figure 8   Illustration of incident, reflected and trans-
            mitted wave.

dent wave by phase or amplitude or where T approaches zero
while R approaches unity, and a substantial reflection of
the incident wave occurs.

According to Snell's law

$$\frac{\sin \varphi_1}{\sin \varphi_2} = 1 + \Delta n$$

a small variation results of the direction of propagation of
the transmitted wave with respect to the incident one, but
this is negligible enough to be omitted.

Two cases have to be considered: incidence from n_1 and incidence from n_2 with φ_2 = angle of incidence which yields total
reflection at $\varphi_2 \geqslant \varphi_{2\,tot}$ ($\varphi_{2\,tot}$ = angle of total reflection)
and $\varphi_1 = \pi/2$ or

$$\varepsilon_{2\,tot} = \frac{\pi}{2} - \varphi_{2\,tot} \simeq \sqrt{2\Delta n}$$

where ε_2 represents the corresponding glancing angle of the
incident wave. In the practical case of $n = 1 + 300 \cdot 10^{-6}$ the
glancing angle ε_2 lies in the region of grazing incidence
when $\Delta N = 10$; and a separate consideration of the case of total reflection is then not necessary.

3.1 Abrupt transition

The reflection and transition coefficients are found from
Fresnel's formula. For both directions of polarization
(E parallel and perpendicular with respect to the plane of

incidence) we get only a noticeable reflection in the case of grazing incidence (5' to 20'). In the case of total reflection (Figure 9) there also occurs a phase shift.

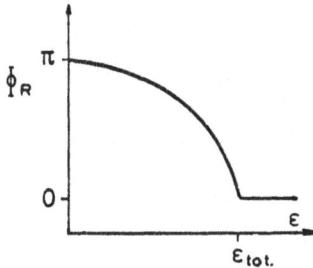

Figure 9   Phase Φ of the coefficient of reflection for abrupt transition (sketch).

3.2   Continuous transition

This case is considered using the unsymmetric Epstein profile (Δz = width of transition zone)

$$n^2(z) = 1 + 2 \Delta n \; \frac{e^{z/\Delta z}}{1 + e^{z/\Delta z}} \; .$$

Considering the amplitude of the reflected wave we obtain the same result as in section 3.1, that is a noticeable reflection occurs at grazing incidence only. However, already at a non-grazing incidence a phase shift occurs in the transmitted wave (Figures 10 and 11).

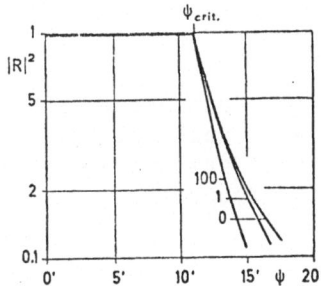

Figure 10   Coefficient of reflection $|R|^2$ against inclination angle ψ for Δz = 0; 1λ and 100λ, ΔN = 10; λ : wavelength of incident wave in vacuum. ψ corresponds to $\mathcal{E}_2$ in text, $\psi_{crit} = \mathcal{E}_{2 \, tot}$

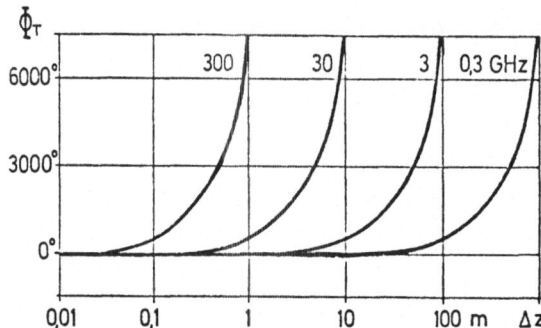

Figure 11   Phase of transmission coefficient Φ_T versus
             thickness of the layer Δz for perpendicular
             incidence and ΔN = 20.

3.3 Ionospheric layer

A layer in the shape of a symmetric Epstein profile (Figure 1) due to an ionospheric discontinuity

$$n^2(z) = 1 + 2\Delta n \frac{4\, e^{z/\Delta z}}{\left(1 + e^{z/\Delta z}\right)^2}$$

formed by a free electron concentration is assumed to exist. Reflection results at grazing incidence and a phase shift even at non-grazing incidence, similar as in 3.2.

3.4 Common properties of boundaries

Assuming the real case to be closer to the continuous transition , are to be expected reflection at boundaries for grazing incidence and a phase shift as well depending on the width of the transition region and on the frequency.

4. EVALUATION OF EXPERIMENTAL DATA

4.1 Ionosphere

The structure of the ionosphere is explored mainly by electromagnetic remote sensing methods. From effects such as aurora reflection, travelling disturbances and ionospheric scintillation it can be concluded that boundaries exist similar to those in the troposphere (Moore 1951, Booker 1958, Lange-Hesse 1967, Harnischmacher and Rawer 1968, Albrecht 1971, Hartmann 1971).

4.2 Troposphere

Weather conditions associated with strong inversion layering are the quasi-stationary high pressure areas which, contrary to intermediate highs, have a great extension in the height. The evidence of reflection processes due to embedded discontinuities can be shown by several experimental facts observed at various frequencies;

a) Study of the reflection at elevated and at ground inversions at 0.5 GHz and 0.09 GHz,

b) evaluation of satellite signals at 0.14 GHz in the presence of weather fronts and during strong elevated discontinuities,

c) the passage of fronts through scatter links and

d) attenuation measurements at 35 GHz.

4.21 Study of the reflection at elevated and at ground inversions at 0.5 GHz and 0.09 GHz. Field-strength measurements of several 0.5 GHz and 0.09 GHz transhorizon links have been correlated simultaneously and independently with vertical profiles of the refractive index. Excellent correlation has been found between field-strength enhancements and the presence of layerlike discontinuities. The field-strength increase depends on height and intensity of the discontinuity (Fengler 1964). Similar results were found by Lane and Sollum 1965, Lane 1969, Thayer 1971, Stone 1958.

4.22 Evaluation of satellite signals at 0.14 GHz in the presence of weather fronts and strong elevated discontinuities. Anticipation and delay of radio satellite rise and set time at 0.14 GHz were evaluated with respect to weather conditions. It was found that pronounced effects were due to a strongly layered tropospheric structure, such as ground inversions and severe weather fronts, in particular cold fronts. A layered tropospheric structure leads to a delay of the satellite signal. Therefore it has to be assumed that reflection processes at boundaries of refractive-index discontinuities, which may be inclined, deflect the propagation path in such a way, that the receiving point becomes located in a shadow zone. This structure exists in association with strong high pressure systems with a large vertical extension. Anti-

cipation of satellite rise time is related to ground inversions and severe weather fronts, again in particular cold
fronts. This effect has to be interpreted as a ducting effect which leads in this case to an enhancement of the field
strength (Fengler 1971).

4.23 The passage of fronts through scatter links. The passage of fronts, in particular cold fronts through scatter
links is accompanied by a decrease of field strength and a
tendency of increase of side scatter depending on the age
of the front. This result was found at 0.14 GHz and 0.09 GHz
(Albrecht 1968 and 1970). It was also noted by Stone 1958
at 3 GHz and was characterized by unusually high fading rates
reaching very deep nulls (Pavlasek 1975).

4.24 Attenuation measurements at 35 GHz. Observations at
35 GHz using the sun as source at various elevation angles
show that there is a strong dependence of attenuation on the
simultaneously measured humidity (Altshuler et al 1968, Telford and Altshuler 1975), see Figure 12. This good correla-

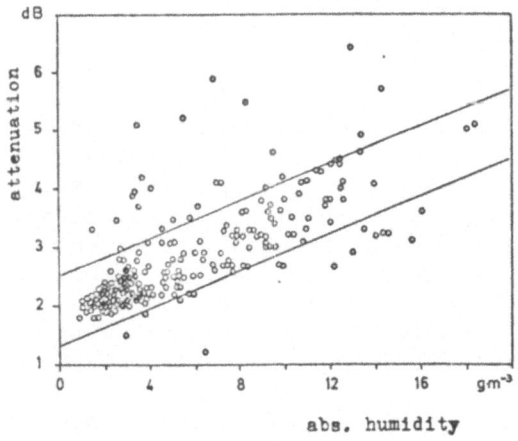

abs. humidity

Figure 12   Attenuation [dB] against absolute humidity $\left[g \cdot m^{-3}\right]$
             elevation angle of the sun 5°.

tion can be interpreted as energy absorption by humidity.
Cases with rain were excluded from this consideration. The
statistical evaluation shows that about 2o % of all 230 individual measurements at an elevation angle of 5° have a
large deviation from the regression line. It was found that
77 % of these runaways with larger attenuation related to the
absolute humidity are in these cases when the rays from the

sun pass through cold fronts. The rest of the runaways (23%) have a smaller attenuation related to the absolute humidity. These cases are associated with strong high pressure areas, i.e. with weather situations which have a tendency to produce pronounced elevated refractive-index discontinuities.

5. CONCLUSIONS

Embedded wide-spread discontinuites which are present about 10 to 30 % of time (depending on season and geographic location) have various kinds of influence on electromagnetic wave propagation in the atmosphere. This is experimentally proved for a wide range of frequencies. These influences can be interpreted in first approximation as reflection processes at atmospheric boundaries.

Of course, other propagation processes need to be considered as well because boundary effects only have been assumed so far.

Real boundaries which exist in the atmospheric structure have arbitrary shape and are irregular. Diffuse reflection or scattering must thereforeresult depending on the ratio between wavelength and dimensions of the irregularity which so far has only been considered in terms of the width of the transition zone.

The presence of precipitation due to fronts can give rise to additional absorption or scatter. Therefore the interpretation of the phenomena in terms of reflection processes is only an initial approximation restricted to planer boundaries, which in reality do not exist, and only to specific frequency ranges depending on the dimension of roughness of the boundaries. Therefore the generalized interpretation of the above mentioned experimental data is not possible. Nevertheless the salient fact remains that significant events occur at these embedded discontinuities whose detailed structure is as yet unknown.

The simplified reflection process model can only be applicable for one element of the surface whereas the total effect of the boundary must be taken as a superposition of all individual elements which give different contributions.

6. IMPLICATIONS TO RADAR SYSTEMS

Wide-spread embedded discontinuities cause several effects when considered in terms of reflection processes alone. The

theoretical treatment leads to deflection at low elevation
angles or to phase disturbances if the boundaries or the
transition zone are inhomogeneous or moving .

Such a deflection results in a change in the angle of arrival which differs from the true direction of the target. It
seems that cold fronts and strongly pronounced elevated layers as well as thick ground based layers produce substantial
effects. As experiments show they can give rise to radar
holes or to an extension of the range in association with
signal distortion as well as an enhancement of the radar
image. Multipath propagation effects lead to further additional erroneous signals.

Another problem associated with embedded refractive-index
discontinuities is that of phase variation. As long as these
variations are uniform over the whole wavefront, they only
have an effect on the signal velocity. If the phase variation
changes rapidly in space then time variable signal distortion will occur as well. This occurs in the case of the discontinuities discussed here which are in motion or are fluctuating and are non-uniform as must be assumed to be the
case. Clutters will be the result. The net result will be
the formation of clutter on the radar display screen and
these effects can be expected to occur at any angle of incidence.

Summarizing there are two dangerous weather situations that
can give rise to severe disturbances: upcoming cold fronts
and strong high pressure areas.

REFERENCES

1.   H.J. Albrecht, Theoretical analysis of medium-dependent
      fluctuations with tropospheric scatter links and compa-
      rision with new experimental data including side-scatter
      characteristic, AGARD-CP-37, Ref. 16, 1968.
2.   H.J. Albrecht, Daily and hourly forecast of tropospheric
      propagation parameters, AGARD-CP-70-71, Ref.45, 1970.
3.   H.J. Albrecht, Prediction of ionospheric scintillation
      for satellite signals in communication and navigation,
      XIX Convegno Internationale delle Communicazioni, Ge-
      nova, 1971.
4.   E.E. Altshuler, V.J. Falcone, Jr., K.N. Wulfsberg,
      Atmospheric effects on propagation at millimeter wave-
      lengths, IEEE spectrum, July 1968.
5.   B.R. Bean and E.J. Dutton, Radio Meteorology, Dover Pub-
      lications, Inc., New York, 1968.

6. H.G. Booker, The use of radio stars to study irregular refraction of radio waves in the ionosphere, Proc.IRE 46, 1958.

7. C. Fengler, The phase of a plane electromagnetic wave transmitting a wide-spread atmospheric discontinuity, AGARD-CP-107, Ref. 3, 1973.

8. C. and G. Fengler, V.H.F. and u.h.f. field-strength variations corresponding to river beds, Proc. IEE, 122,1 , 1975.

9. G. Fengler, The influence of inversions on UHF-propagation over land, Proc. 1964 World Conf. on Radio Meteorology, Boulder,Col., 1964.

10. G. Fengler, Tropospheric effects on the propagation of VHF-satellite signals, Kleinheubacher Berichte, 14, Darmstadt, 1971.

11. E. Harnischmacher and K. Rawer, A simple description of ionospheric drifts in the E-region as obtained by a fading method at Breisach, in: Winds and Turbulence in Stratosphere, Mesosphere and Ionosphere, North-Holland Publishing Company, Amsterdam, 1968.

12. G.K. Hartmann, A brief review of scintillation studies. Journal of Scientific & Industrial Res., 30, 1971.

13. H. Jeske, Die Ausbreitung elektromagnetischer Wellen im cm- bis m-Band über dem Meer unter besonderer Berücksichtigung der meteorologischen Bedingungen in der maritimen Grenzschicht. Hamburger Geophys. Einzelschr. No.6,Cram, de Gruyter & Co, Hamburg, 1965.

14. J.A. Lane and P.W. Sollum, V.H.F. transmission over distances of 140 and 300 km, Proc. IEE, Vol.112, 2, 1965.

15. J.A. Lane, Some aspects of the fine structure of elevated layers in the troposphere, Radio Science, Vol.4, 12, 1969.

16. G. Lange-Hesse, Polarlicht als Rückstrahler ultrakurzer Wellen, Kleinheubacher Berichte,12, Darmstadt, 1967.

17. R.K. Moore, A VHF propagation phenomenon associated with aurora, J.Geophys. Res.,56, 1951.

18. T. Pavlasek, Private communication, McGill University, Montreal, 1975.

19. Kh.P. Pogosyan, The air envelope of the earth, Israel Program for Scientific Translations, Jerusalem 1965.

20. S.A. Stone, Beyond-the-horizon propagation at microwave frequency, Thesis, Dept. of Electrical Engineering, McGill University, Montreal, 1958.

21. L.E. Telford and E.E. Altshuler, Atmospheric attenuation at 15 and 35 GHz for slant paths near the horizon, Digest URSI-Meeting, Urbana, Ill., 1975.

22. G.D. Thayer, Reflections from elevated layers in transhorizon propagation. AGARD-CP-70, Ref. 16,1971.

USEFUL METEOROLOGICAL PARAMETERS FOR PREDICTING SIGNAL BEHAVIOR

Dennis W. Thomson

Department of Meteorology, The Pennsylvania State
University, University Park, Pennsylvania   16802

ABSTRACT. Progress in planetary boundary layer modeling and
measurement techniques suggests that variables describing meteorological processes rather than atmospheric structure ought to be
emphasized for analysis of EM signal propagation. Meteorological
analysis of atmospheric structure related to the interpretation
of EM signals has traditionally been performed using parameters
such as the thermodynamic variables temperature, pressure and
humidity. In particular, the optical refractive index, n, and
radio refractivity, N, space and time-varient fields have been
related to mean vertical gradients, turbulent fluctuations at a
point or in a small volume and changes with respect to time (typically over periods of hours) of the "mean" values. Definition
of the "structure constants" used to describe root-mean-square
turbulent fluctuations of a parameter depends upon assumptions
regarding the fundamental nature of turbulence. Furthermore, n
and N as dependent variables depend in a complicated fashion on
meteorological condition dependent combinations of their respective independent variables.

Recently developed FM-CW radar and acoustic sodar indirect
sounding techniques and new sophisticated models for simulating
the structure and behavior of the atmosphere's planetary boundary
layer have greatly improved our understanding of atmospheric
structure and processes at scales ranging from a few meters to
several kilometers. Observational and theoretical results suggest that interpretation of received EM signals may be facilitated by using measurements and models which provide output dependent upon atmospheric processes and changes thereof rather than,
e.g., one-dimensional vertical profiles of structure. Vertical
time sections obtained using radars and sodars clearly illustrate

Jeske (ed.), Atmospheric Effects on Radar Target Identification and Imaging, 335–356.
All Rights Reserved. Copyright © 1976 by D. Reidel Publishing Company, Dordrecht-Holland.

the dynamic, time varying properties of, e.g., layer features, especially those perturbed by acoustic gravity waves, and the "intermittant" behavior of the turbulent atmosphere. Physical-numerical models have been developed using such observations which predict dynamic variables such as the space-time varying Richardson number and the fluxes of mass, momentum, heat and moisture. With proper adaption, these models could be used to interpret and predict the quality and variations of refracted and scattered EM signals.

1. REVIEW OF TRADITIONAL ANALYSIS TECHNIQUES

Historically, meteorological studies of atmospheric structure have been concerned with the analysis of "weathering dominating" features. Problems such as interpretation of the evolution of major storm systems and the characteristics of dramatic singularities like tornadoes and large thunderstorms have occupied both theoreticians and experimentalists for many decades--and still today fundamental questions are unanswered. Early interest in the smaller scale atmospheric phenomena, processes which we now know influence most directly the propagation of EM waves, was stimulated by scientists interested in, e.g., diffusion and agricultural meteorological problems.

Actually, with the exception of a few special research studies a tremendous "spectral gap" existed until about 1960 in our understanding of the details of atmospheric structure in the domain between the "synoptic-scale" features ($\lambda > 300$ km) which could be studied using radiosondes at 12 hour intervals and hourly surface weather data, and those lowermost atmospheric features which could be continuously examined using instrumented towers and special turbulence sensing equipment on or near the ground. Although the studies of synoptic-scale features were elegant (see, e.g. Danielson, 1959; Saucier, 1955; Petterson, 1956), they were, unfortunately, of little assistance to the radiometeorologist. Research at the U.S. National Bureau of Standards (and other locations) in the 1950's and 60's pretty well defined both the applications and serious limitations of most "synoptic" data for radiometeorological purposes (Bean and Dutton, 1966). Figure 1 is representative of the space-time analysis which can be performed using archived synoptic data. Cross sections such as the one shown were used for interpreting RAKE troposcatter measurements (Birkemeier et al., 1969).

Interpretation of small-scale atmospheric turbulence on the other hand was limited with few exceptions to surface and tower measurements. Thus, although many properties of the atmosphere's lowermost surface layer were well established, it was not possible to specify in other than an average sense the temporal and spatial characteristics of the atmosphere from about 100 m to 2 km.

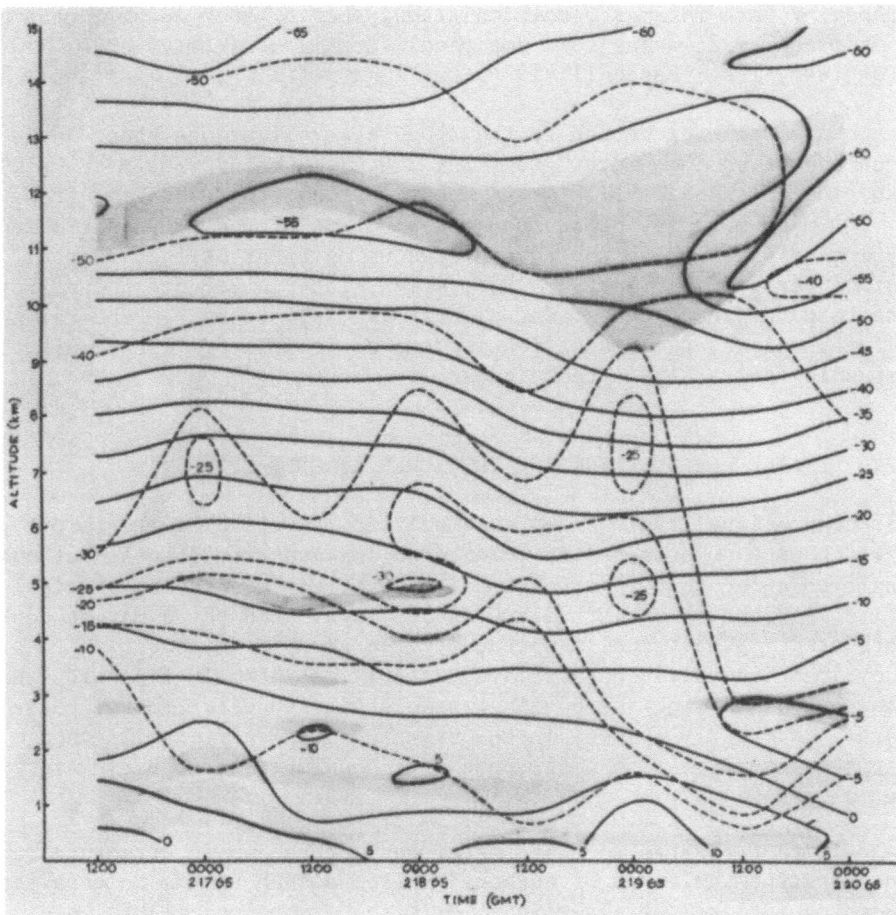

Figure 1.   Vertical time section of temperature and wind field at
             the mid-path of a Rake troposcatter link.   Radiosondes
             at 12 hour intervals from 1200 GMT, 2/16/65 to 0000 GMT
             2/20/65.   Isotherms, crosspath wind speed and inversions
             are solid and dashed lines, and stippled areas, respec-
             tively.   Tropo signals were found to be most effectively
             scattered in the inversion layers.

Descriptions of atmospheric structure and predictions of its
behavior were for most practical requirements, such as pollutant
diffusion, made using semi-empirical formalisms (Lumley and
Panofsky, 1964; Pasquill, 1974).

A notable exception to the above generalizations about inade-
quate data on the "meter" ($3 < \lambda < 300$ m), "urban" (300 m $< \lambda < 30$
km) and "mesoscales" ($30 < \lambda < 300$ km) at altitudes ranging from
$\simeq 100$ m to 5 km is, of course, the substantial body of radar meteorological data which was accumulated using radar as the indirect
atmospheric probe (Battan, 1959; Atlas, 1964). Because the effects
of rain on propagating radar signals are well known, in the balance
of this paper, the discussion is limited to that of "clear" air
sounding and analysis techniques.

2. DETAILED MEASUREMENTS OF THE LOWER ATMOSPHERE

If conventional radiosondes are modified by replacing the baro-
switch with a motor-driven commutator and are then launched at one-
half to three hour intervals, quite detailed "mesoscale" vertical
time sections of the structure of the lower atmosphere may be constructed. A nocturnal temperature inversion, development of the
daytime planetary boundary layer, nearly isothermal, deep elevated
layers and warming and cooling associated with weak frontal pass-
ages are clearly evident in the example shown in figure 2 (used
in the interpretation of troposcatter Doppler shifts, Birkemeier
et al., 1968).

Direct observations of "meter" or "urban" scale structure re-
quire either an array of tethered balloons such as the Universität
Hamburg (Stilke, et al., 1968), British Meteorological Office -
Cardington (Readings, 1975), or National Center for Atmospheric
Research (Morris et al., 1975) systems, or an instrumented air-
craft.

A suitably instrumented aircraft (or helicopter) is certainly
the more flexible system. However, they are expensive (several
hundred dollars/hour) to operate and, furthermore, for "vertical
profiles" still cover a several km^2 area since the ratio of horizontal to vertical velocity is typically 15 or more to one.
Figure 3 shows temperature, refractivity and relative humidity
profiles, respectively, made with an instrumented aircraft (Thomson,
1968). They are representative of the character and quality of
contemporary, digitally logged airborne measurements. For radiometeorological studies, aircraft data are of little value unless
they are carefully analyzed for the feature or process of interest.
Figure 4 summarizes some of the refractive properties of meter
scale layers observed using an aircraft in the common volume of
a troposcatter link (Thomson, 1968). It has been clearly

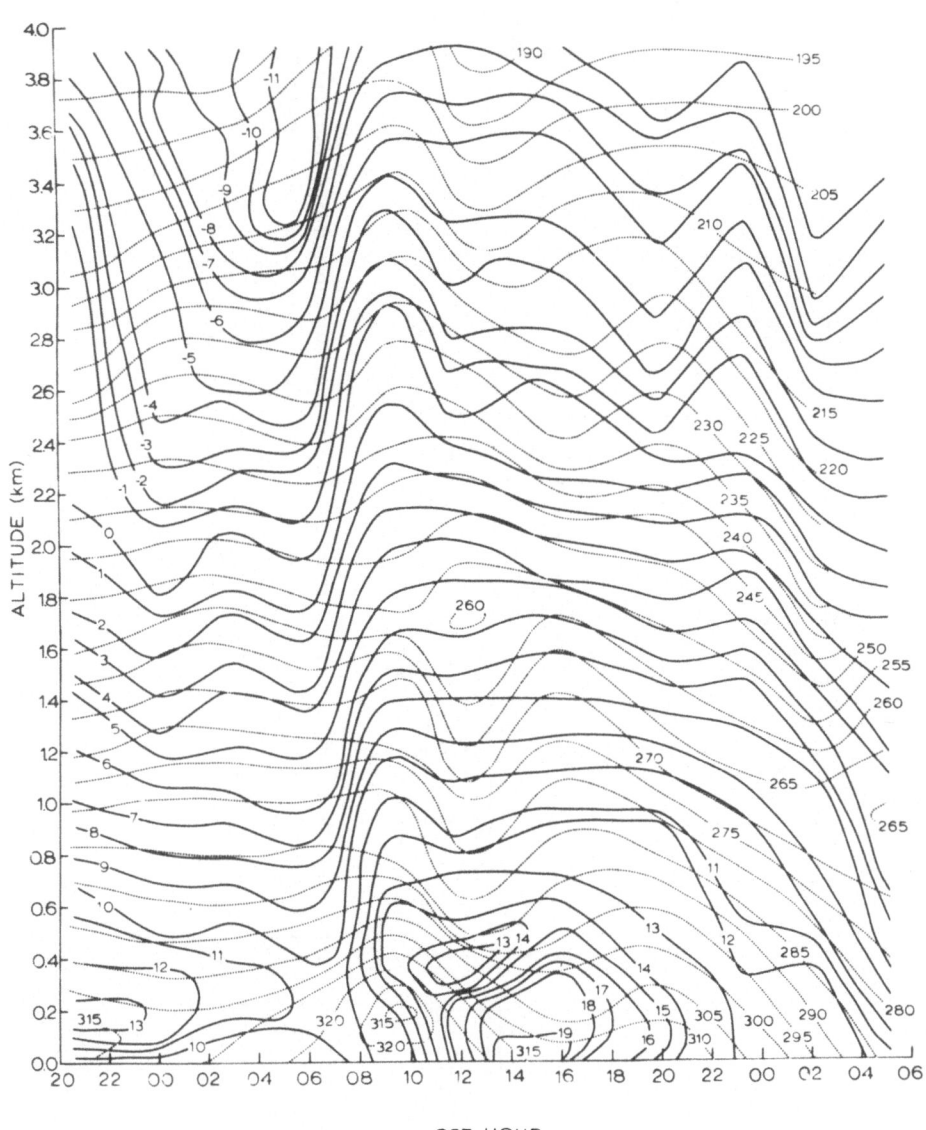

Figure 2. Vertical time section of isotherms and radio refractivity isolines for a 34 hour period measured in the common volume of a phase coherent troposcatter link. Modified radiosonde balloons were launched at 3 to 6 hour intervals. Refractivity was computed from the reduced meteorological data.

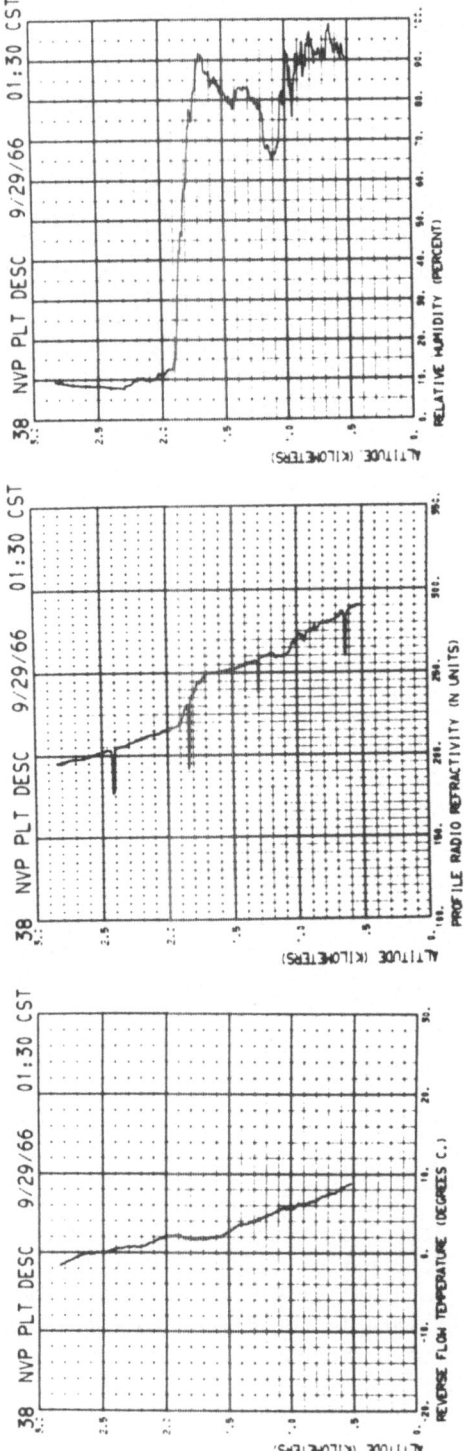

Figure 3. Vertical profiles of temperature, radio refractivity and relative humidity observed using a refractometer and high speed digital recording equipped research aircraft. The aircraft descent from 601:00 to 608:50 CST was through a layer at the top of the planetary boundary layer in the vicinity, according to synoptic analysis, of a weak cold front. "Gliches" on refractometer profile indicate range resets.

established that such layers are associated with radio ducting
and holes and can contribute to anomalous propagation.   However,
in order to analyze propagation problems such as synthetic apera-
ture radar (SAR) image degradation, one should probably concentrate on analysis of "statistical" summaries of turbulent processes rather than quasi-stationary small-scale structural features.

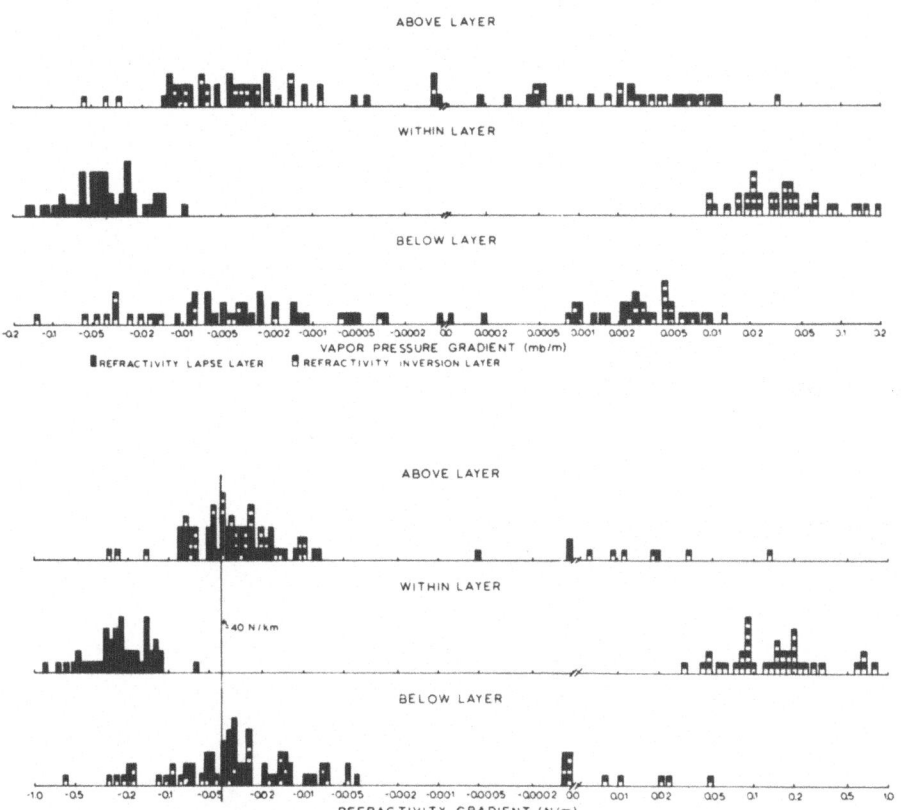

Figure 4.   Distributions of refractivity and water vapor pressure
            gradients within and above and below "meter-scale"
            refractivity lapse (sharply decreasing) and inversion
            (sharply increasing) layers. Gradients within the
            refractive layers (ranging from 2 to 60 m in depth)
            exceeded normal atmospheric values by about an order
            of magnitude. Whereas refractivity gradients within
            the atmosphere depended upon combined water vapor
            pressure and temperature gradients, layer gradients
            were dominated by water vapor gradients.

In the past few years significant progress has been made in airborne measurements research (see Atmos. Tech. No. 1, 1973 and Fall, 1975). The use of inertial navigation systems has greatly improved wind and field turbulence measurements (Axford, 1968; Lenschow, 1975). "Pressure-normalization" schemes (used in figures 5 and 6) have aided the interpretation of urban and mesoscale

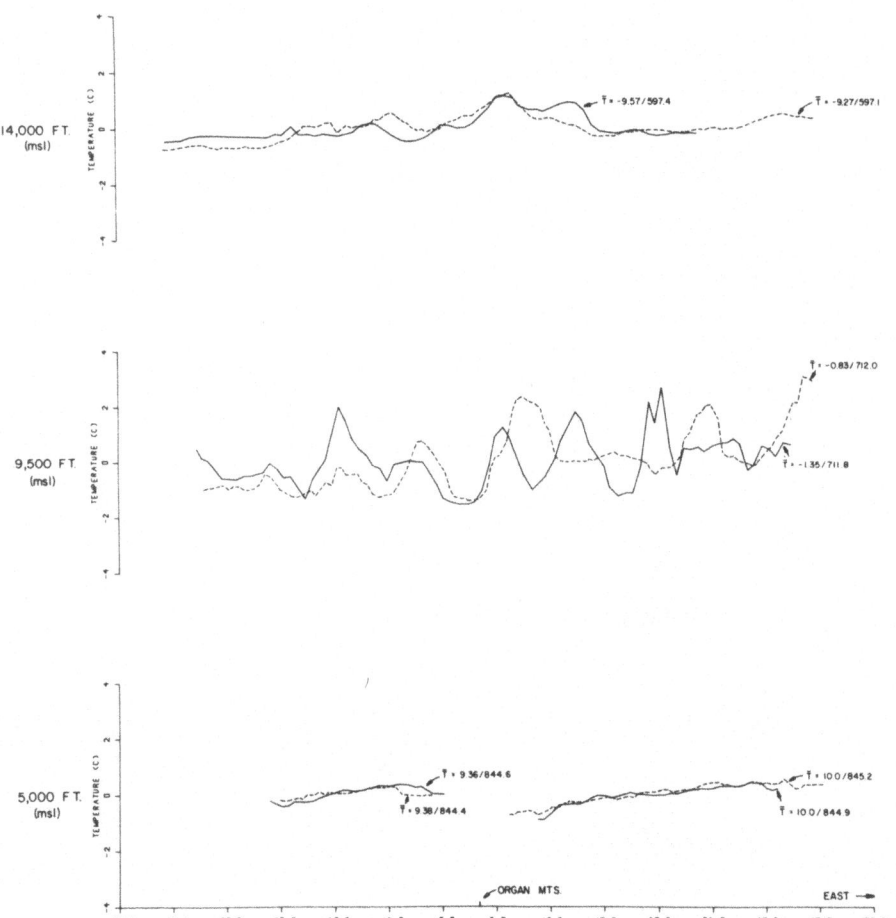

Figure 5.   Horizontal temperature variations measured using a research aircraft at selected flight levels in the vicinity of the Organ Mountains in New Mexico. To correct for aircraft altitude variations, observed mean vertical gradients were used to adjust observations to each mean flight level. The large amplitude temperature "waves" at the 9,500 ft level indicate the presence of an inversion across which strong wind shear was also observed. Patches of moderate to severe clear air turbulence were encountered within the layer.

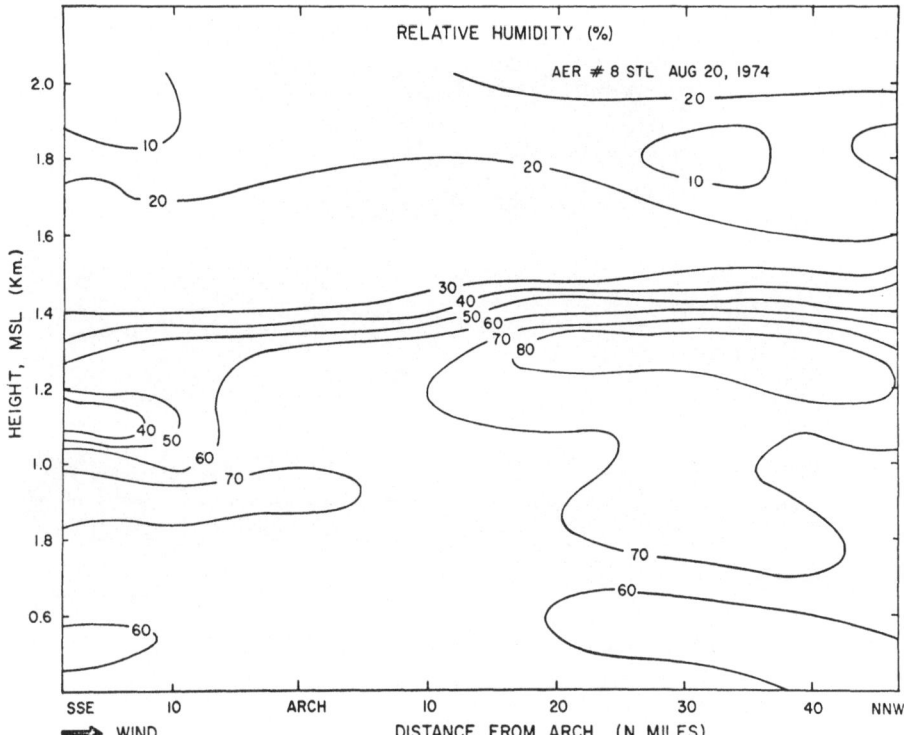

Figure 6. Horizontal cross section of relative humidity field
           measured using a research aircraft in the St. Louis,
           Mo. vicinity. The high gradient region near 1.4 km
           corresponds to the top of the mixing layer. At that
           level the concentration (particles/cc) of atmospheric
           particulate dropped nearly two orders of magnitude.
           Within the mixing layer, concentrations were essen-
           tially independent of height.

features (Mamane et al., 1975; Godowitch, 1976). All of the above
analysis techniques may be applied to parameters such as the radio
refractivity N (x, y, z, t) and C_N^2 ($\vec{r}$, x, y, z, t), the root-mean-
square refractivity fluctuations on the scale of interest (e.g.
λ/2) and as a function of position and time. Techniques for
optimizing SAR data processing on the basis of observed or predicted
C_N^2 vertical time or cross sections should be studied.

3. CONTEMPORARY INDIRECT SOUNDING SYSTEMS

Optical lidar and acoustic sodar systems have in less than a decade
revolutionized our perception of atmospheric structure and processes
in the lowest several kilometers. For example, vertical time section displays of atmospheric aerosol obtained using lidar (figure 7)

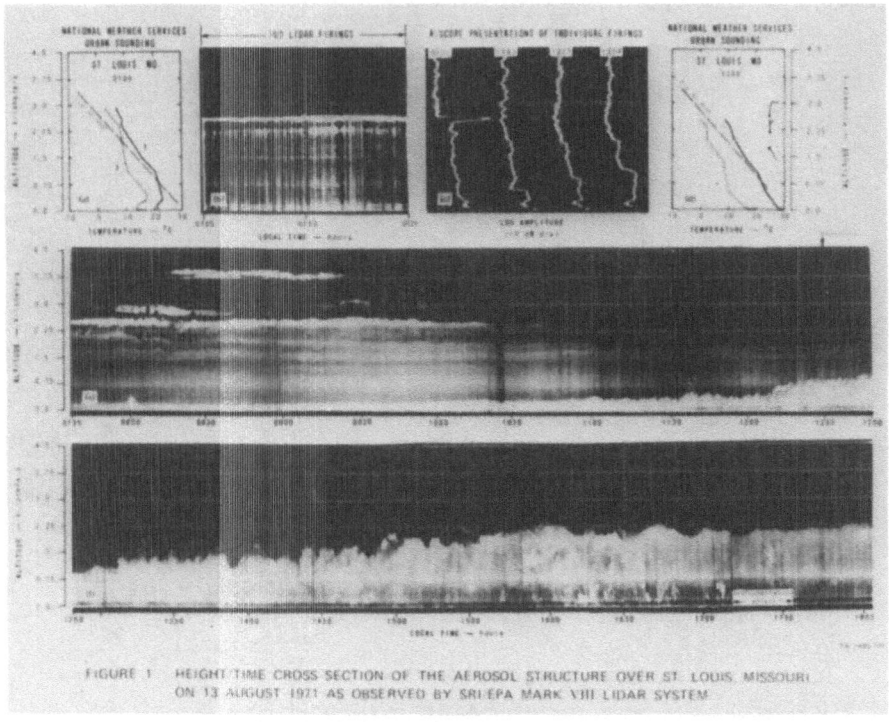

FIGURE 1   HEIGHT/TIME CROSS SECTION OF THE AEROSOL STRUCTURE OVER ST. LOUIS, MISSOURI
ON 13 AUGUST 1971 AS OBSERVED BY SRI/EPA MARK VIII LIDAR SYSTEM

Figure 7.   Application of lidar for monitoring planetary boundary
layer structure.   (Courtesy of R. H. Collis of Stanford
Research Institute, Menlo Park, CA).

greatly facilitate the interpretation of ongoing processes such as
growth of the planetary boundary layer even if we do not yet understand all the details of the scattering of light by atmospheric
aerosol (Reagan and Thomson, 1976) and using sodar (figure 8) we
are now able to monitor the detailed time-height history of turbulent temperature and velocity fluctuations (Little, 1968; 1972;
Derr, 1972; Hall, 1972; Thomson, 1975). Most importantly, the
sophisticated indirect probing techniques such as sodar, lidar
and FM-CW radar are greatly facilitating the development of numerical and theoretical models which can be used even with minimal
data to make highly reliable predictions of turbulent conditions
and the location of certain structural features in the lower
atmosphere.

4. APPLICATIONS OF MINOCOMPUTER TECHNOLOGY

Techniques are now being developed for using mini-computers to
service meteorological data circuits and to provide real-time
analysis capabilities at any location receiving standard coded

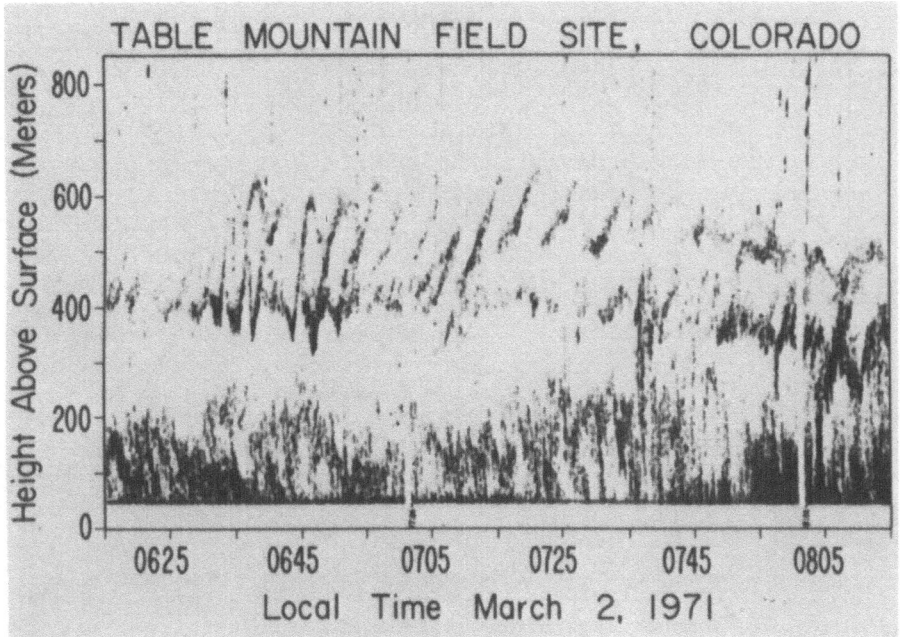

Figure 8. Breaking waves observed using sodar in nocturnal
planetary boundary layer. (Courtesy of F. F. Hall,
Wave Propagation Laboratory, NOAA, Boulder, CO.)

data. One of the most promising techniques for analysis of meteorological fields is the Hermitian interpolation scheme developed
by Shapiro and Hastings (1973). This technique has been programmed
at Penn State University for a PDP-11 minicomputer and is now
used on a daily basis for weather analysis and prediction (Cahir
and Norman, 1975; Norman and Cahir, 1976). Figures 9, 10, and 11
demonstrate analysis techniques now in use which with very little
additional work could be adapted for specific radiometeorological
applications.

Incoming surface and upper air observations are stored on
a disc. They are available at the push of a button to generate on
a video display thermodynamic charts and profiles such as the Skew
T-ln P soundings shown in figure 9. If the Hermitian interpolation
scheme is then applied in a line, an isentropic cross section as
the one in figure 10 may be generated. Presently, regression techniques are being used for analysis of the water vapor field. The
water vapor field shown (figure 11), which was obtained using a
fourth order scheme, could probably be greatly improved if higher
resolution data were available. Since the radio refractive index
is most sensitive in the lower atmosphere to vapor pressure variations, synoptic relative humidity data would be of limited use for

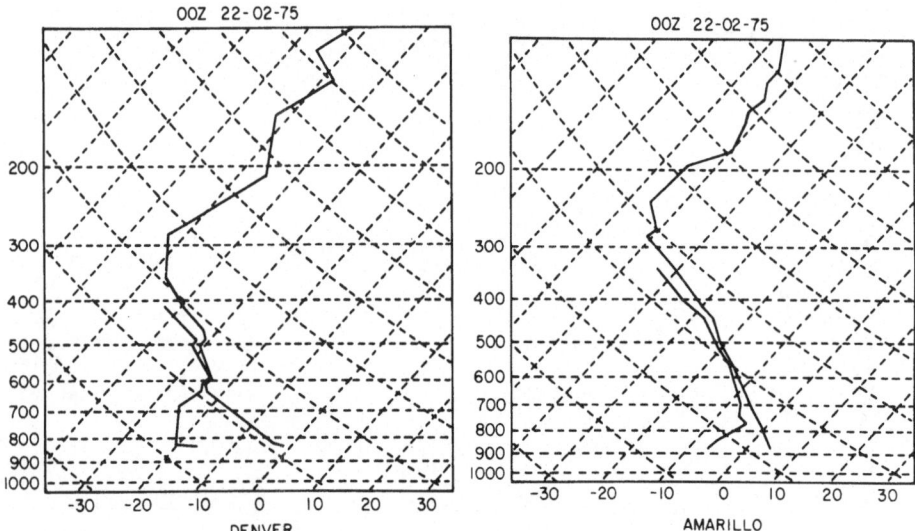

Figure 9. Skews T-ln P soundings generated on the video display
terminal of a minicomputer which services high speed
weather data lines at Penn State University.

radiometeorological analyses. In spite of present limitations
the potential power of such techniques is impressive. Figure
12 shows an interpolated sounding generated from the isentropic
and relative humidity cross sections. We are now experimenting
with the use of these techniques for generating, e.g. refractivity profiles at arbitrary, specified locations, and with
higher resolution research measurements such as the soundings
shown in figure 2.

The minicomputers may also be used for a variety of real-time
special purpose predictions: for the meteorologist such factors
as minimum temperatures or mixing layer depths, for the radio-
engineer refractive bending or duct properties.

5. PREDICTIVE MODELS FOR ATMOSPHERIC TURBULENCE AND STRUCTURE

In general the parameters describing the properties of the lower
atmosphere which are of meteorological interest are those related
to the production and dissipation of atmospheric turbulence in
widely differing atmospheric conditions. Analysis of the turbulent conditions requires knowledge of the (ref. Table I) thermodynamic variables (pressure, temperature and vapor pressure) and
dynamic variables such as the vector wind and its fluctuations as
a function of position and time. Complete solutions to the full
set of equations of turbulent motions have not yet been, and may
never be, obtained. However, using carefully formulated

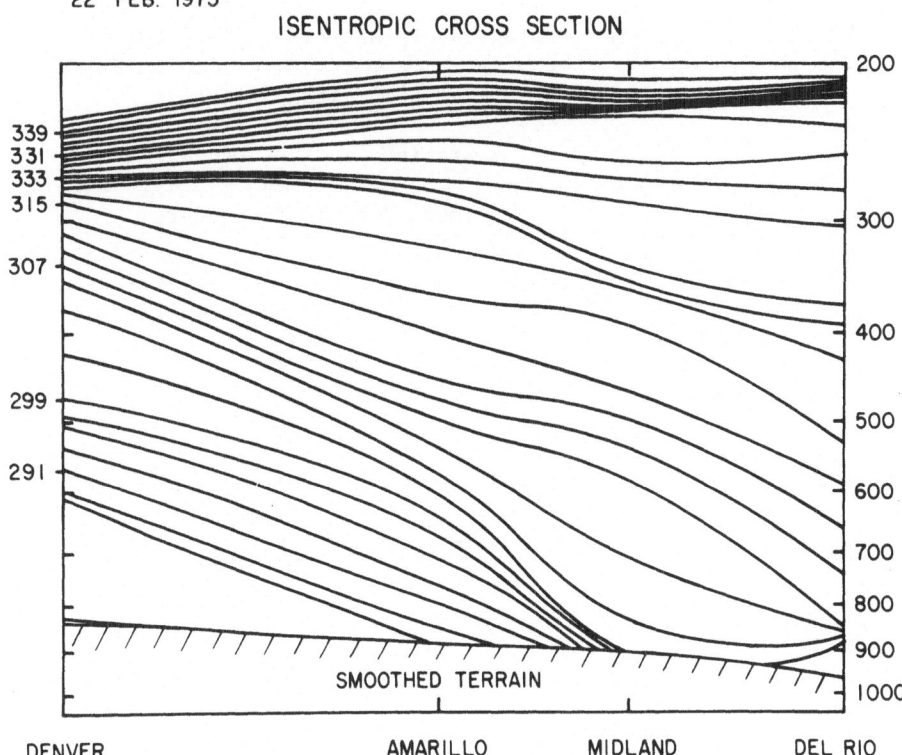

22 FEB. 1975

ISENTROPIC CROSS SECTION

Figure 10.   Isentropic cross section computed and displayed using
             a mini-computer at Penn State University. The
             forecaster-computer operator may select up to seven
             stations from the national network and generate a
             variety of such analyses to aid in formulating his
             forecast for a given city or region.

approximations, substantial progress in modeling many planetary
boundary layer processes has been achieved. The serious reader
can develop an appreciation for both the state-of-the art in
planetary boundary layer modeling and the difficulty of research
in this area by reading a paper such as Deardorff's (1974). Fortunately, for many purposes the radiometeorological variables of
interest such as the vertical refractive index gradient or even
the root-mean-square refractivity fluctuations may be regarded as
passive "contaminants" and thus the relatively simpler "parameteri-
zation" schemes may be used to describe and predict atmospheric
predictions.

    Models such as those devised by Tennekes (1973; 1974) and
Wyngaard (1974) for convective planetary boundary layers, if they
were adapted for radiometeorological applications, would probably

TABLE 1

"Clear Air" Radiometeorological Parameters of Interest

Radio Variables	Meteorological Variables	
	Static	Dynamic
I. "System" parameters	Thermodynamic Parameters	
Power	$P(z)$	$P(x, y, z, t)$
Wavelength	$T(z)$	$T(x, y, z, t)$
Pulse Repetition	$e(z)$	$e(x, y, z, t)$
Frequency		
Beamwidth		
Geometry		
II. "Atmospheric" Parameters		
$N(r, t)$	$N(e, T, p)$	$N(e, p, T, t)$
$\dfrac{\partial N}{\partial z}$	$\dfrac{\partial N}{\partial z}(z)$	$\dfrac{\partial N}{\partial z}(x, y, z, t)$
		$\vec{V}(x, y, z, t)$
		$\dfrac{\partial \vec{V}}{\partial z}(x, y, z, t)$
	$\dfrac{\partial T}{\partial z}$ or $\dfrac{\partial \theta}{\partial z}$	$Ri = \dfrac{g}{\theta}\dfrac{\partial\theta/\partial z}{(\partial v/\partial z)^2}$
$C_N^2(r, t)$		$C_T^2(x, y, z, t)$ $C_e^2(x, y, z, t)$ $C_{\vec{v}}^2(x, y, z, t)$ $C_N^2(x, y, z, t)$
		Momentum flux Heat flux Moisture flux Dissipation of turbulent energy
Extinction Coefficient (r, t) Backscattering Cross Section (r, t) Spectrum of Received Signal		

Figure 11.   Relative humidity field generated and displayed using
a minicomputer which services weather data lines at
Penn State University. Resolution is limited by the
available relative humidity sounding data.

suffice for specification of those turbulent atmospheric properties
which are often responsible for radar image degradation. For example, on the basis of the profiles of atmospheric variables shown
in figure 13, it should be possible to specify the vertical profile
of refractive index fluctuations. A model such as Tennekes's
(figure 14) can be used to predict the depth of a turbulent mixing
layer. Using such a model one could, for example, specify the
maximum altitude at which an SAR equipped aircraft being used for
terrain mapping could fly without having the radar beam traverse
the high refractive index gradient layer normally present at the
top of the mixing layer (note again fig. 3).

Analyses and modeling of the lower atmosphere in stable (e.g.
most nighttime) conditions is not nearly so advanced. To no small
extent this is the consequence of the lack until recently of even
marginally adequate observational data.

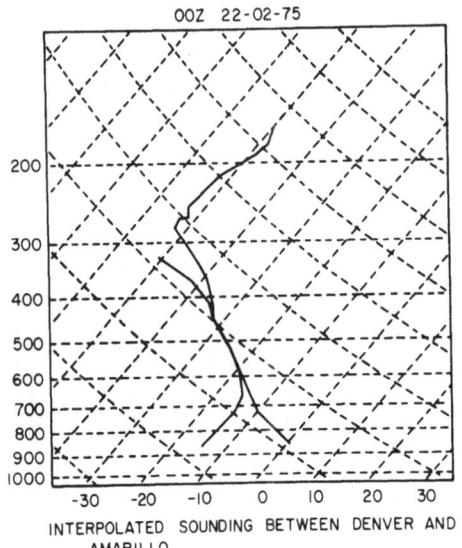

Figure 12. An "artificial" sounding generated for a requested
            location from a minicomputer processed isentropic
            cross section and a relative humidity field. Layer
            of 100% relative humidity was verified using satel-
            lite cloud photographs.

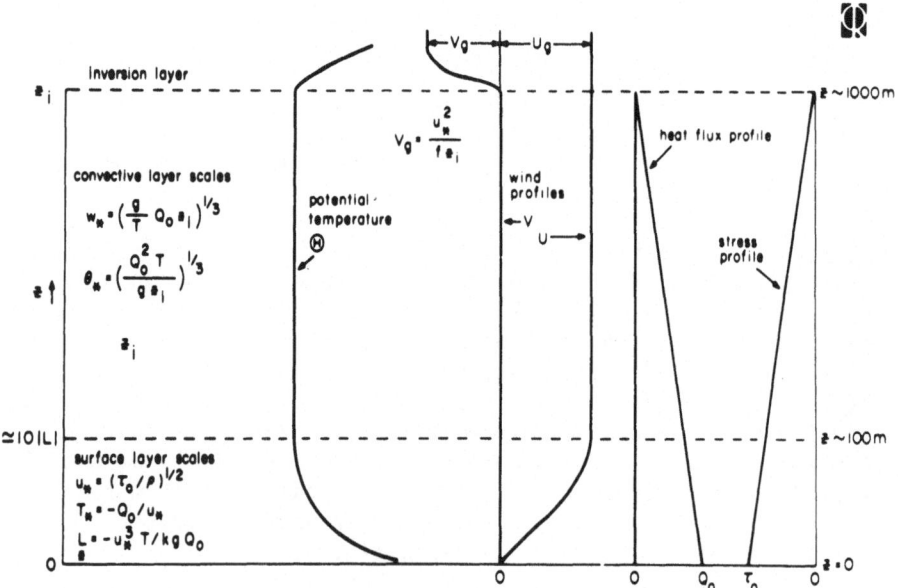

Figure 13. Vertical profiles of potential temperature, wind,
            heat flux and stress for a convectively unstable
            planetary boundary layer. (Courtesy of J.C. Wyngaard,
            Wave Propagation Laboratory, NOAA, Boulder, CO.)

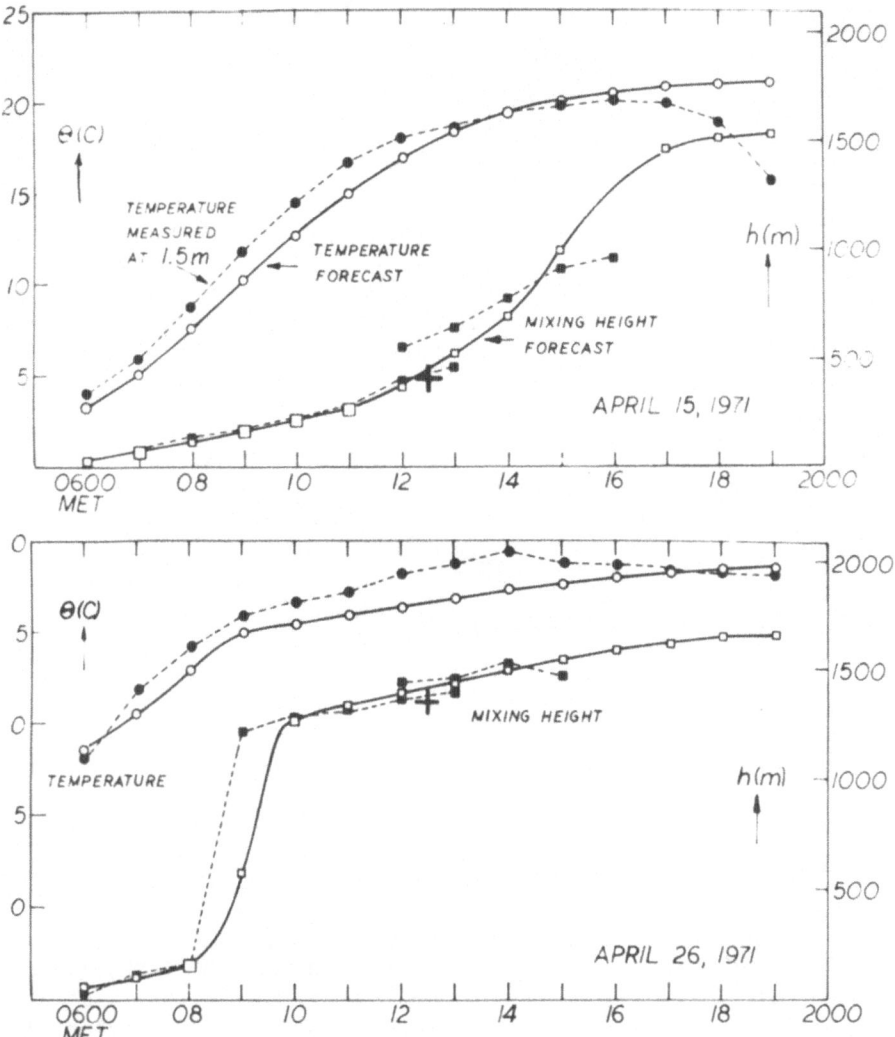

Figure 14. Mixing height and temperature forecasts made using the
Tennekes inversion rise model.

Acoustic sounder, sodar signals which are most sensitive to
turbulent temperature and velocity fluctuations probably now provide our best observations of lower atmospheric structure in stable
conditions. The availability of such data has stimulated meteorologists such as Blackadar (1976) and Wyngaard (1975) to develop
models for stable conditions. Some success has been achieved in
predicting, e.g., rapid changes in the character of surface-based
inversions and nighttime wind profiles (figure 15). It is already

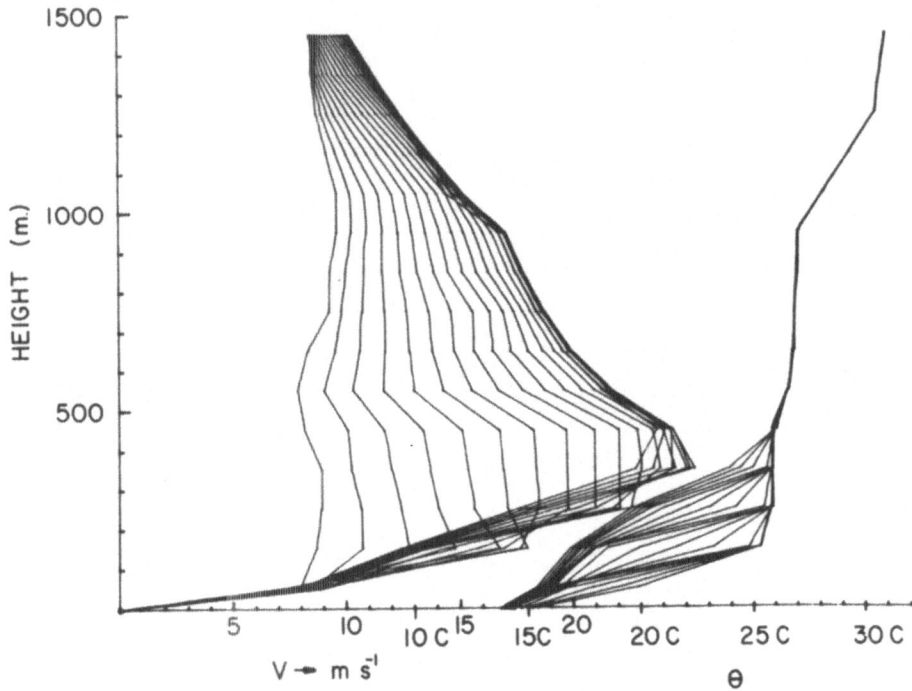

Figure 15.   Wind and temperature profiles predicted at one hour
              intervals for nocturnal planetary boundary layer using
              the Blackadar model. "Episodes" in which the wind
              and temperature profiles change significantly in a
              short period of time are evident.

possible to include predictions of meteorological parameters such
as the eddy diffusivity. At least in principle there should be no
difficulty in expanding existing models to include radio meteorological parameters. However examination of a typical nighttime
sodar record (figure 16) shows that much research remains to be
done. Note, in particular, the changes in depth, altitude and
occurrence of the dark bands corresponding to regions of enhanced
turbulent temperature fluctuations. The time varying "band" structure evident in the sodar returns is the consequence of the same
processes which control the characteristics of duct and certain
types of anomalous propagation. Thus, even now, although sophisticated predictive models for stable atmospheres are not yet
available, the addition of sodar measurements could greatly facilitate studies of radar signals propagated through the lower
atmosphere.

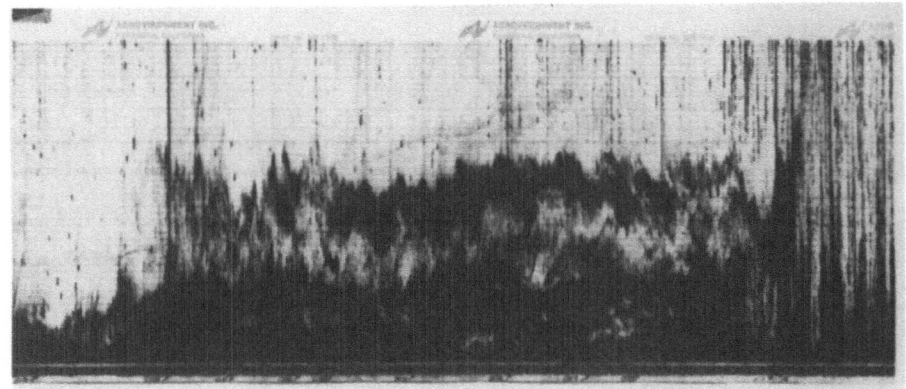

Figure 16. A typical sodar time section of the nocturnal and
early morning planetary boundary layer at University
Park, PA. Dark zones correspond to layers of enhanced
turbulent temperature fluctuations in the stable atmosphere. Sudden transitions in the height, thickness
and character of elevated layers are evident.
Timescale: 14 hours, vertical scale, 500 m.

ACKNOWLEDGEMENTS

The author is indebted to his colleagues at The Pennsylvania State
University, Messrs. Blackadar, Cahir, Lottes, Norman and Tennekes
and to R. Collis of Stanford Research Institute, and F. Hall and
J. Wyngaard of WPL-NOAA for providing useful examples from their
ongoing research. Ms. L. Groover's assistance in preparing the
manuscript is also gratefully acknowledged.

REFERENCES

Atmospheric Technology, National Center for Atmospheric Research,
    Boulder, Co., 1, 1973.

Atmospheric Technology, National Center for Atmospheric Research,
    Boulder, Co., Fall, 1975.

Axford, D. N., On the accuracy of wind measurements using an iner-
    tial platform in an aircraft, and an example of a measurement
    of the vertical mesostructure of the atmosphere, J. Appl.
    Meteor., 7, 645-666, 1968.

Atlas, D., Advances in radar meteorology, in: Advances in
    Geophysics, ed. by H. E. Landsberg and J. Van Meighem, New
    York, 1974.

Battan, L. F., Radar Meteorology, University of Chicago Press, Chicago, Ill., 1959.

Blackadar, A. K., Modeling the nocturnal boundary layer, submitted to Third Symposium on Atmospheric Turbulence, Diffusion and Air Quality, Oct. 19-22, Raleigh, N.C., 1976.

Bean, B. R. and E. J. Dutton, Radio Meteorology, U.S. Government Printing Office, Washington, D.C., 1966.

Birkemeier, W. P., P. F. Duvoisin, A. B. Fontaine, and D. W. Thomson, Indirect atmospheric measurements utilizing RAKE Tropospheric Scatter Techniques - Part II: Radiometeorological interpretation of RAKE channel-sounding observations, Proc. of the IEEE, 57, 4, 552-559, 1969.

Birkemeier, W. P., H. S. Merrill, Jr., D. H. Sargeant, D. W. Thomson, C. M. Beamer, and G. T. Bergemann, Observation of wind-produced doppler shifts in tropospheric scatter propagation, Radio Science, 3, 4, 309-317, 1968.

Cahir, J. and J. M. Norman, A forecaster-initiated real-time cross section analysis for AFOS, Report to the National Weather Service, Contract #5-35290, 1975.

Danielson, E. F., The laminar structure of the atmosphere and its relation to the concept of a tropopause, Arch. Met. Geoph. Biokl., A, 11, 293-332, 1959.

Deardorff, J. W., Three-dimensional numerical study of the height and mean structure of a heated planetary boundary layer, Boundary-Layer Meteorology, 7, 81-106, 1973.

Derr, V., Ed., Remote Sensing of the Troposphere, NOAA, University of Colorado, Boulder, Co., 1972.

Godowitch, J., Case studies of aircraft and helicopter temperature measurements over St. Louis, Mo., M.S. Thesis, The Pennsylvania State University, 1976.

Hall, F. F., Jr., Temperature and wind structure studies by acoustic echo-sounding, in Remote Sensing of the Troposphere, ed. by V. E. Derr, Boulder, Co., 1972.

Lenschow, D. H., Two examples of planetary boundary layer modification over the Great Lakes, J. Atmos. Sci., 30, 568-581, 1973.

Lenschow, D. H., The use of aircraft for probing the atmospheric boundary layer, Atmos. Tech., 7, Boulder, Co., 1975.

Little, C. G., Acoustic methods for the remote probing of the lower atmosphere, Proc. of the IEEE, 57, 4, 571-578, 1969.

Little, C. G., Status of remote sensing of the troposphere, Bull. Am. Meteor. Soc., 53, 10, 936-949, 1972.

Lumley, J. L., and H. A. Panofsky, The Structure of Atmospheric Turbulence, Interscience Publishers, New York, 1964.

Mamane, Y., and J. Pena, Aerosol measurements over St. Louis: Some preliminary results, Paper #75-58.4, 68th Annual Mtg., APCA, Boston, 1975.

Morris, A. L., D. B. Call, and R. B. McBeth, A small tethered balloon sounding system, Am. Meteor. Soc., 56, 9, 964-969, 1975.

Norman, J. M. and J. Cahir, Article in preparation for submission to Bull. of the Am. Meteor. Soc., 1976.

Pasquill, F., Atmospheric Diffusion, Halsted Press, New York, 1974.

Petterssen, S., Weather Analysis and Forecasting, Vol. I, Motion and Motion Systems, McGraw-Hill, New York, 1956.

Readings, C. J., The use of a tethered kite-balloon to probe the lowest levels of the atmosphere, Atmos. Tech., Fall, 1975, Boulder, Co., 1975.

Saucier, W. J., Principles of Meteorological Analysis, University of Chicago Press, Chicago, Ill., 1955.

Shapiro, M. A. and J. T. Hastings, Objective cross section analyses by Hermite polynomial interpolation on isentropic surfaces, J. Appl. Meteo., 12, 753-765, 1973.

Tennekes, H., A model for the dynamics of the inversion above a convective boundary layer, J. Atmos. Sci., 30, 4, 558-567, 1973.

Tennekes, H. and A. P. van Ulden, Short-term forecases of temperature and mixing height on sunny days, in: Symposium on Atmospheric Diffusion and Air Pollution (preprints), San Diego, Ca., 1974.

Thomson, D. W., Airborne measurements of tropospheric structure relating to transhorizon microwave propagation, Ph.D. Thesis, University of Wisconsin, Madison, Wi., 1968.

Thomson, D. W., Acdar Meteorology: the application and interpretation of atmospheric acoustic sounding measurements, 3rd Symposium on Meteorological Observations and Instrumentation, Washington, D.C., 1975.

Stilke, G., K. Mollnhauer, and L. Jahnke, Three-channel-radiosonde for continuous recording of temperature, humidity and air-pressure, Meteor Forschungsergebnisse, B, 1, 54-63, 1967.

Wyngaard, J. C., Modeling the atmospheric boundary layer, in: Advances in Geophysics, ed. by H. E. Landsberg and J. Van Meighem, New York, 1974.

Wyngaard, J. C., Modeling the planetary boundary layer-extension to the stable case, accepted for publication in Boundary Layer Meteor., 1976.

NEW DEVELOPMENTS IN THE DETECTION OF ATMOSPHERIC STRUCTURES

Juergen H. Richter and Douglas R. Jensen

Naval Electronics Laboratory Center, San Diego, CA 92152

INTRODUCTION

The assessment of atmospheric influences on radar target identification and imaging requires a detailed knowledge of the atmosphere's behavior and structure. Otherwise, atmospheric phenomena may be improperly interpreted or remain unrecognized. For example, the dispute in the early 1960's whether "radar angels" were caused by insects, birds or atmospheric refractive structures stemmed largely from an inability to view these unexplained echoes with high range resolution. The need to provide a continuous, high resolution measurement capability for atmospheric structures led to two developments in the late 1960's. One, the FM-CW (frequency modulated, continuous wave) radar (Richter, 1969) accomplished a two order of magnitude improvement in range resolution for microwave remote sensing radars; the other used acoustic energy for atmospheric echo sounding (McAllister, 1968). This paper reviews these two new developments in the detection of atmospheric structures.

FM-CW RADAR

As long as microwave radars have existed, problems have been encountered by non-standard refractive index distributions in the lower atmosphere. Examples are ducting phenomena which may alter detection range envelopes and introduce angular errors or false targets directly or indirectly produced by refractive anomalies. But it was not until the early sixties that

Jeske (ed.), Atmospheric Effects on Radar Target Identification and Imaging, 357–374.
All Rights Reserved. Copyright © 1976 by D. Reidel Publishing Company, Dordrecht-Holland.

substantial investigations were published which used microwave radars themselves to probe the refractive structure in the troposphere (Lane and Meadows, 1963, Atlas et al., 1965; Ottersten and Eklund, 1965; Fehlhaber and Grosskopf, 1964). These investigations demonstrated the feasibility of using microwave radars for sensing refractive structures in the troposphere and provided a remote sensing capability of atmospheric refractive layers. In most of these investigations radars were used which originally were designed for other purposes. The range resolution realized in these efforts was in the order of 100 m. It soon became apparent that supplementary, more detailed observations were needed to answer some fundamental questions concerning atmospheric dynamical processes and microstructure. Accordingly, a high range resolution radar was designed and built at the Naval Electronics Laboratory Center in San Diego.

Design considerations

Fluctuations in the refractive index present in the clear atmosphere may produce radar echoes. These echoes are distributed rather than point targets. Assuming no atmospheric attenuation the monostatic radar equation for distributed targets is (Atlas, 1964)

$$\overline{P}_R = 0.0177 \, P_T \, A_e \, h \, \eta \, H^{-2} \tag{1}$$

with $\overline{P}_R$ = average received power, P_T = transmitted power, A_e = effective antenna area, h = range resolution, η = backscatter cross section per unit volume (reflectivity), H = range of echo region.

The properties of the refractive index field may be expressed in terms of a structure function with a structure constant C_n^2. This structure constant is a measure of the strength of the refractivity fluctuations. The radar reflectivity is related to the structure constant by (Ottersten, 1969)

$$\eta = 0.38 \, C_n^2 \, \lambda^{-1/3} \tag{2}$$

where λ is the radar wavelength. Equation (1) can be used to estimate the power and sensitivity requirements of a radar to detect certain reflectivity values. Assuming an effective antenna area of 2.5 m^2 and a range resolution of 1 m, a ratio between transmitted and received power of 194 dB is required for the detection of reflectivities of 10^{-14} cm^{-1} at a height of 1 km.

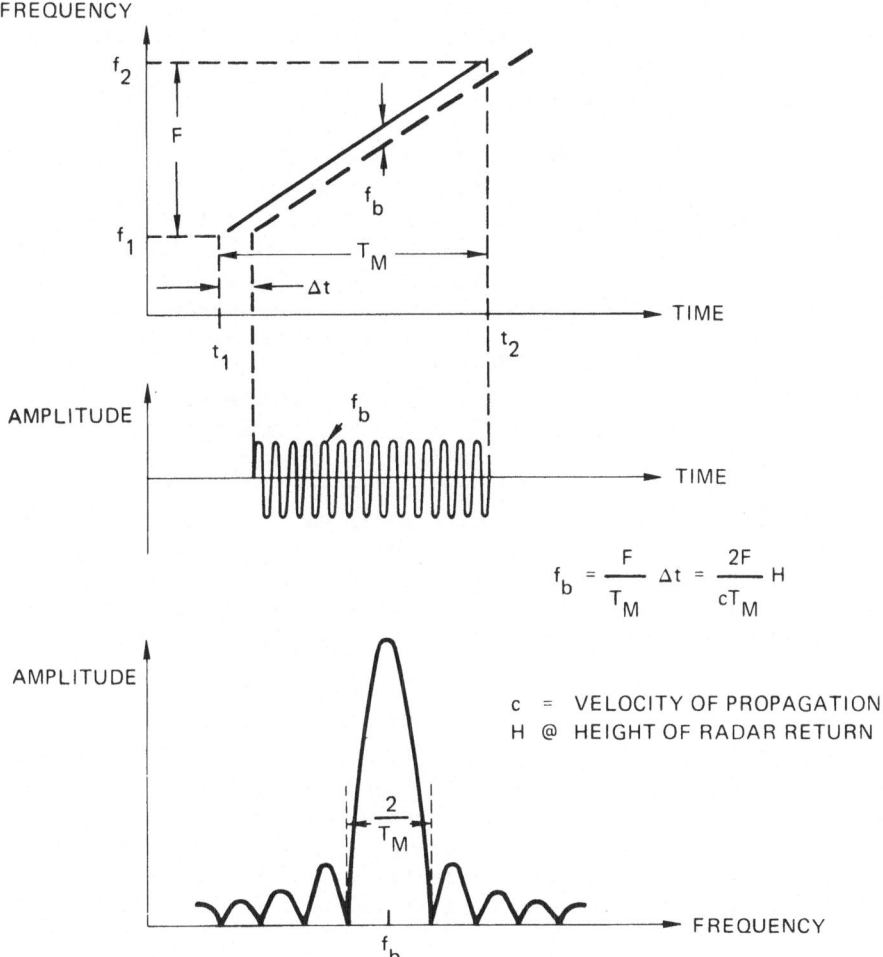

Figure 1. Principle of FM-CW radar using linear frequency modulation.

A high range resolution radar can provide insight into atmospheric microstructure only if it also has small angular antenna beam dimensions. For practical antenna sizes, this requires observations close to the antenna itself. This restricts the maximum antenna size to avoid measurements of radar reflectivities in the near field of the antenna. Observations of very weak echoes necessitate extremely low antenna side lobes. Otherwise ground clutter will override the very weak atmospheric echoes. Experience has shown that antenna side lobes must be about 50 dB below the main lobe for clutter-free observations. Side lobe suppressions of this magnitude are difficult to achieve in the vicinity of the main lobe. Therefore, the antenna has to be pointed vertically or near vertically. How much the antenna can be tilted away from the vertical depends on the magnitude of the side lobe suppression and the terrain.

The previous discussions have shown that high resolution observations of atmospheric structure require a near vertical pointing narrow beam antenna. The radar to be used with such an antenna must permit the reception of signals close to the antenna for two reasons. First, the angular dimensions of the beam must be sufficiently small throughout the observation range of interest to maintain a small measurement volume. This condition is best fulfilled close to the antenna. Second, refractivity structures of interest for microwaves are found frequently in the first kilometer of the atmosphere. It was this requirement for a close minimum range that was considered difficult to meet with a pulse compression radar (an uncompressed high resolution pulse radar was eliminated because of the extremely high peak power such a radar would require). The search for a modulation principle which would permit observations at close ranges and with high range resolution led to a continuous wave radar having linear frequency modulation (FM-CW radar).

In order to discuss the features and requirements of an FM-CW radar the principle of operation is recalled briefly in figure 1. The upper part shows a linear ramp (solid line) which is transmitted during T_M and a reflected ramp (dashed line) which is delayed by Δt^M according to the distance of the reflecting object. These two ramps are mixed instantaneously which yields a difference or beat frequency f_b. The middle part of figure 1 shows this beat frequency and the lower part its frequency spectrum. In the case of multiple targets, reflected waves will arrive at different time intervals and cause different beat frequencies which are superimposed. A spectrum analysis of the beat frequencies allows the targets to be resolved according to their range. The amplitudes of the different beat frequencies are a measure of the reflection coefficients of the targets. For this kind of radar range resolution means the ability to separate adjacent spectra of the kind shown in the lower part of figure 1. The minimum resolvable distance h is given by

$$h = c/kF \qquad\qquad\qquad (3)$$

where F is the frequency excursion, c is the velocity of propagation and k is a constant whose value depends on the criterion used to define "separation" of two spectra (e.g., k = 1 for $\Delta f = 2/T_M$ and k = 2 for $\Delta f = 1/T_M$ where Δf is the frequency separation between two spectra). Equation (3) is essentially the same formula as that for pulse radars and it shows that resolution is only dependent on the frequency excursion or bandwidth of the transmitted signal.

The frequency of the radar was chosen to be around 3 GHz for the following reasons. The reflectivity-wavelength dependence of the scattering volume given by equation (2) is proportional to $\lambda^{-1/3}$. While this wavelength dependence is small, it suggests using a small wavelength provided it does not approach the limiting microscale l_m of the refractivity perturbations. Atlas et al., (1966) therefore suggest an optimum wavelength of about five times the limiting microscale. In weakly turbulent conditions l_m is of the order of 1 to 2 cm; so λ of 5 to 10 cm seemed optimum. The optimum antenna size is determined by a compromise between good angular resolution and acceptable near field ranges. For quantitative observations of reflectivities within 100 m from the antenna and maximum horizontal beam dimensions of 150 m at 1 km from the antenna, an antenna diameter (for a parabolic reflector) of 2-3 m appears to be optimum for 10 cm wavelength. The larger angular beam dimension compared to the range resolution is only acceptable for a near vertically pointing antenna because the atmospheric variability is much smaller in the horizontal than in the vertical direction.

YIG (yttrium iron garnet) tuned transistor oscillators were found to have the necessary linearity required to achieve 1 m range resolution. In the frequency range of 2.8 - 3.0 GHz deviations from straight line tuning of only $\pm$ 17 kHz have been achieved. The modulation of the YIG tuned transistor oscillators is achieved by digitally generated ramps (Richter et al., 1972).

The spectrum analysis of the beat frequencies has to be performed in real time and is being accomplished with an analyzer which time compresses the signal and performs a scanning analysis on the speeded up and stored signal. The 10 kHz analysis band has a 20 Hz bandwidth and can be positioned anywhere from 0-1 MHz. It takes 50 msec to fill the memory and this time determines the modulation time T_M. During T_M the memory is updated and the analysis starts at the end of T_M when the memory is put into "hold". The input signal is time compressed by a factor of 500 which enables the scanning spectrum analysis to be completed in 50 msec. After this time a new modulation cycle starts. The signal to be analyzed is multiplied by a sine squared weighting function which suppresses the side lobes of the sinx/x spectrum shown in figure 1. The resulting spectrum is somewhat broadened and the resolution bandwidth is about 30 Hz.

The antennas of the fixed, vertically pointing sounder (Richter, 1969) consist of two parabolic dishes, 3 m in diameter, with waveguide feeds. In order to achieve good side lobe suppression the antennas were located in pits which is an arrangement that has been used successfully before (Fehlhaber, Grosskopf, 1964). The insides of the pits are lines with microwave absorbers to suppress reflections. The antennas are adjustable within a limited degree ($\pm 2.5^{\circ}$) in order to optimize the common volume for a given height. The antennas of a mobile version (Richter et al., 1972) consist of two 1.83 m parabolic dishes mounted on a steerable antenna pedestal. The antennas are surrounded by shrouds which are lined with microwave absorbers. The isolation for both antenna arrangements exceeds 105 dB over the frequency range of 2.8-3.0 GHz. It is, in principle, possible to operate with a single antenna for transmitting and receiving using either circular polarization or a circulator. The practical problem is to achieve good and constant isolation over the entire frequency band.

The performance characteristics of the fixed and the mobile sounder are compiled in table 1. The minimum detectable signal was determined using a delay line with a delay time equivalent to that of a target at a height of 167 m. The receiving antenna was disconnected for this measurement. When connected, the baseline noise on the A-scope, presumably due to nearby ground clutter echoes, increases the minimum discernible signal (measured with the delay line target at the fixed range of 167 m) by some 10 dB. However, typical atmospheric echoes undulate in range and so they can commonly be discerned at lower signal levels than indicated by calibration measurements with the antenna connected. Thus, -150 dB minimum detectable signal is still thought to be a reasonable estimate.

Parameter	Fixed Sounder	Mobile Sounder
	Value	
Power	100 W	200 W
Center frequency	2.9 GHz	2.9 GHz
Range resolution	1 m max	1 m max
Sweep duration	50 msec	50 msec
Receiver noise figure	5 dB	5 dB
Minimum detectable signal	-150 dBm	-150 dBm
Antenna gain	35 dB	33 dB
Antenna beamwidth	2.5°	4.0°
Isolation between antennas	105 dB	105 dB

Table 1. Performance characteristics of the fixed and mobile FM-CW radar sounders.

The sensitivity of the fixed radar (Stratman et al., 1971) for a point target is given by

$$\sigma_{min} = 3.1 \cdot 10^{-6} r^4 \tag{4}$$

where σ_{min} is the minimum detectable cross section in cm^2 and r the distance of the target in km. Typical insects with cross sections of 10^{-3} cm^2 (Hardy and Katz, 1969) can, therefore, cause dot echoes 25 dB above noise at 1 km height.

For distributed targets the minimum detectable reflectivity (Atlas, 1964) η_{min} in cm^{-1} is

$$\eta_{min} = 5.7 \cdot 10^{-15} r^2 / h \tag{5}$$

where r is the distance in km and h the range resolution in m. Typical values of η for 10 cm wavelength are $10^{-16} - 10^{-14}$ cm^{-1} as given by Hardy et al., (1966), Atlas and Hardy (1966), and Hardy and Katz (1969).

FM-CW radar observations

Figure 2 shows an example of a scanning display obtained with the mobile sounder. The sounder scanned from east to west, started at 25° off the vertical and covered a total of 65° in 135 sec. The scanning display in the left-hand side of the figure represents a true spatial picture assuming the atmosphere was frozen during the scanning period (the concentric lines are radial distance markers for the radar). A layer structure in the visually clear air appears at a height of 515 m with a vertical thickness of approximately 30 m. A radiosonde sounding balloon was launched at 0926 PST and the resulting vertical profiles of temperature, relative humidity, refractive index and wind are plotted on the right-hand side of the figure at the same height scale as the radar picture. Good agreement is found between the gradient in the refractive index profile and the radar returns. The radar apparently detects refractivity fluctuations at the interface of cooler, moist air below 515 m and warmer, dryer air above. The small undulations of the layer between 0-250 m east of the radar are probably wave motions induced by the wind shear measured in this height region.

Strauch et al. (1975) recently demonstrated the ability to obtain doppler information from distributed targets using an FM-CW radar. Their initial measurements were confined to observations of rain and snow. However, it looks feasible and promising to achieve doppler information in clear air conditions. In this case, the FM-CW radar could be used as an

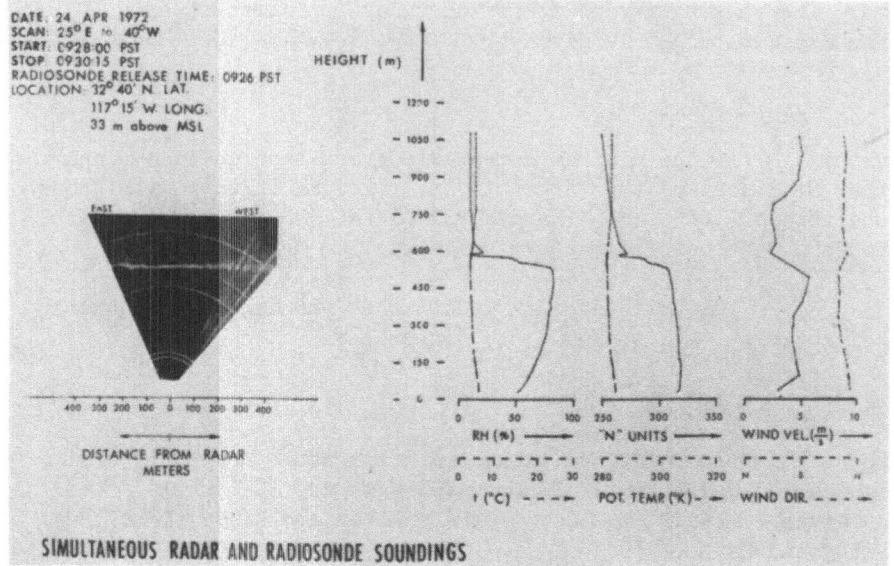

SIMULTANEOUS RADAR AND RADIOSONDE SOUNDINGS

Figure 2. Scanning FM-CW radar display and simultaneous radiosonde sounding. An atmospheric scattering layer in the visually clear air appears at 515 m coinciding with a steep gradient in the refractive index profile.

all weather remote wind sensor.

ACOUSTIC ECHO SOUNDING

The backscatter cross section per unit volume for acoustic energy is dependent on both temperature and wind velocity fluctuations. It may be expressed as a function of the scattering angle θ (Little, 1969)

$$\sigma_{(\theta)} = 0.055 \ \lambda^{-1/3} \cos^2 \theta \ [\frac{C_v^2}{C^2} \cos^2 \frac{\theta}{2} + 0.13 \frac{C_T^2}{T^2}] \ (\sin \frac{\theta}{2})^{-11/3} \tag{6}$$

where λ is the acoustic wavelength, C_v the velocity structure function constant, C_T the temperature structure function constant, c the velocity of propagation, and T the temperature. In the monostatic or backscatter mode ($\theta = 180°$), equation (6) reduces to

$$\sigma_{(\theta = 180°)} = 0.0072 \ C_T^2 \ \lambda^{-1/3} \ T^{-2} \tag{7}$$

i.e. the echo signals are only a function of temperature fluctuations. The acoustic echo sounder equation for this case may be expressed as

$$P_R = P_T \; A \; h \; \sigma \; H^{-2} \; e^{-2\alpha H} \tag{8}$$

where P_R is the received acoustic power, P_T the radiated acoustic power, A the collecting area of the receiving antenna, h the pulse length in space, H the range to the scattering volume and α the average attenuation rate to the scattering volume at range H.

The acoustic sounder used in the following examples is a commercially built unit (manufacturer: Xonics, Inc.). For transmitting, it uses a three by three loudspeaker array (Altec 288C drivers) which at 2 kHz has a 6.5° half power beam width. The electrical input power to the transmitting array is 1350 W which is converted into acoustic power with 27% efficiency. Pulse lengths are selectable 10, 50, and 200 msec and the pulse repetition frequency is variable. The receiver located adjacent to the transmitter consists of a cardioid, condenser-type microphone mounted at the focal point of a 1.2 m parabolic reflector. Both transmitter and receiver are surrounded by shrouds which are lined with acoustic absorbers.

Acoustic sounder observations

Figure 3 shows two examples recorded with the vertically pointing acoustic sounder. The observations on 12 June 1974 show a strong echo layer between 500-600 m and a somewhat weaker layer between 600-650 m. These layered echo structures indicate thermal stability and the strong layer probably coincides with the base of the temperature inversion found commonly in the San Diego coastal area where the observations were taken. The plume-like vertical structures below the layers are probably an indication of thermally unstable conditions. The strong echoes at and below 200 m may be caused by orographically induced mixing (the sounder is located on top of a 120 m ridge). The lower part of figure 3, recorded on 27 February 1975, also shows a layer structure between 280-330 m which again probably traces the base of the temperature inversion. The two observations were paired to illustrate the influence of range resolution on the ability to observe atmospheric microstructure. The observations on 12 June 1974 were taken with a 200 msec pulse length (equivalent to 34 m range resolution) while the 27 February 1975 were taken with a 10 msec pulse length (equivalent to 1.7 m range resolution). If the lower part of figure 3 had been recorded with the same course range resolution as the upper part, the layer would have shown almost no undulations in height and would have appeared as steady as the layers in the upper part of figure 3.

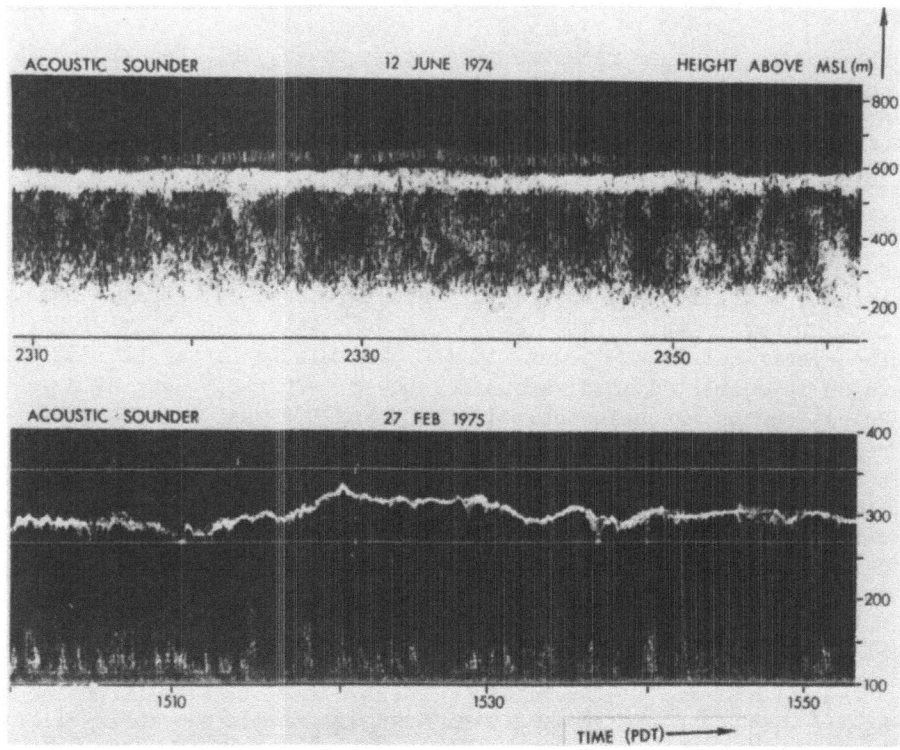

Figure 3. Atmospheric structures sensed by acoustic echo sounding. The range resolution for the upper part was 34 m, for the lower part 1.7 m.

SIMULTANEOUS FM-CW RADAR AND ACOUSTIC SOUNDER OBSERVATIONS

In the foregoing, it has been demonstrated that both an FM-CW radar and an acoustic sounder are capable of observing atmospheric structures with high range resolution. The microwave radar is primarily sensitive to moisture fluctuations while the acoustic sounder senses temperature fluctuations. This sensitivity to different atmospheric parameters should produce different pictures for processes in which moisture and temperature fluctuations are uncorrelated. Therefore, simultaneous observations with the two sounding techniques should produce complementary information.

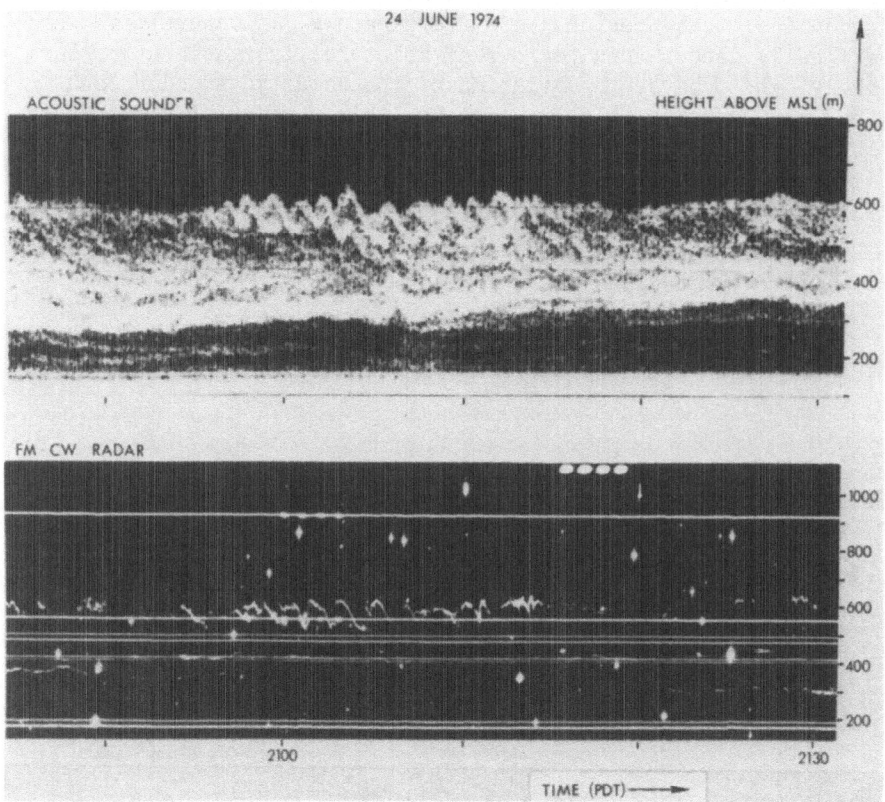

Figure 4. Simultaneous acoustic sounder an   FM-CW   radar
observations of breaking waves.

Figure 4 is an example of simultaneous acoustic sounder and
FM-CW radar observations.   The predominant feature in both
records   are   breaking   waves   (Kelvin-Helmholtz   or   K-H
instabilities) in the height region between 500-630 m caused by
wind shear within thermally stable layers. A radiosonde taken
at 1700 PDT (24 June 1974) at Montgomery Field (MYF) in San
Diego shows stable temperature lapse rates up to 800 m and the
base of a strong temperature inversion at 420 m. The 0500 PDT
MYF radiosonde on 25 June 1974 shows thermal stability up to 400
m and a temperature inversion above.   Both soundings show
directional and velocity shear in the wind profiles at the
height region of the observed K-H instability.   Only weak
stratified layer echoes are visible on the FM-CW radar record
below the K-H instability. Both the K-H instability echoes and
the layers below appear to be   much   thinner   than   the

corresponding features on the acoustic sounder. This may be an
indication that regions of temperature fluctuations are
vertically deeper than regions of moisture fluctuations or, more
likely, the coarser resolution of the acoustic sounder used in
these measurements falsely expands thin regions of temperature
fluctuations. The absence of low level plume-like echoes in the
first 300 m (like the echoes in figure 3) in the acoustic
sounder record coincide with calm surface winds. This supports
the hypothesis that this type of low level plume-like echo may
be an indication of orographically induced mechanical mixing.

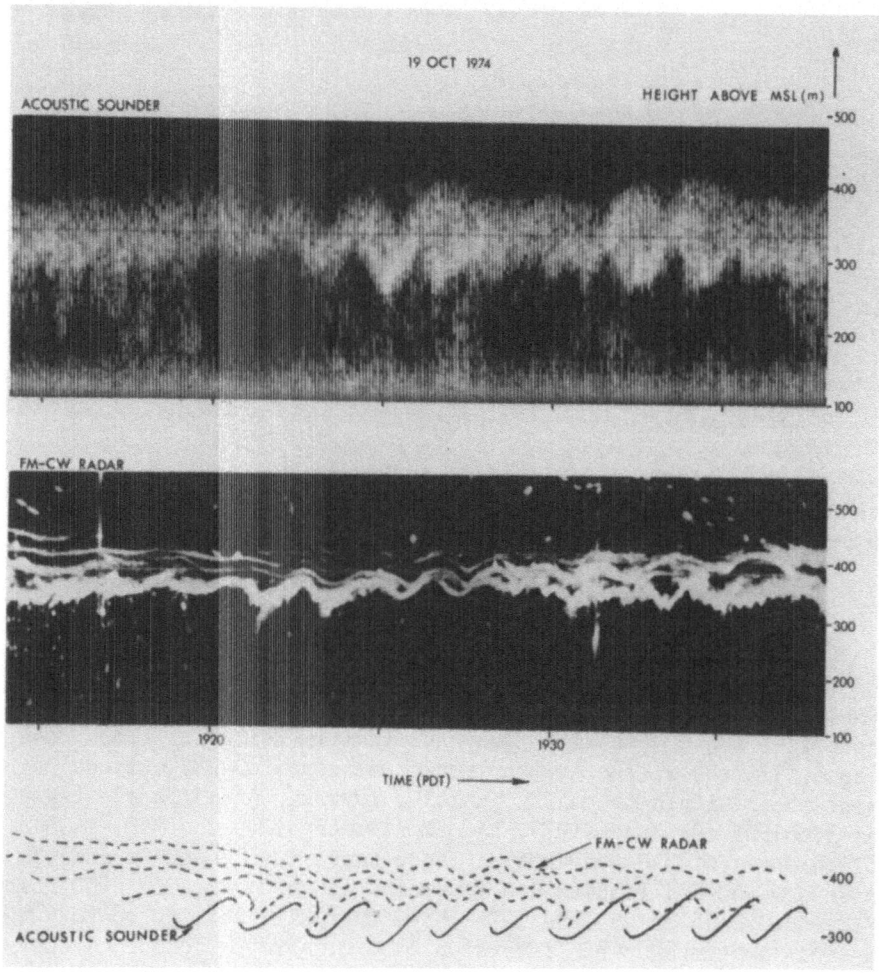

Figure 5. Simultaneous acoustic sounder and FM-CW radar
observations of wave structure.

Another example of a K-H instability observed by the acoustic sounder in San Diego is shown in the upper part of figure 5. During this observation the acoustic sounder was transmitting a 200 msec pulse. The long pulse width produced the poor range resolution seen in figure 5. The corresponding FM-CW radar record shows multiple undulating echo layers for the same time period but no clear indication of breaking waves. Careful examination of the acoustic sounder record shows that the center of the K-H instability lies below the undulating echo regions of the FM-CW radar record. The bottom part of figure 5 is an outline of the predominant echo layers seen by both sounders (FM-CW radar features are traced by dashed lines, the K-H wave from the acoustic sounder record by solid lines) with the same height scale. The lowest of the FM-CW radar echo layers shows downward streamers or plume ejections (e.g. at 1921 PDT). This kind of streamer·has been observed at numerous occasions with the FM-CW radar. Gossard et al., (1973) have suggested that streamers observed on trapped gravity waves may be caused by self-induced reduction of Richardson's number together with wind shear in the medium. The observations in figure 5 suggest an alternative explanation. Entrainment of air from above into the breaking portion of the K-H instability may account for the streamers in the FM-CW radar record in figure 5. The K-H instability at the critical level was apparently not sensed by the FM-CW radar because of insufficient moisture fluctuations. Apparently humidity gradients with refractive index fluctuations were present immediately above a dynamically unstable region outlined by temperature fluctuations on the acoustic sounder. The K-H waves were generated in the dynamically unstable region and their vertical development was sufficient to influence overlying regions and entrain moisture into the troughs of the waves.

Further examples of simultaneous FM-CW radar and acoustic echo sounder observations are given by Richter et al. (1974) and Richter and Jensen (1975). Chadwick and Little (1973) made a quantitative comparison of the sensitivities of these two sounding techniques for the sensing of various atmospheric parameters.

ACOUSTIC WIND SENSING

So far only the amplitude of the backscattered signal has been used to sense regions of strong temperature fluctuations. One can, of course, also measure the doppler shift in the return signal and, thereby, obtain the radial velocity component. For any monostatic arrangement ($\theta = 180^\circ$) the contributions of wind velocity fluctuations to the backscatter cross section vanish according to equation (7) and velocities can be measured only in

regions of strong temperature fluctuations. Therefore, a
bistatic antenna arrangement is more advantageous if wind
velocities are to be measured. Such a bistatic antenna
arrangement has been implemented with the previously described
Xonics sounder. The wind measuring portion of the system
consists of three receiving antennas located 400-500 m from the
transmitting antenna roughly 120° apart (terrain constraints
prevented identical separations). The sidescatter receiving
antennas are similar to the monostatic antenna except that their
beam patterns are vertical fan beams covering the 500 m height
interval for which wind measurements are desired. The doppler
information is processed by a minicomputer in real time and the
three wind components are printed on a teletype machine.

The solid lines in figure 6 are examples of horizontal wind
profiles (v_x and V_y) measured with the acoustic wind sounder.
Each data point along the profile represents an average of ten
individual measurements. For comparison purposes, a wind
profile was obtained by tracking a radiosonde released at 2155
PDT (indicated by dashed lines). Acoustically sensed winds and

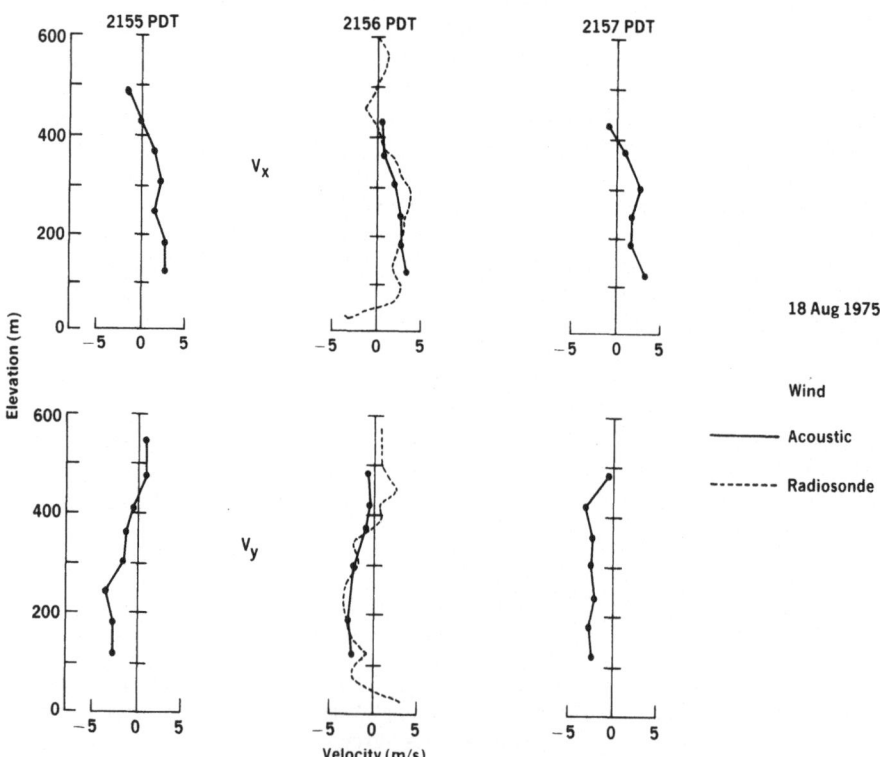

Figure 6. Horizontal wind profiles measured by acoustic methods
(solid line) and by radiosonde tracking (dashed line).

those measured by radiosonde tracking are in reasonable agreement if one considers the spatial and temporal misalignement between the remote and direct sensing techniques.

The present limitations in the wind sensing system are primarily caused by a poor signal to noise ratio. Currently, reliable wind measurements are achievable only under nighttime conditions when both man-made and wind induced noise levels are low. The ranges of measured noise powers and echo intensities for various atmospheric conditions are plotted in figure 7. The solid lines labelled Pr(min) and Pr(max) are echo intensities expected for the described system for very weak and very strong atmospheric temperature and wind fluctuation respectively. (The units of C_m are $K\ m^{-1/3}$, the units of C_v are $m\ sec^{-1}\ m^{-1/3}$.) The ranges of typical noise powers measured during the day and at night with an indication of average and absolute minima and peaks are indicated by lines at the top of figure 7. Figure 7 confirms the experience so far that no reliable daytime wind measurements are feasible with the presently configured system

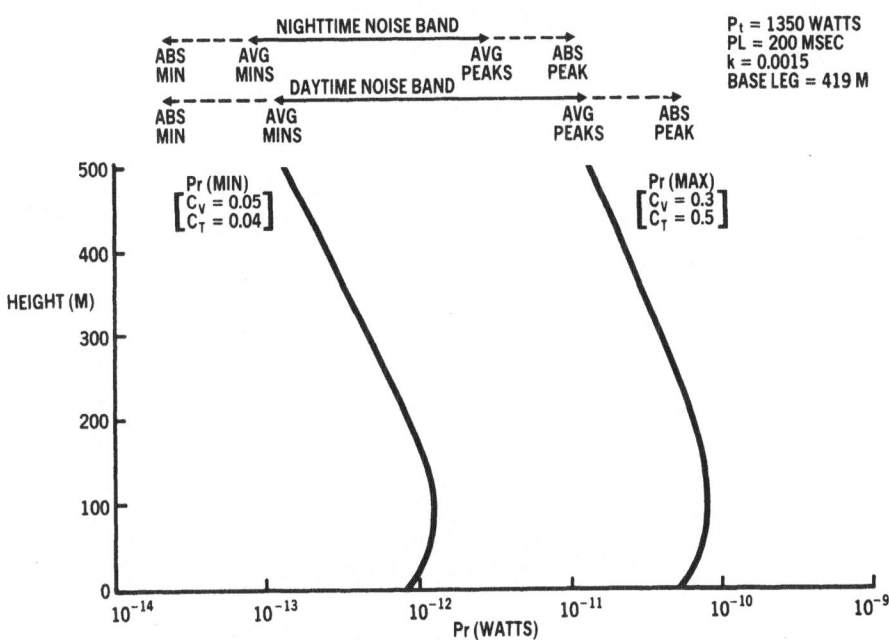

Figure 7. Calculated received maximum and minimum powers (P_r) for an acoustic wind sensing system (see text) and measured day and nighttime noise bands (k is atmospheric attenuation rate in m^{-1}).

and that nighttime measurements are possible only under selected conditions. It is difficult to define exactly these selected condtions. They are probably characterized by strong C_n and C_v values but should not be accompanied by strong surface winds which may contribute to an increase in ambient noise. A significant noise reduction could probably be achieved by using pencil beams rather than fan beams for the bistatic receiving antennas. Pencil beams could either be synthesized by phased transducer arrays or be stacked vertically and electronically switched to track the pulse.

CONCLUSION

The ability to sense atmospheric microstructure is essential for the proper assessment of its effect on electromagnetic signals. Two recently developed ground based remote sensors, the acoustic echo sounder and the FM-CW radar, provide this ability with unprecedented range resolution. Both sounding techniques also have the capability to measure wind profiles using doppler information. Operational reliability of remote wind sensing using FM-CW radar and acoustic sounding techniques remains to be established.

ACKNOWLEDGEMENT

The support of this research by the United States Navy (Naval Electronic System Command, Naval Air Systems Command, and Office of Naval Research) is gratefully acknowledged.

REFERENCES

Atlas, D., (1964), Advances in radar meteorology, Advances in Geophysics, 10, 317-478.

Atlas, D., K. R. Hardy, and K. M. Glover, (1965), Multi-wavelength backscatter from the clear atmosphere, Proc. Intern. Colloqu. Atmospheric Turbulence and Radio Propagation (Ed. Yaglom and Tatarski), Moscow, 269-277.

Atlas, D., and K. R. Hardy, (1966), Radar analysis of the clear atmosphere: Angels, in Proceedings of the 15th General Assembly of URSI, pp. 401-469, International Scientific Radio Union, Munich, Germany.

Atlas, D., K. R. Hardy, and K. Naito, (1966), Optimizing the radar detection of clear air turbulence, J. Appl. Meteorol., 5, 450-460.

Chadwick, R. B. and C. G. Little, (1973), The comparison of sensitivities of atmospheric echo sounders, Remote Sensing of the Environment, 2, 223-234.

Fehlhaber, L. and J. Grosskopf, (1964), Untersuchung der Struktur der Troposphare mit einem Vertikalradar, Nachrichtentechnische Zeitschrift, 17, 503-507.

Gossard, E. E., J. H. Richter, and D. R. Jensen, (1973), Effect of wind shear on atmospheric wave instabilities revealed by FM-CW radar observations, Boundary-Layer Meteor., 4 113-131.

Hardy, K. R., D. Atlas, and K. H. Glover, (1966), Multi-wavelength backscatter from the clear atmosphere, J. Geophys. Res., 71 (6), 1537-1552.

Hardy, K., and I. Katz, (1969), Probing the clear atmosphere with high power, high resolution radars, Proc. IEEE, 57 (4), 468-480.

Lane, J. A., and R. W. Meadows, (1963), Simultaneous radar and refractometer soundings of the troposphere, Nature, 197, 35-36.

Little, C. G., (1969), Acoustic methods for remote probing of the lower atmosphere, Proc. IEEE, 57 (6), 571-578.

McAllister, L. G., (1968), Acoustic sounding of the lower troposphere, J. Atmos. Terr. Phys., 30, 1439-1440.

Ottersten, H. and F. Eklund, (1965), Radar angel activity and its correlation with meteorological parameters, Proc. Intern Colloqu. Atmospheric Turbulence and Radio Propagation (E. Yaglom and Tatarski), Moscow, 269-277.

Ottersten, H., (1969), Radar backscattering from the turbulent clear atmosphere, Radio Sci., 12, 1251-1255.

Richter, J. H., (1969), High resolution tropospheric radar sounding, Radio Sci., 4 (12), 1261-1268.

Richter, J. H., D. R. Jensen, and M. L. Phares, (1972), Scanning FM-CW radar sounder, Rev. Sci Inst, 43 (11), 1623-1625.

Richter, J. H., D. R. Jensen, V. R. Noonkester, T. G. Konrad, A. Arnold, and J. R. Rowland, (1974), Clear air convection: a close look at its evolution and structure, Geophysical Research Letters, 1 (4), 173-176.

Richter, J. H., and D. R. Jensen, (1975), Simultaneous acoustic sounder and FM-CW radar observations, Preprints, 16th Radar Meteorology Conference, 282-289.

Strauch, R. G., W. C. Campbell, R. B. Chadwick, and K. P. Moran, (1975), FM-CW boundary layer radar with doppler capability, NOAA Technical Report ERL, 329-WPL 39.

Stratman, E., D. Atlas, J. H. Richter, D. R. Jensen, (1971), Sensitivity calibration of a dual-beam vertically pointing FM-CW radar, J. Appl. Met., 10 (6), 1260-1265.

INTEGRATED REFRACTIVE EFFECTS PREDICTION SYSTEM (IREPS)

Herbert V. Hitney and Juergen H. Richter

Naval Electronics Laboratory Center, San Diego, CA 92152

SUMMARY

Non-standard variations in temperature and humidity in the lower part of the atmosphere often produce unexpected refraction of radio waves and possible waveguide-like ducts in which radio energy can be "trapped" and carried to great distances. At times the strength of such anomalous refractive effects can seriously alter the performance of electromagnetic systems. In the case of shipboard surveillance radars, low-flying targets can sometimes be detected at ranges far exceeding the normal radar horizon, while at the same time higher-flying targets can only be detected at less-than-normal ranges. Even though it is generally acknowledged that such refractive effects can create serious problems, there is presently no capability in the operational fleet to solve the problem. NELC's Integrated Refractive Effects Prediction System (IREPS) is being developed to fill this need. A more detailed description of IREPS is given by Hitney and Richter (1976).

The objective of IREPS is to provide an onboard near-real-time assessment of the effects of the lower atmosphere on a wide variety of EM systems. Emphasis is placed on easy-to-interpret displays (e.g., detection range versus target altitude) that can be used by a task force commander in determining optimum platform or sensor placement. The implementation of IREPS will center around an interactive graphic-display terminal supported by a 32 kiloword minicomputer and an on-line 2.5 megaword disk-storage unit. The primary inputs to IREPS will be measurements of on-scene refractivity data of the lower atmosphere collected by either balloon-borne

Jeske (ed.), Atmospheric Effects on Radar Target Identification and Imaging, 375–377.

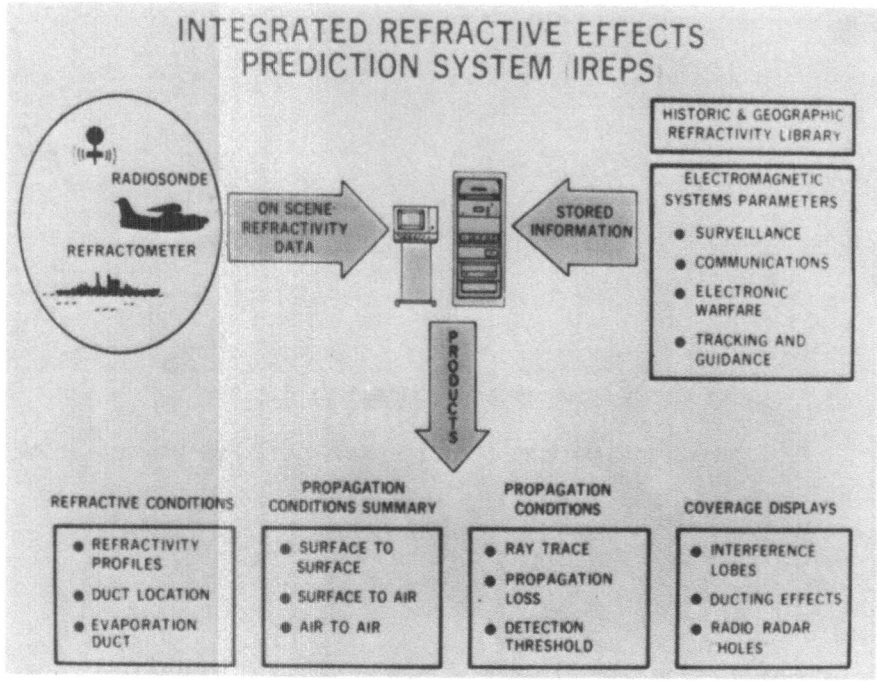

Figure 1. IREPS information flow showing sources of information
         and summarizing various IREPS products.

radiosonde, aircraft-borne microwave refractometer, or
near-surface measurements made aboard ship. IREPS includes
certain stored information consisting of a historic and
geographic refractivity library and a library of EM systems
parameters. The IREPS information flow, as shown schematically
in Figure 1 begins with the on-scene refractivity data being
entered into the system. This information, along with the
stored information, is processed by the computer and one or more
various "products" are displayed on the graphic display
terminal. These products include a display of refractive
conditions that shows the existence, height, and strength of the
evaporation duct. Also a propagation conditions summary can be
produced that gives a system-independent assessment of
conditions for surface to surface, surface to air, and air to
air situations. A ray-trace diagram can be produced from which
coverage can be inferred. Figure 2 is an example of an
IREPS-generated ray-trace diagram for an elevated duct existing
between six and seven thousand feet which shows the extended
coverage that is possible within the duct and the associated
reduced coverage or "hole" above the duct. Also propagation
loss versus range can be displayed for any system whose
parameters have been previously stored. The threshold for

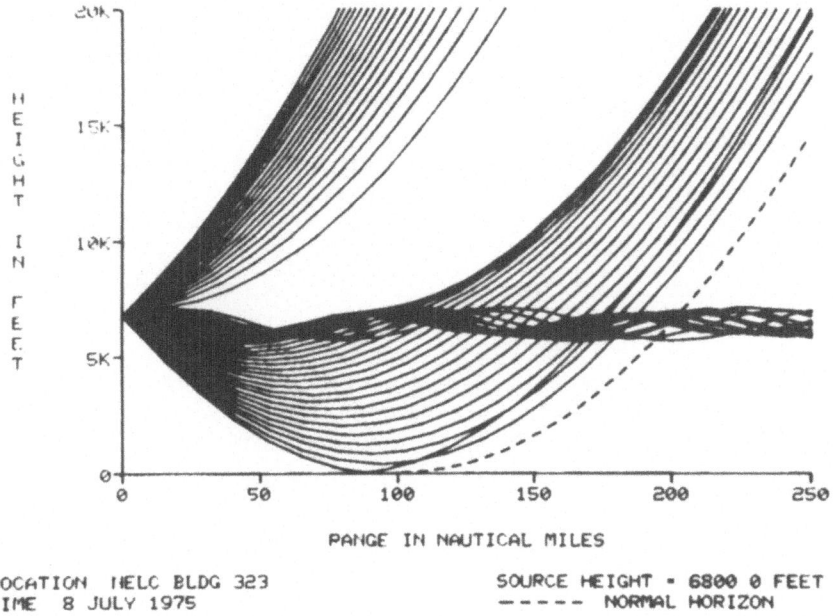

LOCATION NELC BLDG 323   SOURCE HEIGHT = 6800 0 FEET
TIME 8 JULY 1975   ----- NORMAL HORIZON

Figure 2. Ray trace diagram for a sensor located in an elevated
          duct.

detection is superimposed on this display so that actual
detection range for the subject system is apparent. The final
product, and probably the most useful, is the coverage diagram
which shows actual detection range for any stored system as a
function of target or receiver altitude. The characteristic
optical interference lobes as modified by any refractive effects
are displayed as well as any possible ducting effects. The
coverage diagram shows all regions where better-than-normal
detection ranges can be expected as well as any possible radio
or radar "holes".

REFERENCES

Hitney, H. V. and J. H. Richter, "Integrated Refractive Effects
Prediction System (IREPS)", Naval Engineers Journal, April 1976
(to be published)

THE EFFECTS OF PRECIPITATION ON RADAR
TARGET IDENTIFICATION AND IMAGING*

D. B. Hodge

The Ohio State University ElectroScience Laboratory
Department of Electrical Engineering
Columbus, Ohio 43212

ABSTRACT. The properties of precipitation which will influence
radar system design are discussed. The spatial characteristics
of rainfall and the sizes and shapes of raindrops are described. The dielectric behavior of water is combined with
these characteristics to determine the effects of rain on
electromagnetic waves. These effects include: absorption,
scatter, noise emission, phase shift, and depolarization.

******

The treatment of microwave propagation in the troposphere
can be a difficult task even in the case of a clear atmosphere.
The presence of precipitation along a propagation path adds even
further complication to this task. This fact is primarily due
to the extreme spatial and temporal variability of precipitation.
Fortunately, the basic interactions between electromagnetic waves
and precipitation are well understood and may be quantitatively
described with a reasonable degree of certainty. It is the fine
scale spatial and temporal statistics of precipitation which
are not known adequately to permit detailed statistical descriptions of these interactions at the present time.

*This work was supported in part by Grant No. NGR-008-080
between National Aeronautics and Space Administration, Office
of Grants and Research Contracts and The Ohio State University
Research Foundation.

The objective of this tutorial paper is to provide the radar design engineer with enough basic, quantitative information that he may establish realistic bounds on the performance of a given radar system in the presence of precipitation. A great deal of information concerning this topic is available in the literature, and much of this information has been collected and summarized at the recent IUCRM Colloquium on the Fine Scale Structure of Precipitation and Electromagnetic Propagation, Nice, France, October, 1973. The entire proceedings of this meeting, including review papers, are readily available in Reference 1 for further details covering these topics.

The likelihood of the occurence of precipitation is highly variable and may be strongly dependent upon the climatic region of the earth, the season of the year, the time of day, and even, in some cases, the local topography. Nevertheless, one may expect precipitation at many points on the earth's surface more than 5 per cent of the time and, of course, considerably higher percentages of time for the occurrence of precipitation along extended propagation paths. Therefore, if the impact of precipitation on system performance is significant, the reliability of the system may be impaired beyond tolerable limits.

For our purposes, precipitation may be divided into three types: rain, snow, and hail. Rain consists of liquid water drops which may be supercooled if they are carried above the 0°C isotherm by updrafts. These liquid water droplets interact strongly with propagating EM waves in the microwave and millimeter wavelength portions of the spectrum; these interactions include attenuation, phase shift, scattering, depolarization, and noise emission. It must be emphasized that these effects are distinct from the effects of water vapor, i.e., water in its gaseous form, which is present in the clear atmosphere. Snow and hail are normally forms of solid water which do not interact strongly with EM waves in this portion of the spectrum. One exception is, however, the case of melting hail or snow where scattering may be enhanced substantially; this effect leads to the bright band which is often observed near the 0°C isotherm on RHI radar displays. Since the effects of snow and hail are, in general, much less severe than those due to rain, the remainder of this paper will emphasize the impact of rain on radar system performance.

Three distinct types of rain merit discussion in this context: convective, stratiform, and orographic. Convective rain usually occurs in cells, i.e., discrete spatial regions, having horizontal dimensions of only a few kilometers. This

type of rain is generally associated with thunderstorm
activity which can occur due to local convection on hot
summer days or due to the passage of a cold front. These
cells may occur in clusters or in bands with regions of no
rainfall between the cells. The highest rain rates are
generally associated with this type of rain. In contrast,
stratiform rain occurs over widespread regions having dimensions on the order of tens of kilometers. Cellular regions
of higher rain rate may occur embedded in this light rain rate
background; but, nevertheless, the maximum rain rates are
usually considerably lower than those found in convective
storms. Finally, orographic rain is that rainfall which
results from local topographic effects; for example, the
lifting of warm air masses over a mountain range may produce
orographic rain having characteristics which are quite dependent upon the topography and local climatology. Since the
most intense rain rates and, consequently, the most severe
electromagnetic interactions are associated with convective
cells we will specialize further to this case in the following.

It is well known that rain rate probability distributions
vary dramatically over the earth's surface. Nevertheless, it
may be argued that the physical processes leading to precipitation, e.g., convection, are fairly universal and, therefore,
that one may expect the physical properties of precipitation,
cell shape and size, to be reasonably uniform even in different
climatic regions. Thus, if this assumption is correct, one may,
with care, be able to extrapolate some characteristics of rainfall from one region to another with the understanding that the
probability of occurrence may vary drastically.

Convective rain cells typically have horizontal dimensions
of only a few kilometers but may be as large as 10-20 km in
some cases [2]. The vertical extent of these cells may range
up to 20 km; some sample plots of height versus radar reflectivity and equivalent rainfall are shown in Fig. 1.

The cellular nature of such rainfall is exemplified by the
10 GHz PPI display shown in Fig. 2. This figure shows a collection of at least 7 clearly discernible rain cells within a
radius of 12 miles of the radar site at Columbus, Ohio, USA,
on a July day; several cells having nonsymmetrical shapes and
the clear areas between cells can also be noted. Cell orientation as well as orientation between cells tends to show anisotropic characteristics with preferred directions related to
the climatology of the region [3]. The spatial inhomogeneity
of rainfall can be quite significant as demonstrated by Fig. 3;
in this figure two 15 GHz PPI displays of the same rain cell
are presented. The first PPI display shows the shape of the
cell; and the second contour mode PPI display shows the cell

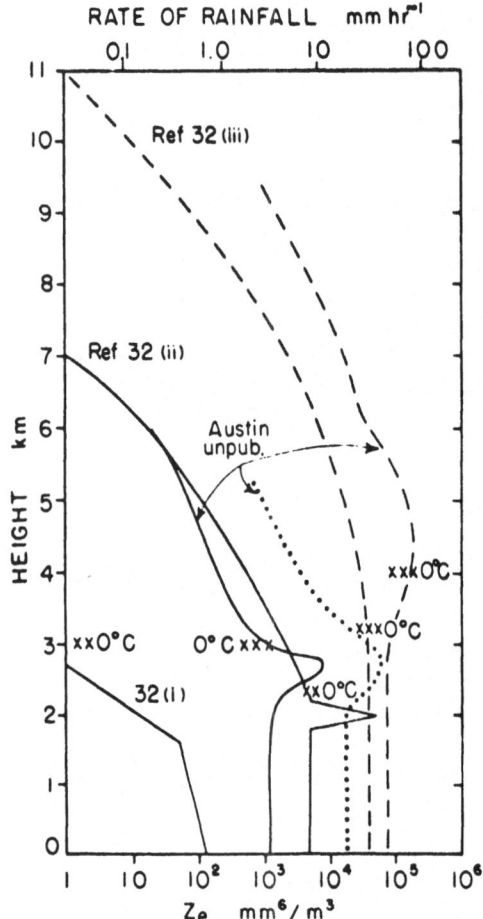

Fig. 1. Height versus radar reflectivity and equivalent
         rainfall rate (Ref. [2]).

with regions of high reflectivity blanked out. Clearly, the
most intense rainfall is occurring only on the far side of the
cell. The vertical inhomogeneity of rain is demonstrated by
Fig. 4; this figure presents a series of vertical 15 GHz
A-scope displays as a raincell passes over the radar site.
Significant reflectivity is observed before the rain actually
reached the ground; and about six minutes later distinct layers
are observed. Finally, the severe attenuation encountered at
15 GHz in heavy rainfall can be noted.

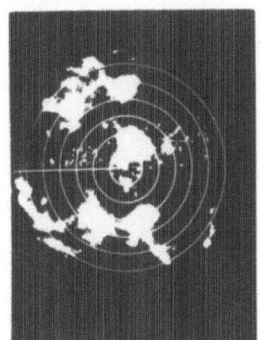

Fig. 2. 10 GHz PPI display showing at least seven clearly
         discernible rain cells in Columbus, Ohio, USA. The
         range rings are at 2 mile intervals. The small
         discrete echoes are ground clutter.

    If one averages horizontal cell dimensions over a number of
cells it is found that the average cell diameter tends to decrease with increasing rain rate, i.e., cells having low rain
rates may be rather large while the effective diameters of cells
having extremely high rain rates tend to approach an asymptotic
value of about 2 Km. The typical lifetime of convective rain
cells is on the order of one-half to one hour; the decaying cell
may be, however, replaced by a new cell in close proximity to
the original cell.

    The water drops that make up rainfall tend to have char-
acteristic shapes as shown in Fig. 5; the smaller drops tend
to be spherical while the large drops flatten in the direction
of motion and the largest drops form a shallow concavity in
the leading surface. It is this lack of symmetry which leads
to polarization dependent attenuation and scattering. The
number of drops per unit diameter per unit volume, $N(D)$, can be
described for our purposes by the Marshall Palmer drop size
distribution [5,6]:

$$N(D) = N_o e^{-\Lambda D} \quad [\text{drops/m}^3 \cdot \text{cm}]$$

where

$$\Lambda = aR^b$$
$$N_o = 8 \times 10^4$$
$$a = 41 \ [\text{cm}^{-1}]$$
$$b = -0.21$$

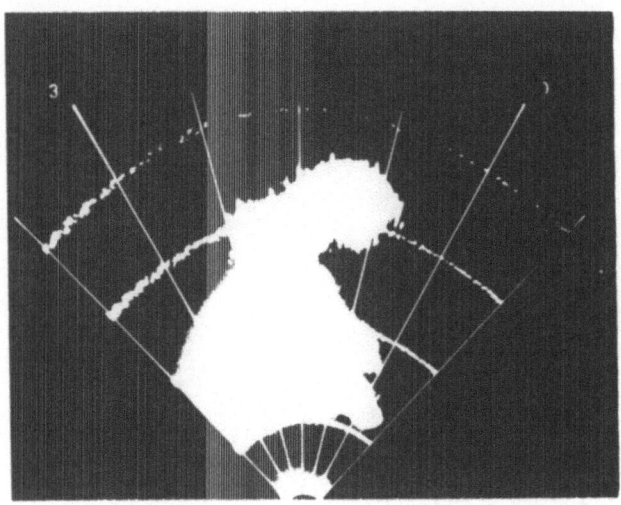

Fig. 3a. 15 GHz PPI display of a single rain cell (2 mile range marks.

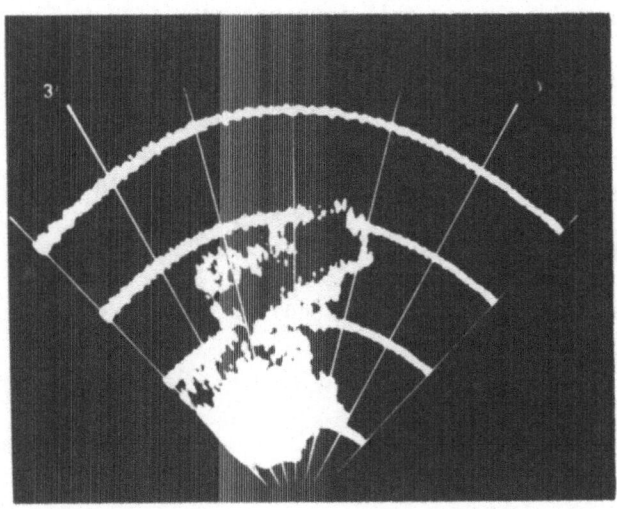

Fig. 3b. PPI display one minute later than Fig. 3a with region of high reflectivity blanked out.

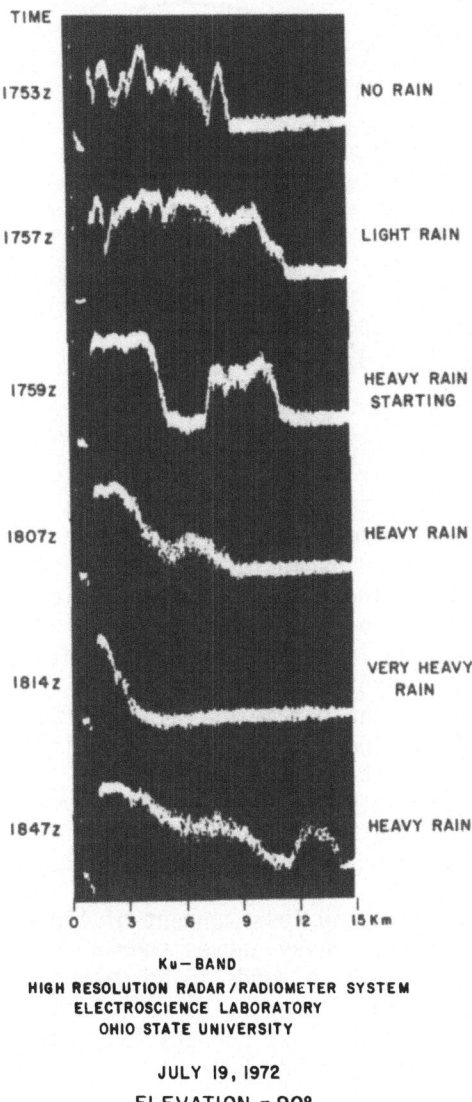

Fig. 4. 15 GHz vertical A-scope displays as rain cell passes
         over radar site.

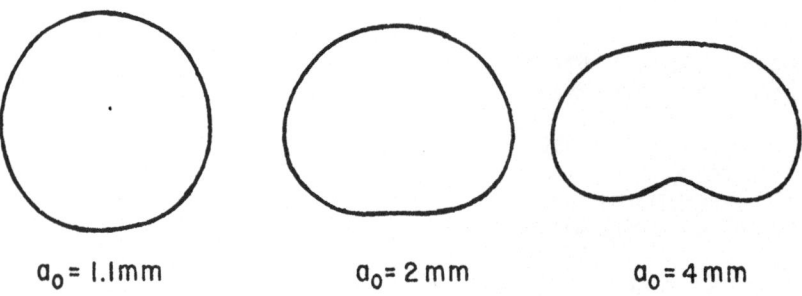

$a_0 = 1.1$ mm $a_0 = 2$ mm $a_0 = 4$ mm

Fig. 5. Characteristic rain drop shapes for drops with
         rotational symmetry (Ref. [4]). (a_0 = radius of
         sphere having equal volume.)

and D is in centimeters and R in mm/hr. This distribution is
shown in Fig. 6 for various rain rates; it can be noted that
drop sizes larger than 5 mm are extremely rare. As the rain
rate increases, the number of large drops increases; and these
electrically large drops tend to dominate the electromagnetic
interactions. It is of interest to note that the liquid water
content of rainfall is normally much less than the equivalent
liquid water content of the water vapor contained in even a
normal dry day as shown in Fig. 7. The parameter rain rate is
used widely in describing precipitation since it is a quantity
which is easily measured at the surface of the earth. This
parameter can, however, be misleading since rain drops may be
carried upward within a rain cell due to strong updrafts;
therefore, this parameter must be used with a degree of caution.

The relative dielectric constant and loss tangent of liquid
water are shown in Figs. 8 and 9 [7]. These curves show a
rather marked temperature dependence which tends to be masked
by other factors when attenuation and scattering calculations
are performed. We also observe that the loss tangent is large
over the frequency range of roughly 10 to 100 GHz.

The presence of these dielectric drops along the propagation path of an electromagnetic wave gives rise to a number of
resultant effects as shown pictorially in Fig. 10. These
effects include; attenuation, dispersive phase shift, depolarization, change in angle of arrival, backscatter, and
noise emission. The attenuation results from both absorption
and scattering loss and may be expressed as

$$\alpha_t = \alpha_a + \alpha_s \ [\text{dB/Km}]$$

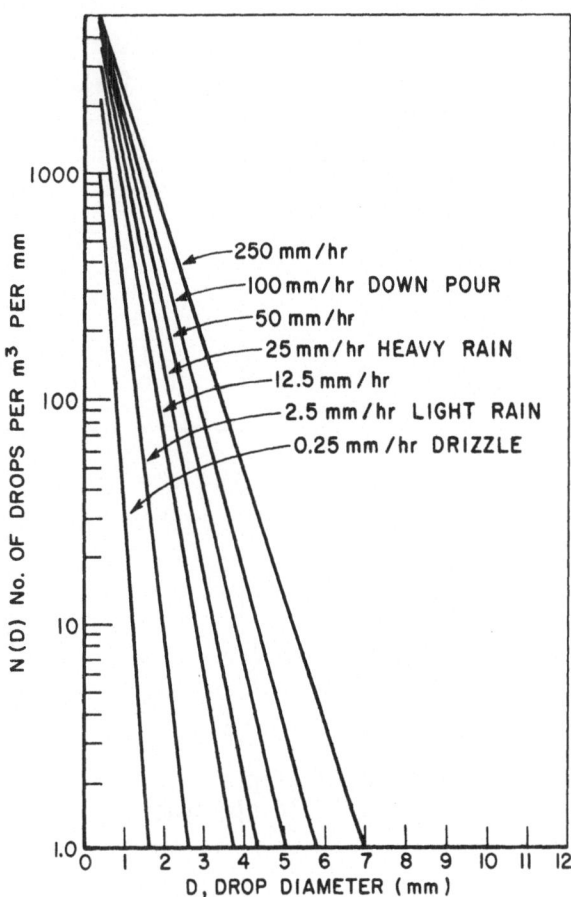

Fig. 6. Drop size distributions.

where

α_t = total attenuation rate

α_a = rate of attenuation due to absorption

α_s = rate of attenuation due to scattering

and either attenuation rate is given by

$$\alpha_i = 4.343 \int_0 Q_i(D)N(D)dD.$$

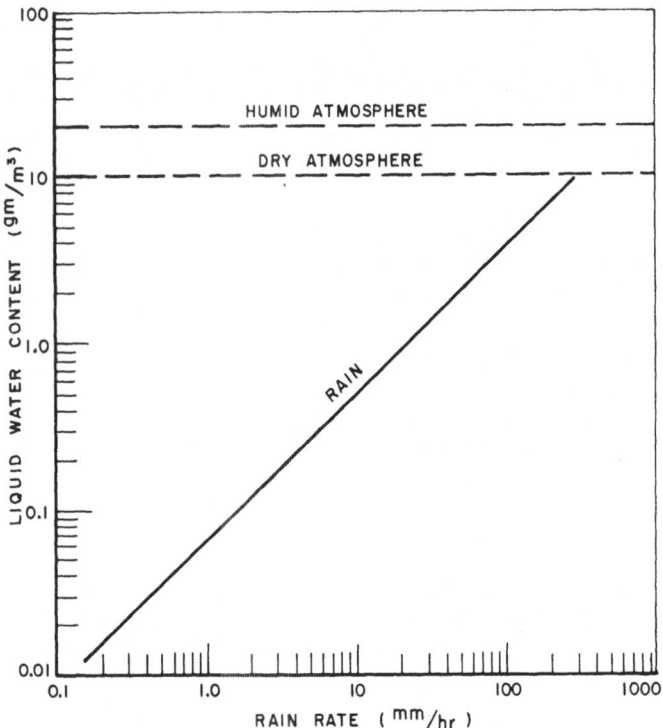

Fig. 7.   Equivalent liquid water content of rainfall and
          water vapor for typical dry and humid atmospheres.

Here, Q_i is either the absorption or total scattering cross
section of a drop of diameter D [8]. These cross sections are,
of course, dependent upon frequency as well as the drop temperature and drop size. Plots of these attenuation rates are given
in Fig. 11 for various rain rates; these calculations were based
upon the Marshall Palmer drop size distribution, Stogryn's
dielectric constant, and the Mie cross sections for spherical,
lossy dielectric drops. For low rain rates and/or low frequencies the dominant effect is that of absorption; however,
for higher frequencies and/or higher rain rates scattering loss
can become significant. This characteristic is emphasized by
the plot of albedo, i.e., the ratio of scatter loss to total
attenuation, shown in Fig. 12; the albedo approaches 0.5 for
high frequencies and high rain rates just as the albedo of a
single large drop approaches 0.5 with increasing size. The
attenuation rate is plotted along with the attenuation rates
due to fog and atmospheric gases in Fig. 13; this figure indicates that attenuation due to rain may be as much as an
order of magnitude higher than that encountered in either fog
or clean air at most frequencies of interest. This rain at-

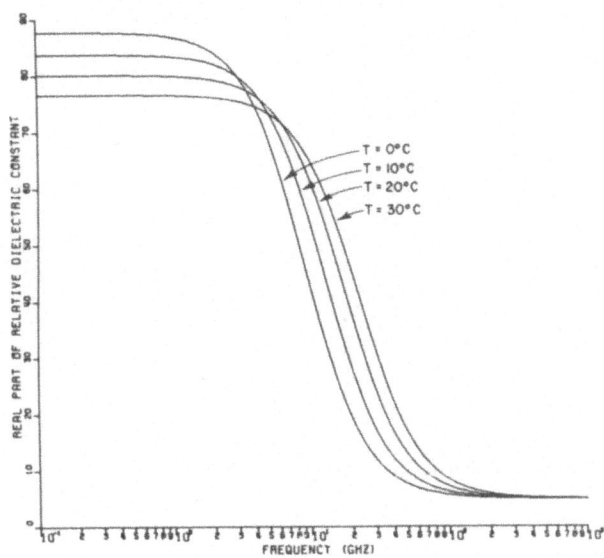

Fig. 8a.   Real part of the relative dielectric constant of
           liquid water.

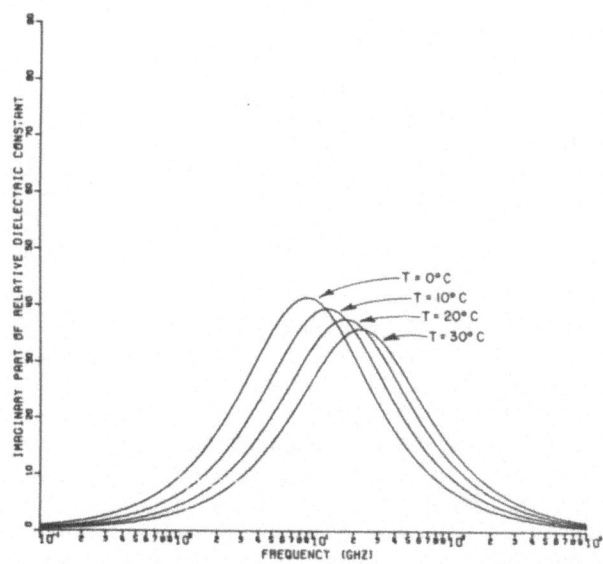

Fig. 8b.   Imaginary part of the relative dielectric constant
           of liquid water.

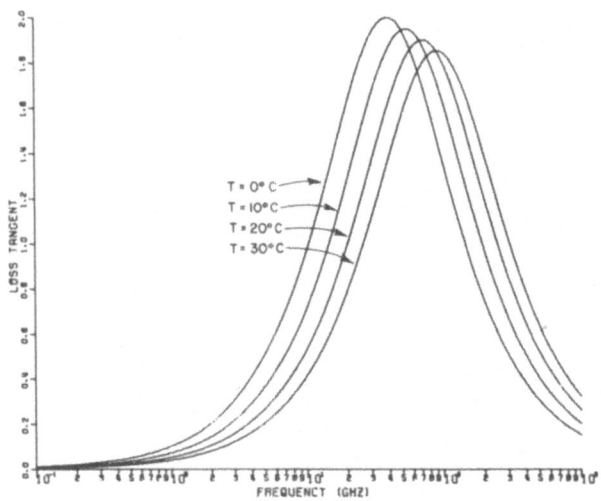

Fig. 9. Loss tangent for liquid water.

tenuation is produced by spatially inhomogeneous rain cells
passing through the propagation path; and, thus, the total
path attenuation may vary rapidly with time. Fading rates
as rapid as 1/8 dB/sec were encountered, for example, on a
15.3 GHz earth-satellite propagation path [10].

Since the rain filled atmosphere is a lossy medium,
thermodynamic arguments indicate that the medium itself must
emit noise. For sufficiently narrow beam antennas, the
observed radiometric noise emission may be related quite
simply to the total path attenuation by

$$T_s = T_m (1-10^{-A/10}) \quad [°K]$$

where T_s is the radiometric sky temperature, T_m is the mean
absorption temperature of the medium, and A is the total path
attenuation expressed in decibels [11]. The measured correlation between radiometric sky temperature and total path
attenuation for a 15.3 GHz earth-space propagation path is shown
in Fig. 14. Note that these radiometric sky temperatures do not
exceed the mean absorption temperature and, thus, do not present
a problem in system design since much higher noise levels are
generated in current receiver front ends. Nevertheless, this
noise emission provides a very useful means of detecting the
presence of rain and indirectly measuring total path attenuation.

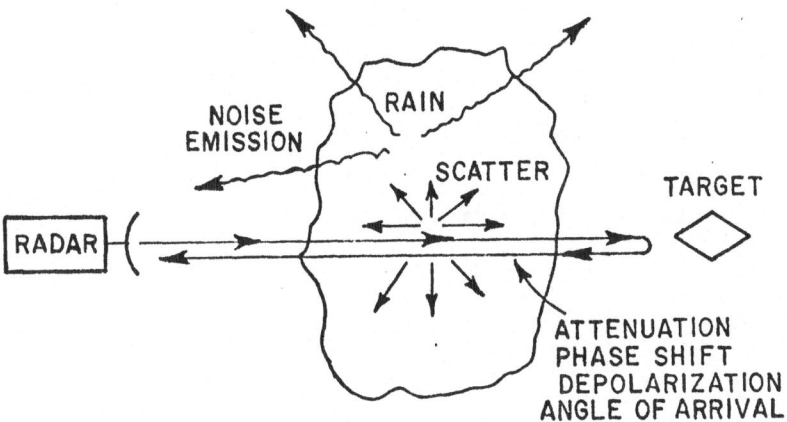

Fig. 10. Interactions between an electromagnetic wave and rain.

It has already been pointed out that scattering produces attenuation; in addition, the energy backscattered to the radar produces a radar return, i.e., clutter. The radar range equation for this clutter case is

$$P_{RCVD} = \frac{\lambda^2 G\, c\, \tau\, P_{TRANS}}{2^7\, \pi\, r^2} \cdot \eta$$

where λ is the wavelength, G is the radar antenna gain, c is the velocity of propagation, τ is the radar pulse length, r is the range to the scattering volume, η is the volumetric radar cross section of the rain, and P_{TRANS} and P_{RCVD} are the transmitted and received powers, respectively [12]. The r^2 range dependence is a consequence of the assumption that the radar resolution cell is filled with rain. The volumetric radar cross section is given by

$$\eta = \int_0^\infty \sigma(D)N(D)dD$$

where σ is the usual radar backscatter cross section of a single rain drop of diameter D; η is plotted in Fig. 15 for various rain rates. The calculation of the volumetric radar cross sections shown in Fig. 15 were based on the same assumptions as those noted earlier for the attenuation calculation shown in Fig. 11. This radar backscatter signal is produced by scattering from many randomly spaced rain drops; consequently, the signal varies randomly about a mean value

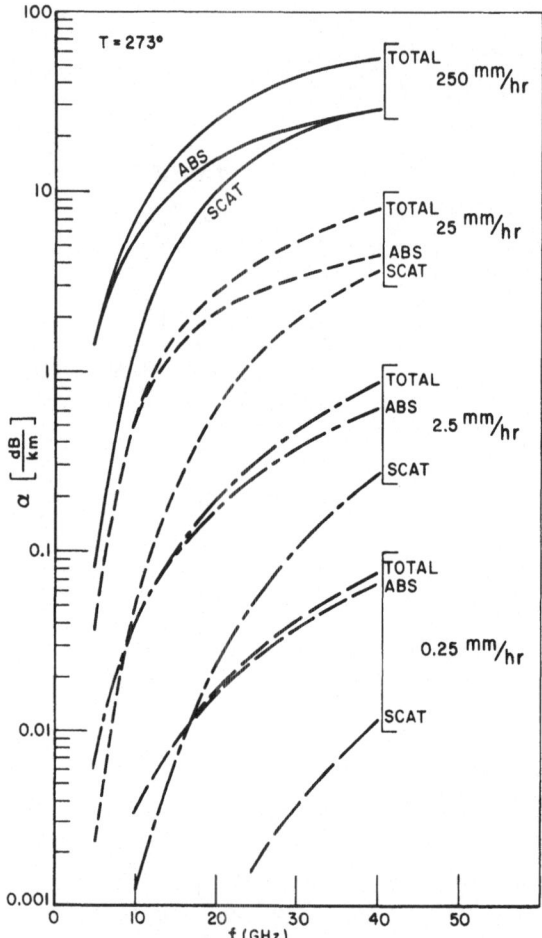

Fig. 11. Attenuation rates in rain.

as the relative positions of the drops change. The time to independence, i.e., the time required for the drops to rearrange
themselves so that an uncorrelated signal is produced, is on the
order of milliseconds to tens of milliseconds for typical radar
frequencies [13].

As a wave propagates through rain it suffers a phase shift
in addition to the attenuation described earlier. This phase
shift may be calculated using the methods of Van du Hulst [14].
Phase shifts calculated in this manner are shown in Fig. 16 [15].
These phase shifts are recognized to be comparable or, in some
cases, less than the phase shifts produced by the atmospheric
gases in clear air [16]. However, it may be that the inhomogeneity of the rainfall within the illuminated region may

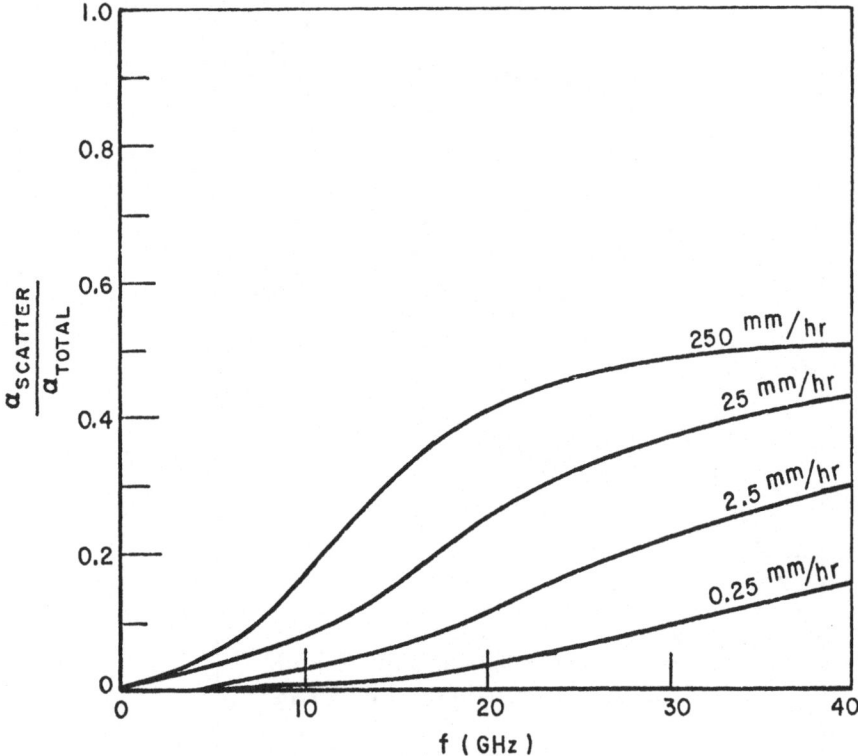

Fig. 12. Albedo as a function of rain rate and frequency.

produce nonuniform phase shifts and attenuation across the beam
which, in turn, may alter the direction of propagation. There
is very little data available at the present time concerning
these effects.

It was noted earlier that nonspherical drop shapes produce
depolarization. These effects are discussed in considerable detail in another paper presented at this meeting [17]. Let it
suffice here to simply indicate the magnitude of this problem
in intense rain; in this case, differential attenuations as
high as 5 dB/Km, differential phase shifts as large as ±20
deg/Km, and cross polarization coupling exceeding -20 dB may
be encountered between orthogonal linear polarizations [18].

The preceding discussion has indicated the nature of the
interactions between propagating electromagnetic waves and rain.
These effects are reasonably well understood and their gross
influence on system performance may be estimated. In most cases,
however, little information is available concerning the likelihood of occurrence of these effects. This is largely due to

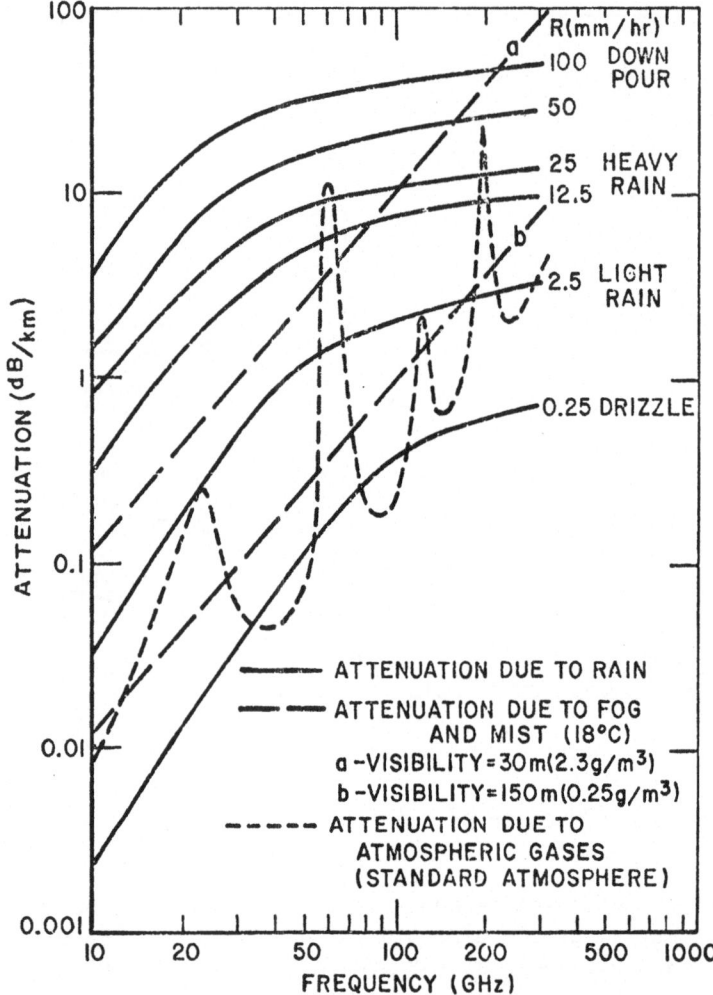

Fig. 13. Comparison of attenuation due to atmospheric gases,
rain, and fog (Ref. [9]).

the lack of data describing the fine scale temporal and
spatial statistics of rainfall and the dependence of these
statistics on climatic region. Current and future communication link experiments will certainly shed light on some of
these questions if care is exercised in the extrapolation of
the results to the radar case. For example, it is necessary to
categorize propagation paths into four classes depending upon
whether a single rain cell or multiple rain cells are likely
to be encountered on the path and whether the path is a
terrestrial path influenced only by horizontal variations in
rainfall or an earth-space path influenced by both horizontal
and vertical spatial rainfall variations.

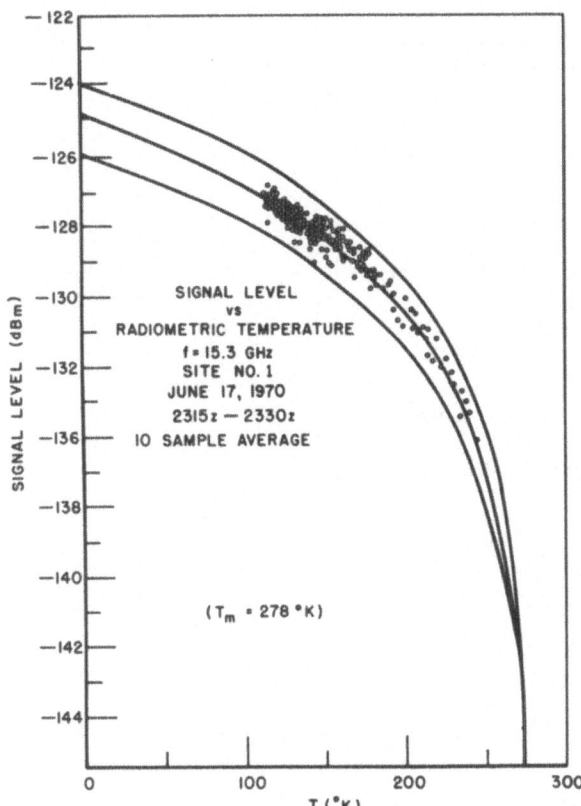

Fig. 14. Comparison of rain attenuation and radiometric sky
          temperature.

REFERENCES

[1] IURCUM Colloquium on the Fine Scale Structure of
     Precipitation and Electromagnetic Propagation, Nice,
     France, October, 1974. Entire proceedings contained in
     Journal de Recherches Atmospheriques, v. 8, No. 1-2, 1974.

[2] "The Structure of Precipitation Systems - A Review,"
     T. W. Harrold and P. M. Austin, Journal de Recherches
     Atmospheriques, v. 8, No. 1-2, p. 41, 1974.

[3] "Radar Measurements of Site-Diversity Improvement During
     Precipitation," J. I. Strickland, Journal de Recherches
     Atmospheriques, v. 8, No. 1-2, p. 451, 1974.

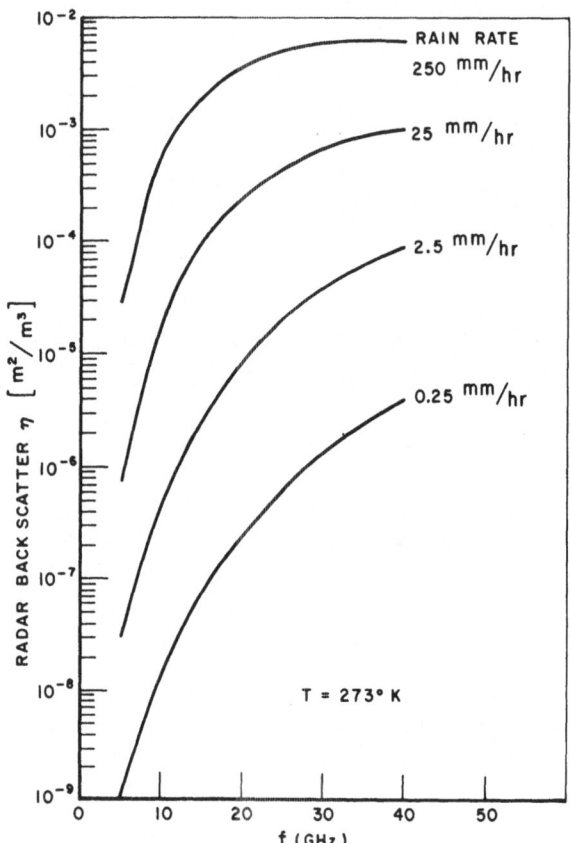

Fig. 15.   Volumetric radar backscatter cross section of rain.

[4] "A Semi-Empirical Determination of the Shape of Cloud
     and Rain Drops," H. R. Pruppacher and R. L. Pitter, Jour.
     Atmospheric Sci., v. 28, p. 86, 1971.

[5] "The Distribution of Raindrops with Size," J. S. Marshall
     and W. M. Palmer, Jour. of Meteorology, v. 5, p. 165, 1948.

[6] "The Cloud Physics of Particle Size Distributions - A
     Review," R. C. Srivastava, Journal de Recherches
     Atmospheriques, v. 8, No. 1-2, p. 23, 1974.

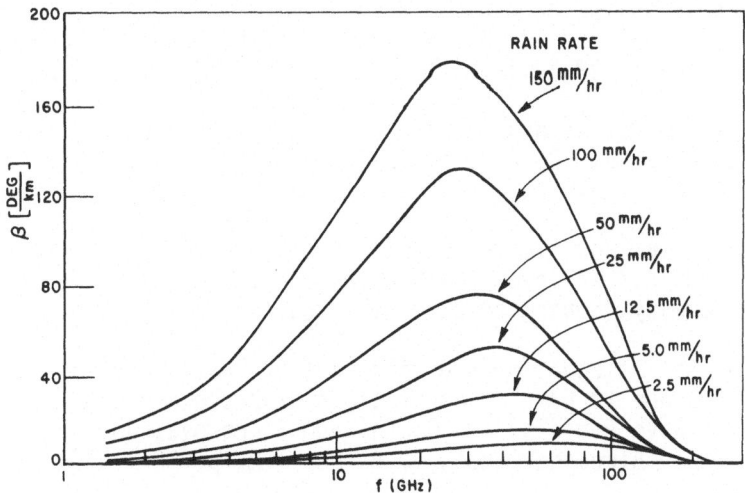

Fig. 16.   Phase shift due to rain.

[7]   "Equations for Calculating the Dielectric Constant of
      Saline Water," A. Strogyn, IEEE Trans., v. MTT-19,
      p. 733, 1971.

[8]   "Propagation of Short Radio Waves," D. E. Kerr, Dover,
      p 445, 1951.

[9]   "Atmospheric Effects of Millimeter Wave Communication
      Channels," E. Mondre, NASA Goddard Space Flight Center,
      Rept. X-733-70-250, March, 1970.

[10] "A 15.3 GHz Satellite-to-Ground Path-Diversity Experi-
      ment Utilizing the ATS-5 Satellite," D. B. Hodge, Radio
      Science, v. 9, p. 1, 1974.

[11] "Rain Attenuation at 15 and 35 GHz," K. N. Wulfsberg and
      E. E. Altshuler, IEEE Trans., v. AP-20, p. 181, 1972.

[12] "Introduction to Radar Systems," M. I. Skolnik, p. 539,
      McGraw-Hill, 1962.

[13] "Advances in Radar Meteorology," D. Atlas, p. 404 in
      "Advances in Geophysics," H. E. Landsberg and J. Van
      Mieghem, Vol. 10, Academic Press, 1964.

[14] "Light Scattering by Small Particles," H. C. Van du
      Hulst, p. 31, Wiley, 1957.

[15] "Computed Transmission through Rain at Microwave and Visible Frequencies," D. E. Setzer, Bell Sys. Tech. Jour., v. 49, p. 1873, 1970.

[16] "Transit-Time Variations in Line-of-Sight Tropospheric Propagation Paths," D. A. Gray, Bell Sys. Tech. Journ., v. 49, p. 1059, 1970.

[17] "Some Polarization Effects for Millimeter Wave Propagation in Rain," W. Vogel, NATO Advanced Study Institute, Goslar, Germany, September 22-October 3, 1975.

[18] "Differential Attenuation and Differential Phase Shift of Radio Waves due to Rain: Calculations at Microwave and Millimeter Wave Regions," T. Oguchi and Y. Hosoya, Jour. de Recherches Atmospheriques, v. 8, No. 1-2, p. 121, 1974.

SOME POLARIZATION EFFECTS FOR MILLIMETER WAVE PROPAGATION IN RAIN

W. J. Vogel, B. M. Fannin and A. W. Straiton
Electrical Engineering Department
The University of Texas at Austin

ABSTRACT

A theoretical investigation of cross polarization generation and radar echo cancellation for the propagation of electromagnetic waves at 17, 30 and 92 GHz through rain is carried out. Both effects are due to canted, nonspherical drops. The canting angles are assumed random with preferred directions. The drop shapes used were determined at UCLA by Pruppacher, Beard and Pitter. They are oblate with increasingly flat and finally concave bottoms as drop size increases. The scattering from single particles is solved by point matching for arbitrary angles of incidence of a plane wave. Propagation through the anisotropic rain medium is handled mathematically with the use of the characteristic polarizations and propagation constants. Cross polarization effects for linear and circular transmit polarization and rain echo cancellations for circular polarization are found for rain rates up to 150 mm/hr and ranges up to 5 km. The cancellation is improved with non-circular radar polarization; however the optimum polarization is critically dependent upon the canting angle distribution.

I. INTRODUCTION

An economical utilization of radio spectrum and equipment, especially on satellites, suggests the transmission of different signals simultaneously with the same frequency band but

Jeske (ed.), Atmospheric Effects on Radar Target Identification and Imaging, 399–423.

separated in polarization. Rain destroys the orthogonality of
these two polarizations and the produced cross talk places a
limit on system performance. Also improved target detection
in rain can be achieved with a circularly polarized radar. This
is due to the property of antennas being blind to the polarization
that is orthogonal to their own and the fact that a sphere reflects
left-hand rotating circular polarization if it is radiated with
right-hand circular polarization (or vice versa). Consequently,
if rain is assumed to consist of spherical drops, a circularly
polarized radar could not see it.

These two aspects of precipitation induced effects on millimeter wave propagation are to be discussed in this paper. Specifically, they are

(a) cross polarization generation and

(b) rain echo cancellation

at 17, 30 and 92 GHz. There would be no cross polarization and
no received radar signal (neglecting multiple scattering effects)
for spherical drops, however rain consisting of canted drops
with non-spherical shape introduces polarization dependent
attenuations and phase shifts which cause significant deviations
from the ideal case.

Previous treatments of cross polarization by Saunders (1),
Thomas (2), Oguchi (3) and Wiley et al (4) and of echo cancellation by Ridenour (5), White (6), Offut (7), Panasiewicz (8),
Wheeler and Badertscher (9) and McCormick and Kendry (10,
11, 12) have left room for improvement on the theory. This
paper uses the equilibrium shapes of falling raindrops measured
at the University of California at Los Angeles (13). The scattering parameters for single drops are calculated by a point
matching method for arbitrary angles of incidence of a plane
wave. Previous treatments assumed all drops to be canted by
the same angle and only around the direction of propagation but
the treatment in this paper allows for a random distribution with
preferred directions of drop orientations. This is facilitated by
the introduction of the characteristic polarizations and propagation constants with the rain assumed to be an anisotropic
spatially-homogeneous medium.

II. SOME PHYSICAL PROPERTIES OF RAIN

The rates of precipitation considered in this study were
.25, 1.25, 2.5, 5.0, 12.5, 25, 50, 100 and 150 mm/hr, and
these cover from very light drizzle to the heaviest showers.
For the ranges to be considered, .5 to 5 km, the rain rates are
assumed to be spatially homogeneous. The Laws and Parsons
drop size distribution as used by Medhurst (14) is adapted here
as representing the number densities of drops in different size
ranges per unit volume.

The values of the radio refractive index for liquid water
used in the computations in this study were

$$n = 7.37 - i2.50 \text{ at } 17 \text{ GHz}$$
$$n = 5.99 - i2.85 \text{ at } 30 \text{ GHz}$$
and $\quad n = 3.62 - i2.10 \text{ at } 92 \text{ GHz}.$

These values were determined from the Debye formula given in
Kerr (15).

Measurements of the shape of water drops falling at terminal velocity in air were performed in a vertical wind tunnel at
the University of California at Los Angeles by Pruppacher and
Beard (13) in 1970 and a semi-empirical model was developed
by Pruppacher and Pitter (16) that could account for the measured
shapes. It is reported that only drops with radii smaller than
about .5 mm can be approximated by oblate spheroids. Drops
between .5 and 2 mm have the shape of an asymmetric oblate
spheriod with increasingly pronounced flat base and drops larger
than 2 mm are reported to develop a concave depression in the
base which is deeper for larger drops and plays a significant role
in the breakage of the largest drops. These drop surfaces can be
represented by cosine series which expresses $r(\theta)$, the drop center to surface distance, in terms of the cosine of integer multiples of the polar angle θ. The series coefficients for some of the
selected drop sizes are given in Table II of reference (13) and
values for the other drops were obtained by interpolation of Beard
and Pruppacher data. The normalized computed drop shapes are
given in Fig. 1 for a_o = .25, 1.25, 1.75, 2.25, 2.75 and 3.5 mm.

Measurements on the canting angle distributions of raindrops in storms of varying intensities are virtually nonexistent.
The only direct measurements of which the authors are aware

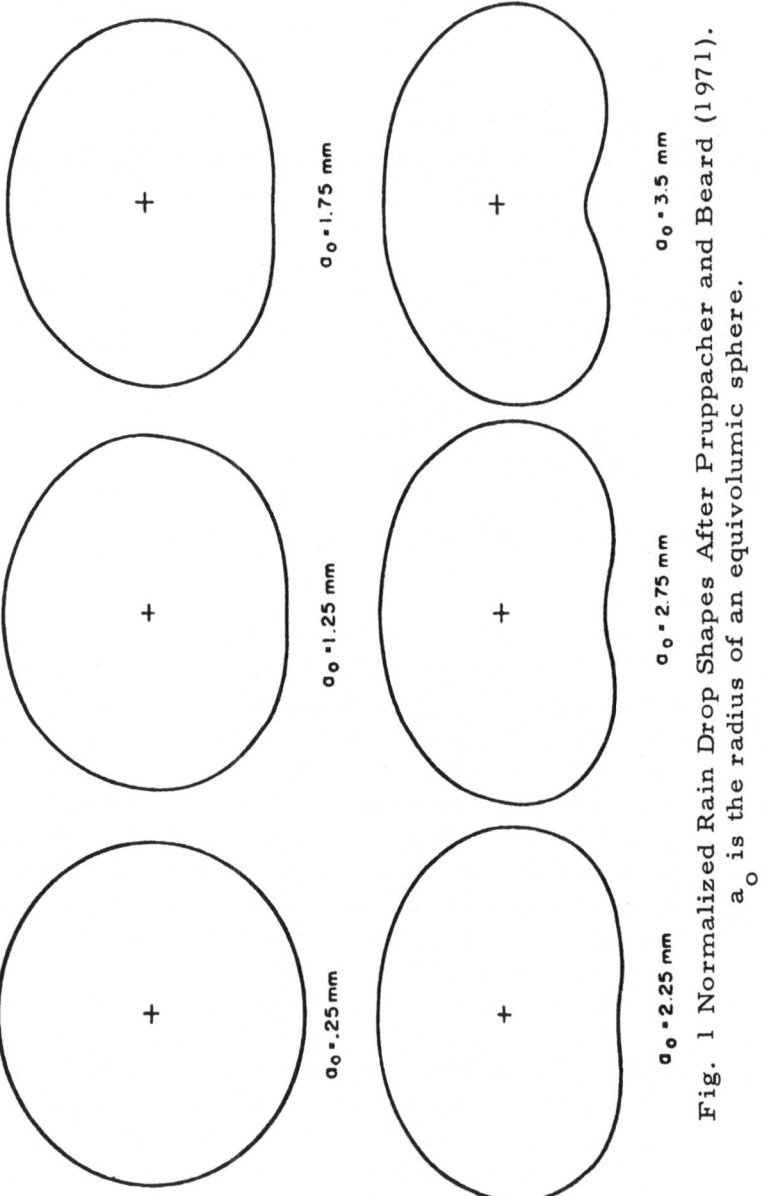

Fig. 1 Normalized Rain Drop Shapes After Pruppacher and Beard (1971). a_o is the radius of an equivolumic sphere.

were reported by Saunders (1) in 1971. Since for Saunder's data angles were deduced from photographs of drops taken with one camera, only the projection of the angles on the film plane could be measured. Since there is no evidence that the canting angle depends strongly on the drop size, these two parameters have been considered to be independent.

Fig. 2 shows the canting angle geometry which defines the angles δ and φ. The z coordinate is taken to be vertically upward and the propagation vector for the incident plane wave is restricted to the x-z plane. Due to the scarcity of data the chosen densities for δ and φ are based on a guess of the possible situation, these densities ($f_1(\delta)$ and $f_2(\varphi)$) being shown in Fig. 2 and described mathematically by binomial functions. As can be seen from the plots, these densities describe a worst case situation in which the drops are predominantly canted in one direction. The computed values of cross polarization for linear polarization are therefore larger than might be measured. For circular polarization the agreement is better because in this case the mean canting angle is less significant. As more measurements become available the canting angle distributions can be deduced with more certainty. The general method of the calculation can be demonstrated anyway. It is assumed that the statistics of the canting angles are the same everywhere in the rain volume.

III. POINT MATCHING SOLUTION TO SCATTERING PROBLEM

The point matching method can be applied, at least theoretically to any reasonable size and shape of raindrop at any frequency. The basic idea is to expand the incident, transmitted and reflected fields into a set of eigenfunctions, the easiest being spherical waves, and then to match the boundary conditions at a finite number of points on the surface of the scatterer. This procedure leads to matrix equations which can be solved for the expansion coefficients. After the field expansion coefficients have been determined they can be used to calculate the quantities which will be of interest here, namely the fields scattered in the forward and backward directions and extinction, scattering, absorption and backscattering cross sections.

Greenberg (17) in 1965 applied the point matching method to solve the scattering of a plane scalar wave incident along the axis of a smooth convex cylindrically symmetric body using

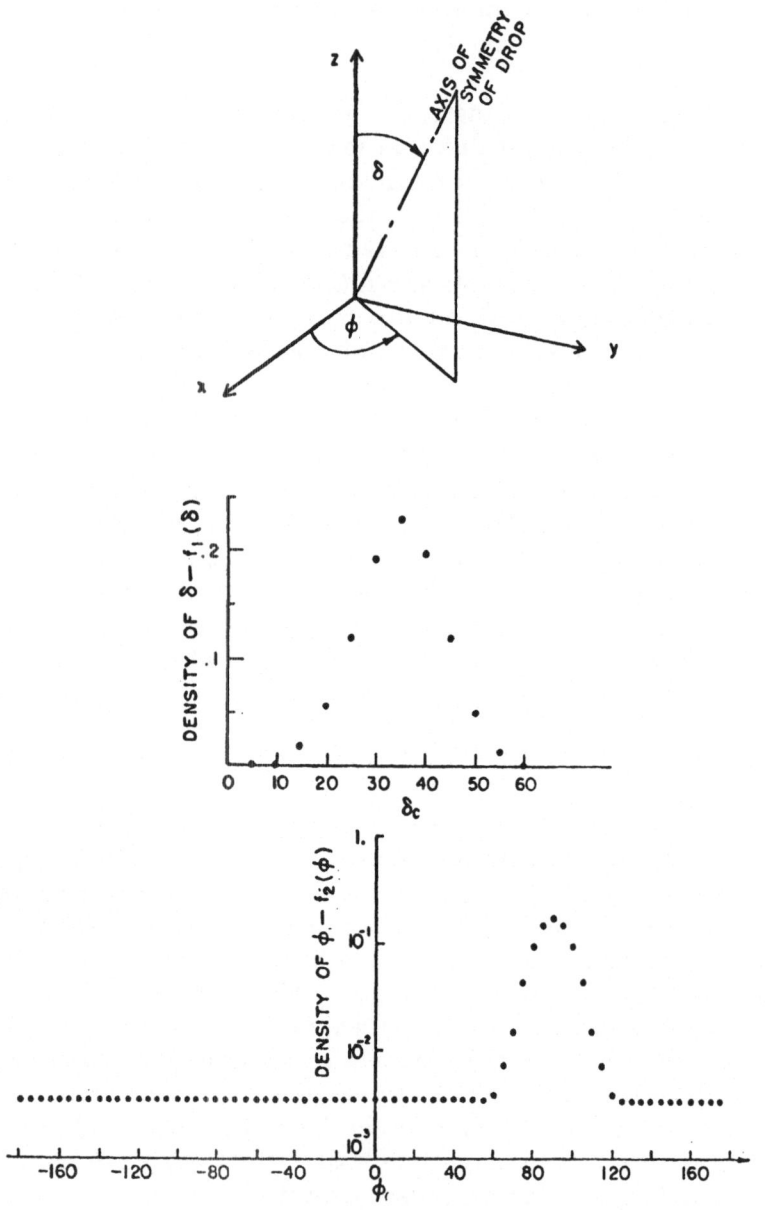

Fig. 2 Canting Geometry and Canting Angle Densities

spherical wave functions.

Oguchi (3), (18) pioneered in applying this method to microwave propagation through rain in 1973. He calculated the forward and backscattered far fields produced by a plane wave incident horizontally on oblate spheriods with complex refractive index.

Morrison and Cross (21) have applied this method combined with least square fitting in 1974.

A Fortran IV program has been written to run on a CDC-6600 computer, which evaluates the scattered field expansion coefficients, field factors and cross sections. Input parameters to this program are the frequency, the refractive index of the scatterer (raindrops), the shape of the drop, and the polarization (vertical or horizontal). For this consideration the drops are taken to be not canted so the axis of symmetry of the drop is in the vertical direction. Thus, due to the symmetry, the scattered fields will not have cross polarized components.

The point matching program was used to calculate the relevant scattering parameters for the 14 dop sizes that appear in the Laws and Parsons' tabulation and for angles of incidence from 0° to 180° in 15° steps for 17, 30 and 92 GHz. When values for other angles were needed they were found by interpolation.

It was found that, in general, the angular dependence of the scattering factors increased with increasing size, and therefore deformation, of the drops and is relatively larger for vertical than for horizontal polarization. This is due to the fact that the electric field in the horizontal polarization "sees" approximately the same linear dimension regardless of the angle of incidence whereas for the vertically polarized field the dimension changes between minimum and maximum radius of the drop.

IV. MEDIUM FORWARD SCATTERING MATRIX FOR RAIN WITH CANTED DROPS

For the case of the plane wave incident upon a region containing densely-packed (on a scale equal to the first Fresnel Zone dimension), randomly-positioned, weak scatterers, Van de Hulst (19) has derived an expression relating the effective refractive index for the medium to the forward scattered far field for the individual scatterers. He deduces that the perturbation imposed on the wave due to a thin slab of thickness dx normal to the path is equal to

$$d\vec{\underset{\sim}{E}} = -j\lambda_o \left(\sum_{\substack{i}} a_i \right) dx\, \vec{\underset{\sim}{E}}_i = -j\, (\Delta k)\, dx\, \vec{\underset{\sim}{E}}_i \qquad (1)$$

u. v.

in which λ_o = free space wavelength, a_i = the forward scattering factor for the i-th particle, the summation is over the particles in a typical unit volume, $\vec{\underset{\sim}{E}}_i$ is the field incident on the slab, and Δk denotes the effective change in propagation constant. Thus eq. (1) serves to relate Δk to the scattering factors for the particles. In this paper $\rightarrow$ and $\sim$ are used to indicate vector and phasor quantities, respectively.

Van de Hulst did not take into accout cross polarization effects but the left equality in eq. (1) can easily be generalized to do this; ie.,

$$d\vec{\underset{\sim}{E}} = \begin{bmatrix} dE^+ \\ \underset{\sim}{} \\ dE^- \\ \underset{\sim}{} \end{bmatrix} = -j\lambda_o\, dx \begin{bmatrix} \Gamma^{++} & \Gamma^{+-} \\ \Gamma^{-+} & \Gamma^{--} \end{bmatrix} \begin{bmatrix} E^+_{\sim i} \\ E^-_{\sim i} \end{bmatrix} \qquad (2)$$

in which the + and - superscripts refer to components in the vertical plane containing the propagation vector and perpendicular to this plane respectively. A $\Gamma^{\alpha\beta}$ is obtained by summing the corresponding forward scattering coefficient for the individual drops in a unit volume. Of course, the summing is accomplished by integrating over the density functions for drop size and the canting angles. The $\Gamma^{\alpha\beta}$'s are thus functions of rain rate and direction of propagation.

The forward scattering coefficients for individual drops were computed, using the point matching method discussed in Section III, only for cases of the incident E-field either in a plane containing the propagation direction and the axis of symmetry of the drop or perpendicular to this plane. Therefore, to carry out the computations to obtain the medium forward scattering coefficients necessitates decomposing the incident E-field vector into components in and perpendicular to the drop-oriented plane of incidence; multiplying each by the forward scattering coefficient for the drop size, polarization and angle of incidence under consideration; and then breaking down these vectors into their horizontal and vertical components before summing over the drop sizes and canting angles. The computational manipulations outlined above are extensive but involve straightforward conversions based on spherical trigonometry.

V. CHARACTERISTIC POLARIZATIONS AND THEIR PROPAGATION PROPERTIES

Once the Γ-matrix, the cumulative or medium forward scattering matrix for horizontal and vertical polarizations, has been determined it is possible, and very desirable, to convert to the characteristic polarizations for this matrix.

A convenient way to describe the polarization of an electromagnetic wave is by use of the complex polarization coefficient presented by Beckman (20). This coefficient, p, is defined by

$$p = \frac{E^+_{\sim}}{E^-_{\sim}} . \tag{3}$$

Thus

$$\vec{E}_{\sim} = \begin{bmatrix} E^+_{\sim} \\ E^-_{\sim} \end{bmatrix} = E^-_{\sim} \begin{bmatrix} p \\ 1 \end{bmatrix} \tag{4}$$

If the two characteristic or eigen vectors for the Γ-matrix are normalized to be of the form

$$\begin{bmatrix} p_1 \\ 1 \end{bmatrix} \quad \text{and} \quad \begin{bmatrix} p_2 \\ 1 \end{bmatrix} \tag{5}$$

then p_1 and p_2 are termed the characteristic polarization coefficients and the vectors in (5) the characteristic polarizations or vectors.

Transforming to a system in which the characteristic polarizations are the base vectors, the Γ-matrix converts to a diagonalized one with the characteristic or eigen values of the Γ-matrix, λ_1 and λ_2, as the elements on the principle diagonal. That is, in this representation there is no cross polarization, the two characteristic polarizations propagate independently. Therefore, in keeping with eq. (1), the scattering medium produces changes in the propagation constants for these two waves by the amounts

$$\Delta k_{1,2} = \lambda_o \lambda_{1,2} = 2\pi\lambda_{1,2}/k_o. \tag{6}$$

The attenuation in dB per unit length and the phase shift relative to free space in degrees per unit length for the characteristic waves are then

$$A_{1,2} = -8.68 \, \mathrm{Im}(\Delta k_{1,2})$$

and $\hspace{10cm}$ (7)

$$\varphi_{1,2} = (180/\pi) \, \mathrm{Re}\{\Delta k_{1,2}\}.$$

One deduces the expressions for the characteristic polarization coefficients and characteristic values to be (20)

$$P_{1,2} = \frac{1}{2\Gamma^{-+}}\left[\Gamma^{++} - \Gamma^{--} \pm \sqrt{(\Gamma^{++} - \Gamma^{--})^2 + 4\Gamma^{+-}\,\Gamma^{-+}}\right] \tag{8}$$

and

$$\lambda_{1,2} = \frac{1}{2}\left\{\Gamma^{++} + \Gamma^{--} \pm \sqrt{(\Gamma^{++} - \Gamma^{--})^2 + 4\Gamma^{+-}\,\Gamma^{-+}}\right\}$$

$$= \Gamma^{-+} P_{1,2} + \Gamma^{--}. \tag{9}$$

To compute the strength and polarization of a wave after it has propagated a distance through rain involves three steps:

(1) Synthesize the initial field with two components having the characteristic polarizations.

(2) Determine the attenuation and phase shift for each by multiplying the corresponding coefficients in eq. (7) by the path length.

(3) Combine the two characteristic waves that result and determine the resultant strength and polarization.

VI. RAIN INDUCED LOSS AND CROSS POLARIZATION

Two waves propagating in the same direction are said to be orthogonal when

$$\vec{\underset{\sim}{E}}_1 \cdot \vec{\underset{\sim}{E}}_2^* = 0, \tag{10a}$$

this condition existing if their polarization coefficients satisfy

$$p_1 p_2^* = -1, \tag{10b}$$

the * denoting complex conjugate.

For waves arriving from a given direction an ideal antenna is "blind" (that is, collects no power from the wave) for a unique polarization and presents the maximum effective area for reception to the orthogonal polarization. The polarization of a receiving antenna is taken to be that of the latter wave, the polarization of the wave it receives most efficiently. A transmitting antenna will be considered to have the polarization of the wave it radiates in the specified direction. With these definitions, an antenna generally has a different polarization when used as a receiving element than when used as a transmitting element. A receiving antenna (t) and transmitting antenna (t_x) are matched if $t = t_x$.

The polarization efficiency, η, of an antenna at receiving an incident wave with polarization p is defined as the ratio of the power collected by the antenna (t) receiving the wave (p) to the power this antenna would collect if the wave polarization matched that of the antenna (p = t), the two waves having the same power densities. Thus

$$\eta = \frac{\left| 1 + pt^* \right|^2}{(1 + |t|^2)(1 + |p|^2)} \tag{11}$$

In the absence of rain $p = t_x$. The steps followed to compute the changes in incident power density and p when rain is inserted in the path were detailed in the previous section.

The expression for α, the dB loss due to rain in the path for propagation between matched antennas $(t_x = t)$ is

$$\alpha = -10 \log\left(\frac{P_R}{P_o}\right) = -10 \log\left(\frac{S_R}{S_o}\right) - 10 \log(\eta_R), \tag{12}$$

R and o subscripts denoting quantities with rain and with only free space along the path, respectively.

Cross polarization, K, is defined as the efficiency with which an ideal antenna orthogonally polarized relative to the transmitting antenna receives a wave after propagation through a rain-filled medium. Thus K, expressed in dB, is given by

$$K = 10 \log \frac{\left| t_{tx} - p \right|^2}{(1 + \left| t_x \right|^2)(1 + \left| p \right|^2)} \cdot \tag{13}$$

Since the elements in the Γ-matrix in eq. (8) are complex, the characteristic polarizations are generally not quite orthogonal. A measure of the deviation from orthogonality is given by K_{12}, the efficiency with which an antenna matched to one characteristic polarization will receive the other; ie.,

$$K_{12} \text{(in dB)} = 10 \log \frac{\left| 1 + p_1 p_2^* \right|^2}{(1 + \left| p_1 \right|^2)(1 + \left| p_2 \right|^2)} \cdot \tag{14}$$

VI. BACK SCATTER CANCELLATION

Radar echo or back scatter cancellation, K, is here defined as the efficiency with which the radar antenna receives the back scattered energy. This definition differs from the one used by McCormick and Hendry (10) in that they normalized by dividing by the efficiency with which an antenna with polarization orthogonal to the one under consideration would receive the wave. Therefore, for good cancellation the two definitions give values very close to each other.

Concepts from the preceding sections are basic to the calculation of K. Considering a single scatterer, the transmitted wave is attenuated and has its polarization changed upon propagation through the rain before reaching the scatterer (as discussed in Section V). Back scattering coefficients, computed using the point matching method as discussed in Section III for the incident E-field in and perpendicular to the drop-oriented plane of incidence, can thus be applied to determine the initial backward-traveling wave. The rain then again modifies the wave strength and polarization on its way back to the receiving antenna. The received power due to the i-th scatterer, ΔW_i, is found using eq. (11) to be

$$\Delta W_i = A S_i \eta_i = \frac{A S_i |1 + p_i t^*|^2}{(1 + |t|^2)(1 + |p_i|^2)} \tag{15}$$

in which A = the effective area of the antenna for matched condition, S_i = power density of the returned wave, p_i = polarization coefficient of the returned wave and η_i = efficiency with which antenna receives the returned wave. S_i and p_i depend on the angle of the radar path relative to the vertical and transmitted polarization as well as the size, shape and canting of the i-th drop.

Since raindrops are randomly distributed in space their radar returns will be incoherent. There are also large numbers of raindrops so for most purposes it is appropriate to conclude that the total received power does not deviate significantly from its average value obtained by adding the power contributed by the individual drops. Again the summation is accomplished by integrating over the density functions characterizing the random features of the raindrops. Considering the quantities in eq. (15) bearing i-subscripts to be functions of drop size (which also establishes drop shape) and the two canting angles; multiplying by the corresponding density functions N(R), $f_1(\delta)$ and $f_2(\varphi)$; and integrating gives

$$K(\text{in dB}) = -10 \log \left\{ \frac{\int\int\int S(R, \delta, \varphi) \eta(R, \delta, \varphi) F(R, \delta, \varphi) \, dR \, d\delta \, d\varphi}{\int\int\int S(R, \delta, \varphi) F(R, \delta, \varphi) \, dR \, d\delta \, d\varphi} \right\} \tag{16}$$

in which $F(R, \delta, \varphi) = N(R) f_1(\delta) f_2(\varphi)$, the joint density function. Since the computation of $S(R, \delta, \varphi)$ and $\eta(R, \delta, \varphi)$ for each choice of (R, δ, φ) is rather time consuming, to approximate the integrals in eq. (16) the ranges of the variables of integration were broken into rather moderate numbers of intervals.

It has been assumed herein that the same antenna, or one with the same polarization, is used for reception as that used for transmission.

To find the polarization affording optimum cancellation for a given direction of propagation, rain rate, and range, a numerical minimization technique was used. For the sake of economy, the drop size giving the largest contribution to the numerator

integral in eq. (16) for a circularly polarized antenna was first
determined and then the transmit polarization to minimize this
integral considering only this one drop size was sought. After
this approximately optimum polarization was determined, the
corresponding cancellation was computed taking into account all
drop sizes.

VIII. RESULTS

Numerical results have been obtained using the techniques
briefly detailed in Sections II-VII for the rain model described in
Section II for 17 GHz, 30 GHz and 92 GHz propagation, at angles
of 0° to 180° relative to vertical in 15° steps, through rain of intensities from 0.25 mm/hr to 150 mm/hr, for ranges of 1.0 km
to 5.0 km, and for various polarizations. Some of the results
are presented below.

A.   Attenuation and Phase Shift
Attenuation and incremental phase shift constants were
computed for vertical and horizontal incident-wave polarizations
for rain with no canting of the drops ($\delta = 0°$). Neither was found
to vary absolutely with polarization and/or angle of incidence, for
a fixed rain rate and frequency, by more than 5% and for most
cases it was considerably less. Therefore, the effect of drop
deformation and tilt is not likely to be very significant except for
those situations in which the observed quantity is characteristi-
cally rather weak.

B.   Characteristic Polarizations
The characteristic polarization coefficients p_1 and p_2,
defined in Section V, were computed for each case. They were
found to be nearly real (linear polarization), the imaginary part
of p typically being two to four orders of magnitude less than the
real part. The characteristic polarizations are closely aligned
with the mean directions of the drop axes and very close to orthogonal. This last fact is well illustrated in Fig. 3 which gives the
spread in the cross polarization of p_1 and p_2, as expressed by
eq. (14), for a number of rain rates.

The differential attenuation and phase shift between p_1 and
p_2, for horizontal propagation are given as functions of rain rate
in Figs. 4 and 5.

The general trend of the differential phase shift at 17 and

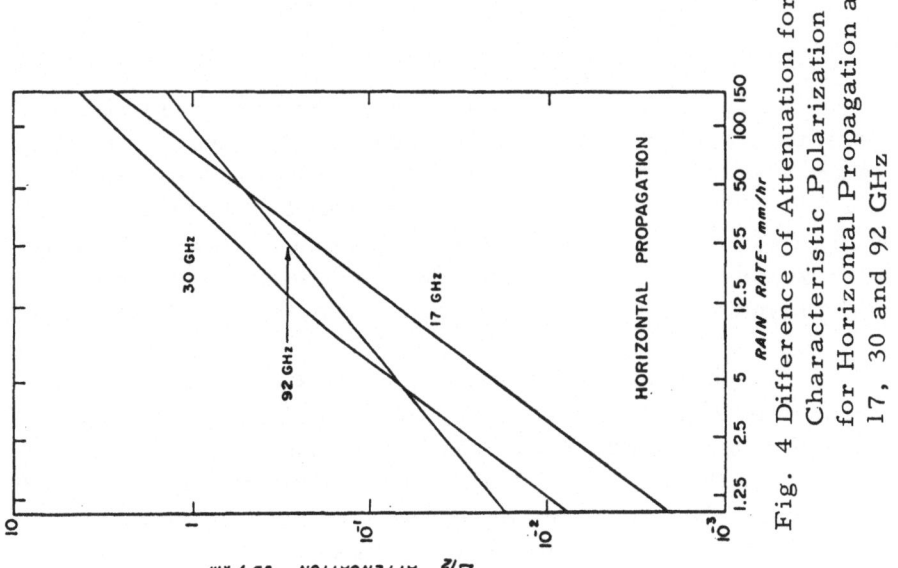

Fig. 4 Difference of Attenuation for Characteristic Polarization for Horizontal Propagation at 17, 30 and 92 GHz

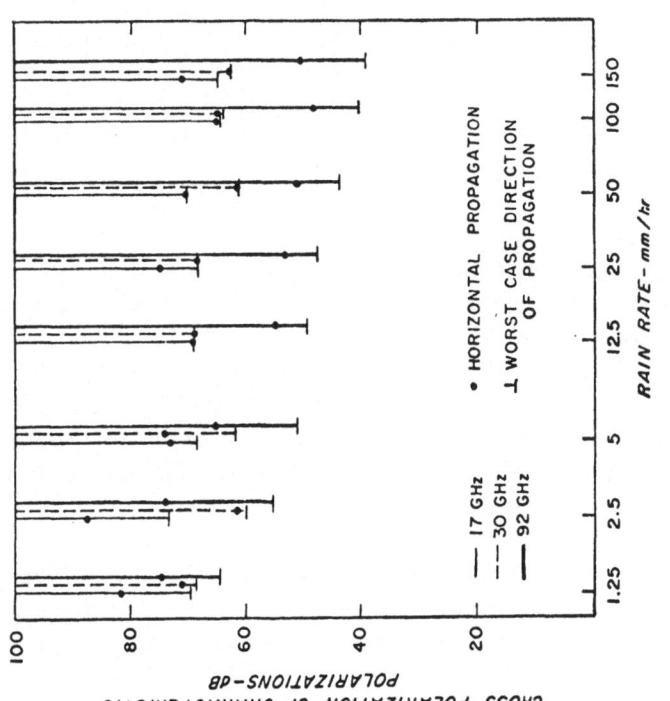

Fig. 3 Comparison of the Orthogonality of the Characteristic Polarization

30 GHz in Fig. 5 is similar to that found by Oguchi for 19.3 and 34.8 GHz for rain with oblate spheriod drops.

C. Cross Polarization

The energy converted from the initial wave to the cross-polarized one upon passage through rain, more specifically the cross polarization as defined by eq. (13), is plotted in Fig. 6 as a function of path length for a rain rate of 50 mm/hr, frequency of 30 GHz, and a horizontal propagation path for several linear polarizations. The strong dependence of the cross polarization on the orientation of the linear polarization is immediately noted.

Achieving the low cross polarizations depicted by the characteristic-polarization curve would not be expected in practice because relative slight deviations in the canting angle distribution would cause the medium to not match the design specifications for the antenna. However, the thought of making a radio link with adaptable (linear) orthogonal polarizations to reduce cross talk is intriguing.

Figs. 7 and 8 show the average cross polarization as a function of the average attenuation for the right hand-left hand circular polarization and horizontal-vertical linear polarization cases. For equal attenuations, the cross polarization decreases as the frequency goes up (for the range of frequencies considered). Also, for this rain model used there is little difference between these two transmit polarizations.

D. Radar Echo Cancellation

Fig. 9 gives the radar rain echo cancellation as given by eq. (16), as a function of the one way attenuation for circular polarization. At 17 and 30 GHz it is approximately constant to attenuations of 3 dB and then decreases logarithmically. The cancellation is better at 92 GHz where even for high attenuations of 40 dB it is still 20 dB. These curves have been derived by taking an average over the range dependency. The error bars give extrema for 1, 2.5 and 5 km range.

The results of cancellation optimization by polarization adaption are found in Tables I, II, and III. At 17 GHz for horizontal propagation it is in the neighborhood of 20 dB until at higher rates a rapid decrease occurs. Optimization for 25 mm/hr at 3.5 mm range improved the cancellation there by 7 dB. The same polarization led to a general improvement over the rain rates and

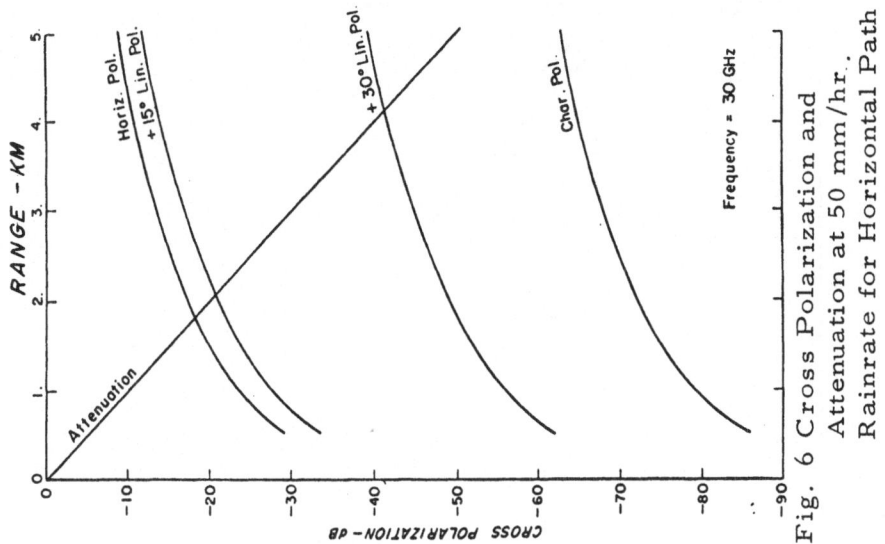

Fig. 6 Cross Polarization and Attenuation at 50 mm/hr. Rainrate for Horizontal Path

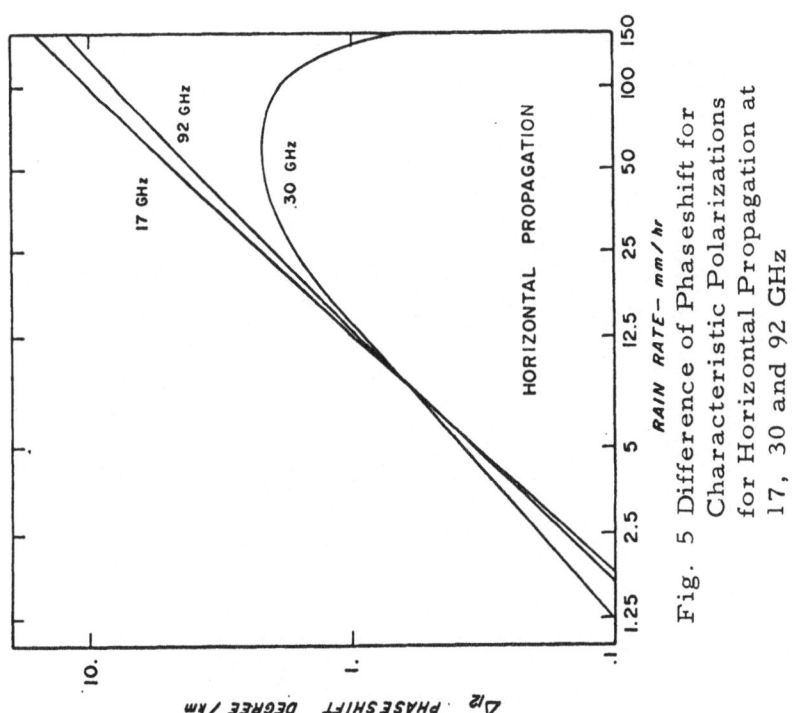

Fig. 5 Difference of Phaseshift for Characteristic Polarizations for Horizontal Propagation at 17, 30 and 92 GHz

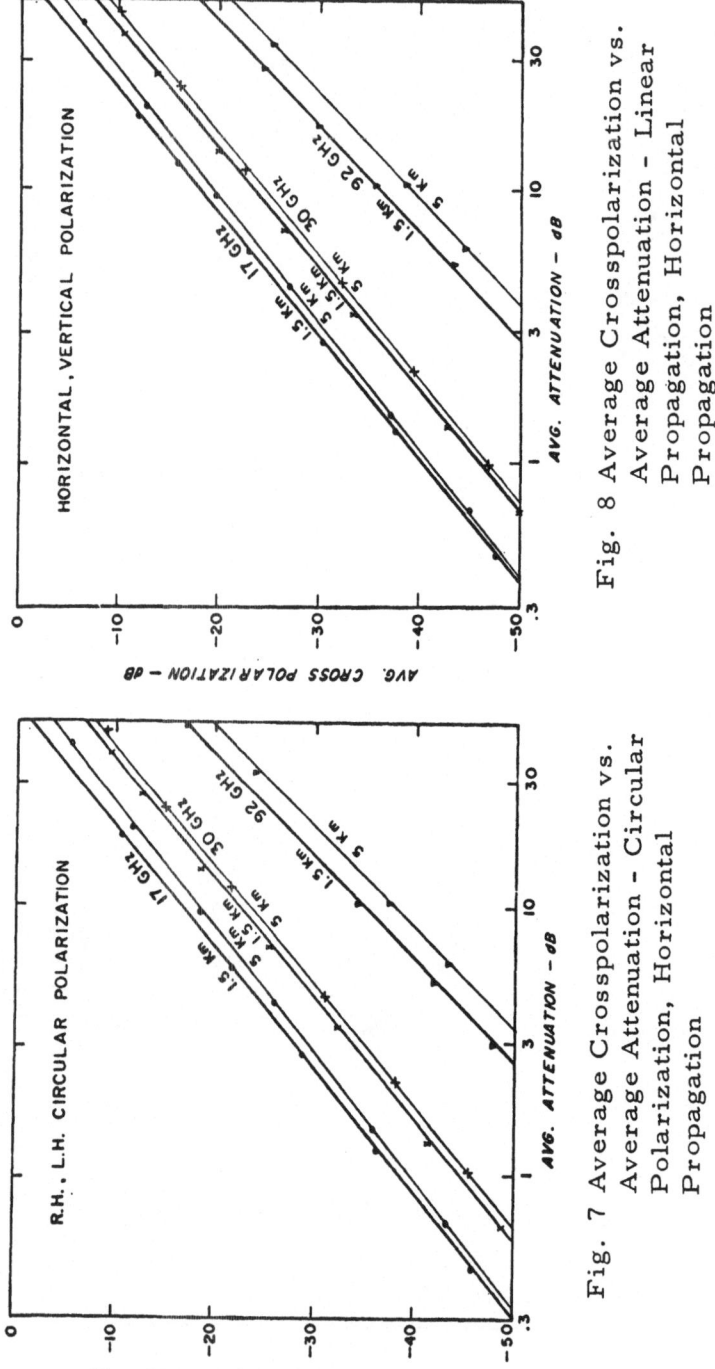

Fig. 8 Average Crosspolarization vs.
Average Attenuation - Linear
Propagation, Horizontal
Propagation

Fig. 7 Average Crosspolarization vs.
Average Attenuation - Circular
Polarization, Horizontal
Propagation

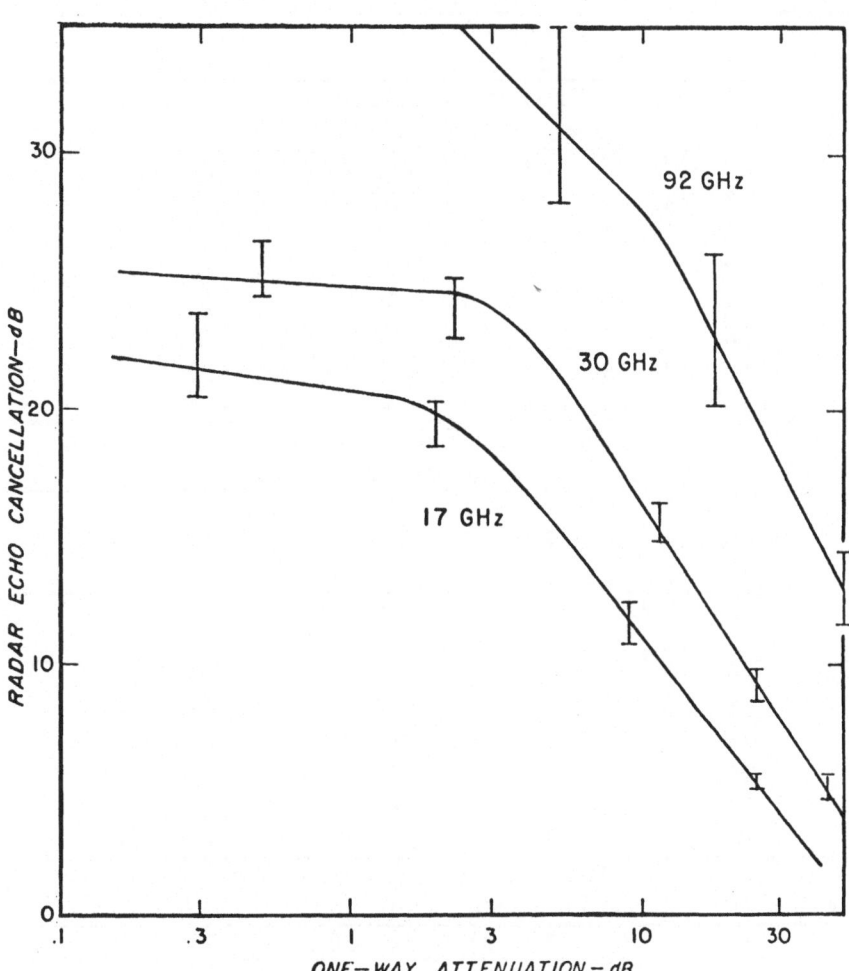

Fig. 9 Radar Echo Cancellation Versus One-Way
Attenuation Circular Polarization

Table I

17 GHz Radar Echo Cancellation - dB

Rain Rate mm/hr	.5	1	1.5	2	2.5	3	3.5	4	4.5	5
R. H. Circular Polarization p = i Horizontal Propagation										
1.25	24	24	24	24	24	24	24	24	24	24
2.5	22	22	22	22	22	22	22	22	22	22
5.0	21	21	21	21	21	21	21	21	20	20
12.5	20	20	20	19	19	19	18	18	18	17
25	19	19	18	17	16	15	14	14	13	12
50	18	16	14	13	11	10	9	8	7	6
100	16	12	9	7	5	4	4	3	3	2
150	13	8	6	4	3	3	2	2	2	3
L. H. Elliptical Polarization p=-.987-i.06 (Optimum for 25 mm/hr at 3.5 km)										
12.5	25	25	25	25	24	24	24	24	24	23
25	24	24	24	23	23	22	21	20	20	19
50	23	23	21	20	18	17	16	15	15	14
R. H. Circular Polarization p = i Angle of Propagation = 150°										
1.25	29	29	29	29	29	29	29	29	29	29
2.5	27	27	27	27	27	27	27	27	27	27
5	26	26	26	26	26	26	26	26	26	26
12.5	25	25	25	25	25	25	25	24	24	23
25	25	25	24	23	23	22	21	20	19	18
50	24	23	21	19	17	16	15	14	13	12
R. H. Elliptical Polarization p=1.00+i.03 Angle of Propagation = 150°, Optimum at 3.5 km										
25	31	31	31	31	31	31	30	30	30	30
R. H. Circular Polarization p = i Angle of Propagation = 0°										
25	27	27	26	25	25	24	23	23	22	21
Transmitter Pol. Horizontal, Receiver Pol. Vertical Angle of Propagation = 0°										
25	34	34	34	34	34	33	33	33	33	33

Table II

30 GHz Radar Echo Cancellation - dB

Rain Rate	Range - km									
mm/hr	.5	1	1.5	2	2.5	3	3.5	4	4.5	5
R. H. Circular Polarization									p = i	
Horizontal Propagation										
1.25	26	26	26	26	26	26	26	26	26	26
2.5	25	25	25	25	25	25	25	25	25	25
5	24	24	24	24	24	24	24	23	23	23
12.5	24	23	23	22	21	20	19	18	17	16
25	23	21	19	17	15	14	13	12	11	10
50	21	17	14	11	9	8	7	6	6	5
100	17	11	8	6	5	5	4	4	4	3
150	13	8	6	5	4	4	3	3	3	3
L. H. Elliptical Polarization p=-.826-i.301										
(Optimum for 25 mm/hr at 3.5 km)										
12.5	17	18	19	20	20	21	22	23	25	26
25	18	20	22	24	17	31	32	30	27	25
50	20	25	30	28	24	21	19	18	17	16
R. H. Circular Polarization									p = 1	
Angle of Propagation = 150°										
12.5	30	30	29	28	27	26	25	24	23	22
25	30	28	26	24	22	20	19	18	17	16
50	28	23	20	17	15	14	12	11	10	9
R. H. Elliptical Polarization p=-1.02+i.09										
Angle of Propagation = 150° Optimum for 25 mm/hr at 3.5 km										
12.5	32	32	33	33	33	34	34	35	35	35
25	32	33	34	35	35	36	36	36	36	35
50	33	34	25	35	34	33	32	31	30	29

Table III

92 GHz Radar Echo Cancellation - dB

Rain Rate mm/hr	.5	1	1.5	2	2.5	3	3.5	4	4.5	5
Range - km										
R. H. Circular Polarization p = 1 Horizontal Propagation										
1.25	37	37	37	38	38	38	37	37	37	36
2.5	35	36	36	35	35	35	34	33	33	32
5	33	34	33	32	31	30	29	28	27	26
12.5	30	30	28	26	24	23	22	20	19	18
25	28	26	23	21	19	17	16	15	14	13
50	25	20	17	15	13	11	10	9	8	7
100	20	15	12	9	8	6	5	5	4	3
150	17	12	9	7	5	4	3	3	2	2
R. H. Elliptical Polarization p=-.745 + i.493 Horizontal Propagation										
12.5	19	20	21	21	22	23	24	26	27	29
25	19	21	23	25	27	30	32	31	29	26
50	21	24	28	30	26	23	21	19	17	16
R. H. Circular Polarization p = i Angle of Propagation = 150°										
21.5	28	28	27	27	26	26	25	24	24	23
25	25	24	24	23	22	21	20	18	19	18
50	21	21	20	18	17	16	15	14	13	12
L. H. Elliptical Polarization p=-1.0 - i.2 Angle of Propagation = 150° Optimum for 25 mm/hr at 3.5 km										
12.5	31	31	32	32	32	33	33	33	34	34
25	29	30	30	31	31	31	32	32	32	32
50	27	28	28	28	28	28	28	28	27	27

distances given. For a 150° angle of propagation an improvement of 9 dB at 3.5 km and 12 dB at 5 km was achieved by changing p from i to 1. + i.03. This polarization is an almost linear one with 45° inclination. It seems surprising to find cancellation for such a case and might be a clue to discrepancies reported between the theory and experiments for radar backscatter measurements in rain. The last two rows in Table I compare cancellation with circular polarization to cancellation with horizontal transmit and vertical receive polarization at a 0° angle of propagation (straight down). It is noted that the orthogonal inear polarizations result in an order of magnitude better cancellation.

At 30 GHz and horizontal propagation optimization for 25 mm/hr at 3.5 km improves that cancellation by two orders of magnitude there. At closer range cancellation is decreased by up to 5 dB. For a 150° angle of propagation a drastic improvement is also noted. For this case the polarization is near linear.

At 92 GHz and horizontal propagation cancellation optimized for 25 mm/hr at 3.5 km is changed from 16 to 32 dB but at .5 km range from 28 to 19 dB. For a 150° angle of propagation no such decrease at close range is found.

The cancellations calculated seem to be in general agreement with those reported in the literature and especially McCormick and Hendry's measurements at 16.5 GHz.

REFERENCES

(1) M. J. Saunders, "Cross polarization at 18 and 30 GHz due rain," IEEE Trans. Antennas Propagat., vol. AP-19, 1971.

(2) D. T. Thomas, "Cross polarization discrimination in microwave radio transmission due to rain," paper presented at 1970 URSI. Published in Radio Science, vol. 6, pp. 833-839, Oct. 1971.

(3) T. Oguchi, "Scattering properties of oblate raindrops and cross polarization of radio waves due to rain: calculations at 19.3 and 34.8 GHz," J. of Radio Research Laboratories, vol. 20, p. 102, 1973.

(4) P. H. Wiley, et al., "The influence of polarization on milli-

meter wave propagation through rain, " interim report no. 1,
NASA Grant NGR-47-004-91, 1973.

(5) L. Ridenour, (ed.), Radar system engineering, MIT Lab.,
Lab. series No. 1, McGraw-Hill, New York, 1947.

(6) W. D. White, "Circular polarization cuts rain clutter, "
Electronics, vol. 27, pp. 158-160, 1954.

(7) W. B. Offut, "A review of circular polarization as a means
of precipitation clutter suppression and examples, " Proc.
National Electronics Conference (Chicago), vol. 11, 1955.

(8) J. J. Panasiewicz, "Enhancement of aircraft return by use
of airborne reflectors and circular polarization, " IRE Con-
vention Record, vol. 4, pt. 8, 1956.

(9) M. S. Wheeler and K. A. Badertscher, "Radar mapping in
heavy rain with orthogonal radar modes at X- and K_A-Band, "
NAECON 70 Record, pp. 72-80, 1970.

(10) A. Hendry and G. C. McCormick, "Polarization properties
of precipitation scattering, " NRL of Canada, Radio and E. E.
Div., vol. 21, 1971.

(11) G. C. McCormick and A. Hendry, "The study of precipitation backscatter at 1.8 cm with a polarization diversity radar, "
14th Radar Meteorology Conference preprints, Tucson,
Arizona, 1970.

(12) G. C. McCormick and A. Hendry, "Results of precipitation
backscatter measurements at 1.8 cm with a polarization diversity radar, " 15th Radar Meteorology Conference preprint,
AMS, Boston, Mass., 1972.

(13) H. R. Pruppacher and K. V. Beard, "A wind tunnel investigation of the internal circulation and shape of water drops
falling at terminal velocity in air, " Quart. J. Royal Meteorol.
Society, vol. 96, 1970.

(14) R. G. Medhurst, "Rainfall attenuation of centimeter waves:
comparison of theory and measurement, " IEEE Trans.
Antennas Propagat., vol. AP-13, 1965.

(15) D. E. Kerr, Propagation of short radio waves," McGraw-Hill, New York, 1951.

(16) H. R. Pruppacher and R. L. Pitter, "A semi-empirical determination of the shape of cloud and rain drops," J. of Atmos. Sci., vol. 28, 1971.

(17) J. M. Greenberg, "Scattering by nonspherical systems," in R. L. Rowell and R. S. Stein, ed., Electromagnetic Scattering, Gordon and Breach, New York, 1965.

(18) T. Oguchi, "Attenuation and phase rotation of radio waves due to rain: calculations at 19.3 and 34.8 GHz," Radio Science, vol. 8, 1973.

(19) H. C. Van de Hulst, Light scattering by small particles, J. Wiley & Sons, New York, 1957.

(20) P. Beckman, The depolarization of electromagnetic waves, Golem Press, Boulder, 1968.

(21) A. Morrison and M. J. Cross, "Scattering of a plane electromagnetic wave by axisymmetric raindrops," BSTJ, vol. 53, no. 6, July-Aug. 1974.

RADAR ECHOES FROM BIRDS AND THEIR EFFECTS ON RADAR PERFORMANCE

Warren L. Flock

University of Colorado, Boulder, Colorado 80309

INTRODUCTION

The first observation of a radar echo from a bird was apparently that made with the 200-MHz XAF radar on the U.S.S. New York off Puerto Rico in 1939. A number of reports of radar echoes due to birds were made subsequently during World War II, and by 1957-58 radar was being used to study bird migration. A good treatment of the subject of radar ornithology has been provided by Eastwood (1967). The concern of this paper, however, is primarily for the effects of birds on radar system performance rather than for the use of radar to study birds.

What is clutter to one person may be the desired signal to another but in most conventional applications of radar birds are a source of clutter. An effort is made here to describe the characteristics of this clutter.

IDENTIFYING BIRD ECHOES

In considering radar echoes from birds, it is necessary first of all to be able to distinguish bird echoes from echoes from other objects. The identification of bird echoes can be made most readily when many birds, or flocks of birds, are flying in a consistent direction for a period of time. A time-exposure photograph of a radar PPI screen then results in a number of parallel streaks or trains of dots of a length compatible with the flight speed of the birds. Figure 1 is an especially clear example of radar echoes from birds, as recorded by a time-exposure photograph. Such photographs have an 180° ambiguity as to the direction the

Jeske (ed.), Atmospheric Effects on Radar Target Identification and Imaging, 425–435.
All Rights Reserved. Copyright © 1976 by D. Reidel Publishing Company, Dordrecht-Holland.

birds are flying, but this ambiguity can be resolved by comparing
successive photographs or visually following individual targets
on the radar screen. Another technique for observing bird movements is to photograph the PPI display with a motion-picture
camera, with one frame for each or alternate rotations, and to
project the film much speeded up so that the bird movements of a
night can be visualized in a few minutes. More often than not
what one sees on radar displays are echoes from flocks or
clusters of birds rather than individual birds, but radars of
sufficient resolution and sensitivity can detect single birds, as
a function of range from the radar. The long range surveillance
radars of the Federal Aviation Administration of the United States,
for example, are capable of detecting a single bird the size of
a gull (more precisely a target having a cross section of $0.01 \, m^2$)
at a distance of 78 km.

When birds are not flying in parallel paths and not maintaining a constant direction, the identification of their echoes is
more difficult. Some knowledge of what birds are in the area
and their habits may then be needed. For example, in winter the
North Platte, Nebraska radar screen may be clean and nearly free
from echoes at night but at the first sign of light in the morning
large bright areas appear on the radar screen as in Fig. 2. The
cause of such echoes is not necessarily obvious from study of the
radar screen alone, but knowledge of the area and of the habits
and occurrence of mallard ducks (Anas platyrhynchos) in the area
indicates that the echoes are due to ducks flying from their
resting areas on water to feed on nearby fields.

Sometimes neither the appearance of the echoes on a PPI
display or a knowledge of the birds in the area is sufficient to
decide if certain echoes are due to birds or not. In such cases
determination of the frequency spectra of the amplitude fluctuations of the echoes may provide the desired identification. In
order to record the amplitude variations, it is necessary for the
radar to dwell for a sufficient time on the target, which condition may be achieved by use of a tracking radar or by aiming the
radar beam in a favorable direction and letting birds fly through
it. A principal feature of the amplitude variations of the echo
from a bird is modulation at the wingbeat rate, which varies inversely with the size of the bird (Houghton, 1972, 1973), an
example being shown in Fig. 3. In this particular case, it was
possible to not only identify the echo as due to a bird but to
identify the bird as to species, on the basis of the very low
wingbeat rate and a knowledge of what birds would be expected in
the area. The wingbeat rate for the great blue heron (Ardea
herodias) of Fig. 3 was about 2.7 Hz as compared with near 7 Hz
for a mallard and 10 Hz or more for small passerine species.

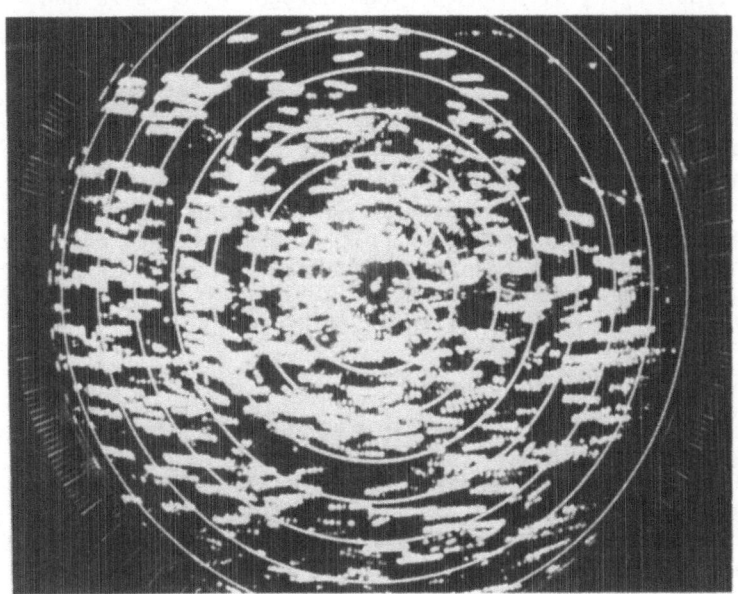

Fig. 1 Eastward migration of birds over Oliktok, Alaska (near Prudhoe Bay): 40 nautical mile (74 km) range, 5 nautical mile (9.3 km) range marks, 5.6-min time exposure, geographic north up, 2005 ADT, 28 May 1972.

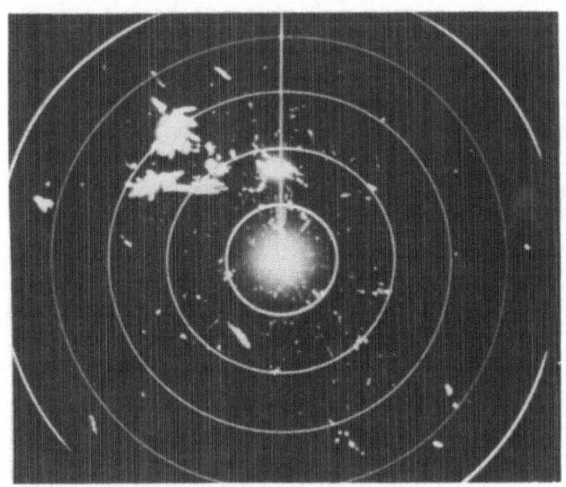

Fig. 2 Mallards leaving resting areas near sunrise in Platte Valley, west of North Platte, Nebraska: 50 nautical mile range, 10 nautical mile range marks, 5-min time exposure, magnetic north up, 0715 CST, 4 Feb. 1968.

In general the "amplitude signature" technique considered
here allows classifying birds into rough categories of size and
type. This approach appears to be useful when a number of birds
are in the beam simultaneously (Fig. 4), as well as when only a
single bird is present. The spectrum of Fig. 4 is the average of
thirty 4-s spectra, each having a frequency resolution of 0.25 Hz.
The lowest peak frequency of about 3.6 Hz is believed to be due to
gulls and the second peak frequency of 6.6 Hz is due to mallards.
The third prominent peak is believed to be the second harmonic
of the mallard fundamental frequency. It is essential to actually
compute the spectrum in the case of echoes from a number of birds
whereas visual examination of the amplitude variations may be
sufficient in the case of a single bird (Flock, 1974).

The doppler spectra of radar echoes from birds have the very
interesting feature depicted in Fig. 5 of a doppler-frequency
smear or spike which repeats at the wingbeat rate (Green and
Flock, 1972). This feature may also be useful for identification
purposes, and the combination of amplitude and doppler spectra or
signatures is potentially more useful than one type alone. Echoes
from birds are sometimes difficult to distinguish from land and
sea clutter or small aircraft and helicopters, but the amplitude
and doppler signatures mentioned can facilitate making the
distinction.

OCCURRENCE AS A FUNCTION OF TIME OF YEAR AND DAY

Bird echoes tend to be most numerous in the northern hemisphere during spring and fall migrations but can be encountered at
any time of the year at temperate or tropical latitudes. In the
lower 48 states of the United States, for example, spring waterfowl migration takes place mostly in late February, March, and
early April. Shorebirds migrate somewhat later in the spring,
and passerine bird migration generally reaches a peak in May
(Fig. 6) and extends until perhaps mid June. After the spring
migration is over, many flying birds are still present as swallows,
swifts, and nighthawks, for example, feed in the air, and other
species such as gulls and alcids fly from nesting to feeding areas.
Some birds, furthermore, do not stay long in their breeding areas.
Some shorebirds, for example, return from breeding areas to lower
latitudes as early as July. Southward shorebird migration continues through August, to be joined by passerine migrations.
Migrating ducks may reach peak numbers in October, with goose
migrations being prominent in November and December. In the winter
ducks, geese, and cranes make daily flights from resting areas to
feeding areas. No time of year is free from bird activity.

In the arctic, it is true, essentially no bird activity is

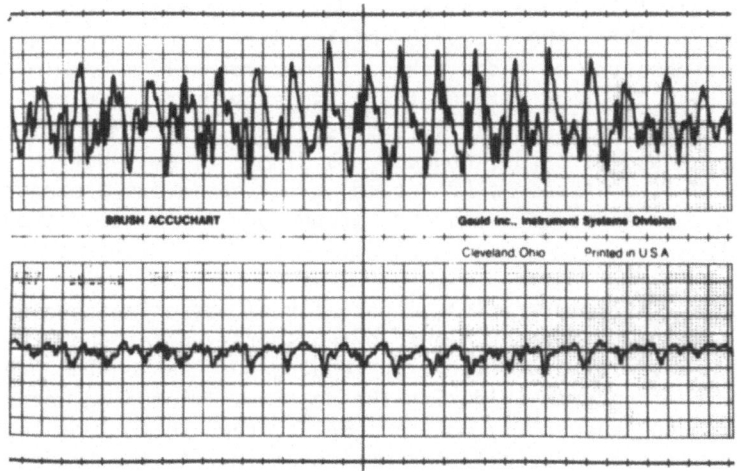

Fig. 3 Variation in amplitude of radar echo from great blue heron, Boulder, Colo., 2015 MDT, 28 May 1973. Horizontal scale: 5 div/s. The upper trace represents the signal after filtering by a 2-200 Hz bandpass filter, and the lower trace is the unfiltered signal.

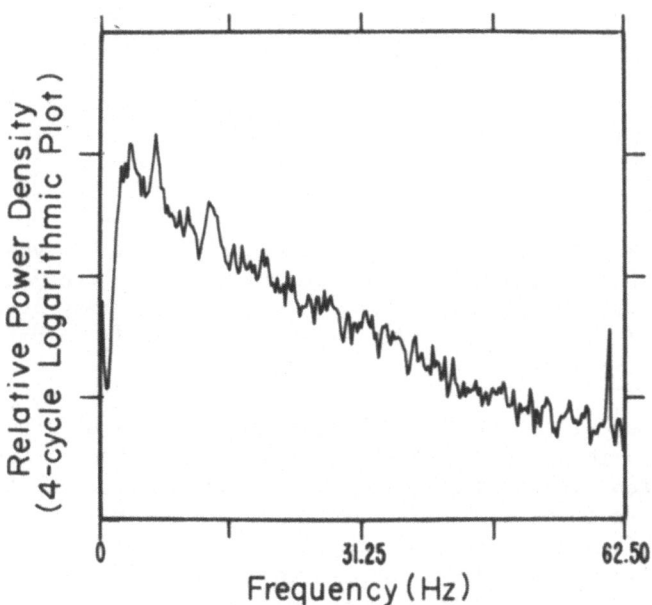

Fig. 4 Spectrum of amplitude variation of radar echo at a time when both gulls and mallards were in the radar beam, Boulder, Colo., 11 Feb. 1973.

encountered in the winter but at Pt. Barrow, Alaska, for example, bird movements take place from early May through early November (Flock, 1973). With the exception of the Arctic in winter, an experienced observer of radar echoes from birds seldom examines a radar screen for long without seeing some evidence of bird activity, though not necessarily extensive activity. Much of the radar clutter due to birds, however, is dismissed by radar personnel in the United States as AP (standing for anomalous propagation, which condition tends to cause more ground and sea clutter than otherwise).

In general bird migration, especially of passerine species, tends to take place more at night, often shortly after sunset, than in the day. Bird activity for feeding purposes on the other hand is more often greater during the daytime. Wintering waterfowl and cranes tend to fly out to feeding areas in the morning, ducks often returning quickly and making another flight towards evening but geese and cranes perhaps staying away from resting areas for a good part of the day.

EFFECTS OF BIRD ECHOES

Assuming that bird echoes are of common occurrence, what effect can they have on radar system performance and how can any possibly adverse effects be eliminated or ameliorated?

A principal adverse effect is that which any clutter signal may have, namely that of obscuring or reducing the ability to detect desired signals. Targets such as aircraft and ships may be difficult to detect amongst bird clutter, and radar return from the land, vegetation, ocean waves, and ocean ice may be obscured by and contaminated by bird echoes, especially if the investigators are not aware of and alert to this possibility. Figure 7 shows a DEW radar screen at Oliktok, Alaska at a time of heavy bird migration. It is obvious that it would be difficult to detect a small aircraft among all the birds.

A related effect is that the echo from a flock of birds may be confused with an echo from some other object. Many a fighter plane has been scrambled to investigate a suspicious echo and found the echo to be caused by a flock of birds. Helicopters and small aircraft can be difficult to distinguish from birds because their velocities and radar cross sections are similar to those of flocks of birds. Sea clutter sometimes has a spiky nature that makes it difficult to distinguish sea and bird clutter. Atmospheric conditions that cause signals to be received from greater ranges than normal tend to cause greater than normal ground and sea clutter and high levels of bird clutter as well, as birds are commonly found at relatively low altitudes. Mixtures of the

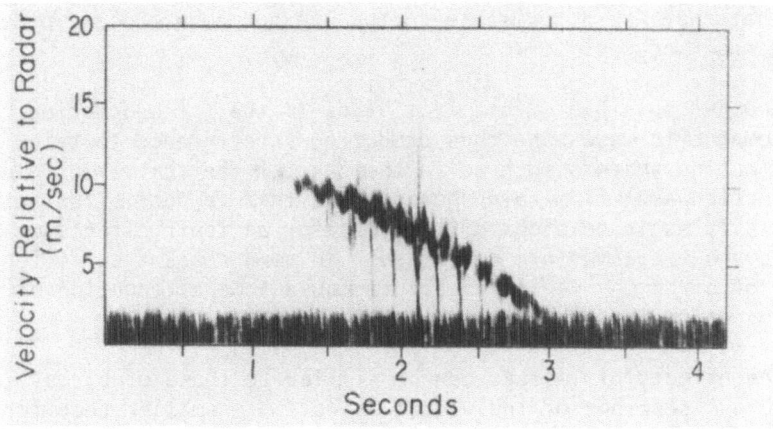

Fig. 5 Doppler spectrum of a shoveler (Spatula clypeata),
Boulder, Colorado.

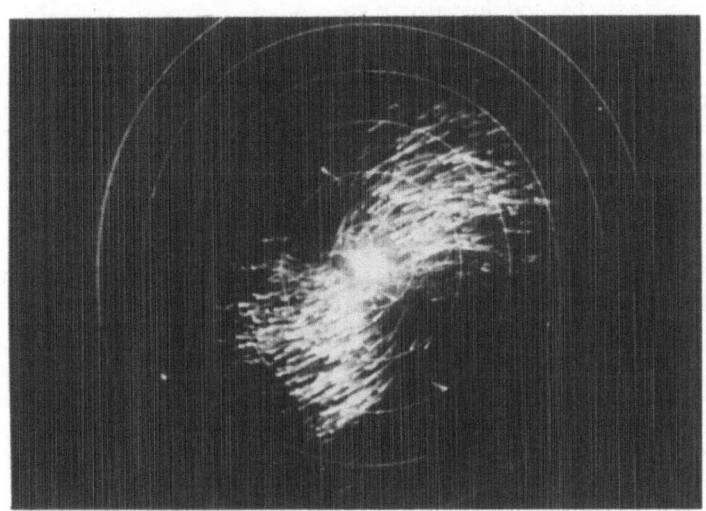

Fig. 6 Passerine bird migration as recorded by scan converter at
Houston Federal Aviation Administration Center, 2326 CDT, 4 May
1968. The MTI-video signals are from the New Orleans, Louisiana
radar and depict a reverse migration to the southwest. (geomagnetic
north up). The memory provided by a storage tube, rather than a
time exposure, is responsible for the formation of the streaks.

different types of clutter can be very confusing and difficult to identify. Although birds commonly fly in the 0 to 2000 m height range, shorebirds may fly at altitudes as high as 6000 m or more, such that they are not visible to the eye but nevertheless provide strong radar echoes.

Another possible effect from birds is that of scattering electromagnetic waves and thus producing interference to tele-communication systems much as in the case of the scatter of elec-tromagnetic waves from rain showers. In this respect birds act essentially as large blobs of water as far as their effects on electromagnetic waves are concerned. In some cases a sufficient number of birds may be in the air to cause some attenuation of microwave signals.

The effects of insects can be similar to those of birds. The radar cross sections of individual insects are smaller than for birds, of course, and insects are usually not strong fliers and tend to move with nearly the same velocity as the wind. The amplitude spectra of insect echoes should contain higher fre-quencies than for birds because of their smaller wings and higher wingbeat rates. Hummingbirds, however, have wingbeat rates in the 30 to 80 Hz range and overlap in wingbeat rates can be expected between insects and hummingbirds.

What are the possible measures for reducing the effects of bird echoes (or insect echoes)? It is commonly suggested that MTI (moving-target-identification) be used to eliminate bird echoes but MTI video is actually often the desirable type for observing and recording bird echoes. MTI video velocity thresh-olds normally permit recording bird echoes except for the birds which have essentially zero radial velocity, as is the case for any kind of moving target. STC (sensitivity-time-control) circuit-ry has been widely used to suppress bird echoes, and it does help but can not accomplish the purpose entirely. Some degree of STC is generally desirable whether one is interested in recording or suppressing bird echoes; a sufficiently strong STC action can suppress bird echoes to some degree. However, the echo from a flock of birds can be stronger than the echo from the largest aircraft, as in the case of Fig. 8 depicting migration across the Bering Strait (Flock, 1972). Thus elimination of all bird echoes on the basis of amplitude alone could well result in the elimination of all echoes of any kind.

If measures such as MTI and STC are not completely effective in combating bird clutter, what approach can be taken? To answer this question requires returning to the subject of the identifica-tion of bird echoes, as before bird echoes can be eliminated they must be identified and distinguished in some way from other echoes.

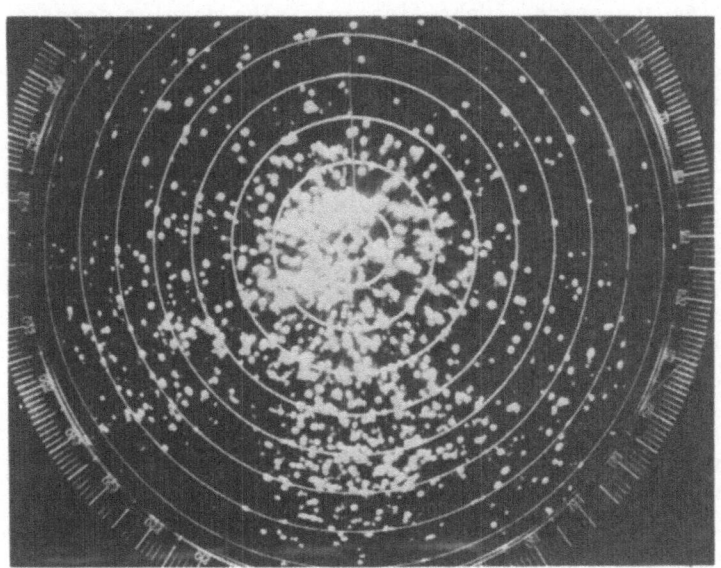

Fig. 7   Intense bird activity at Oliktok, Alaska: 40 nautical
mile range, 5 nautical mile range marks, one-rotation exposure,
geographic north up, 0740 ADT, 2 June 1972.

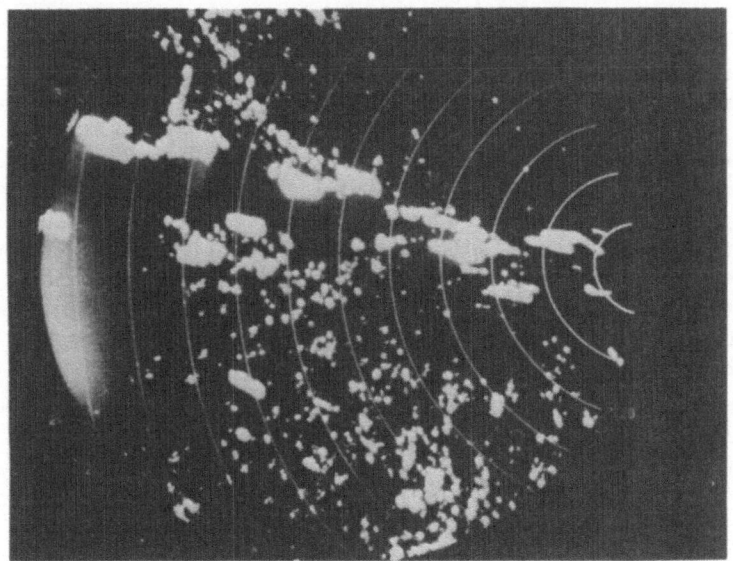

Fig. 8 Large flocks of sandhill cranes (<u>Grus canadensis</u>)
migrating across Bering Strait from Alaska to Siberia, off-
centered PPI display, 5 nautical mile range marks, 5-min time
exposure, geomagnetic north up, 2008 ADT, 10 May 1970.

The spectra of the amplitude fluctuations, which spectra show the fundamental frequencies and harmonics of the wingbeat rates, appear to provide the needed identification. The doppler spectra of echoes from birds may also be useful for this purpose. Obtaining these spectra requires dwelling on the targets for sufficient lengths of time, (ideally for a second or more). von Schlachta (these proceedings), however, has shown that the ratio, $\sigma/\bar{a}$, where σ is standard deviation and $\bar{a}$ is mean value, is often a useful criterion for distinguishing echoes. This ratio can be obtained by a continuously rotating antenna which dwells on a particular target for only a short period of time. Obtaining the spectra is desirable whenever feasible, that is, whenever it is possible to dwell on or track targets for sufficient lengths of time, but such a mode of operation is not always possible. Both the spectral and $\sigma/\bar{a}$ techniques have virtue, each being particularly applicable under certain conditions. The two techniques are not mutually exclusive, however, and an ideal procedure would be to obtain both types of data. Such data plus analysis of the movement and densities of echoes could be the basis for a sophisticated data processing system for automatically recognizing and eliminating bird echoes.

It should be pointed out, however, that radar echoes from birds contain information which can be of value in helping to minimize the hazard of collisions between birds and aircraft and for environmental monitoring purposes. Thus systems for processing radar echoes should in most cases at least allow the option of retaining and recording echoes from birds.

An effort has been made here to stress that radar echoes from birds are common and widespread and that they can affect radar system performance, but some caution is advisable lest the number of birds present at a given time be over-emphasized. Upon examining a radar PPI screen that is virtually obliterated by bird echoes, which condition occurs not infrequently and may be serious in its own right, one might well assume that the air is full of birds. But the density of birds required to produce the effect is not necessarily as high as might be assumed, especially in the case of a surveillance radar which has a relatively large vertical beamwidth and which therefore records bird echoes from over a range of heights. An observer might well step outside and search for birds with binoculars and see nothing at all at a time when the scope is full of birds.

CONCLUSION

Radar echoes from birds are a common and widespread phenomenon and a major cause of radar clutter. A few bird echoes on an otherwise clean radar screen can be readily recognized and accounted for, but the use of sophisticated data processing systems may be required to effectively combat severe clutter from birds. The utility of bird echoes should be recognized, however, and the option of retaining bird echoes should generally be provided.

Although considerable information about radar echoes from birds is available in the literature, many persons are still unaware of radar echoes from birds or consider them to be of infrequent occurrence, much as was the case when Bonham and Blake (1956) wrote their early very interesting paper on the subject. They pointed out that "the odd thing about this phenomenon is that it has apparently gone unrecognized by many who have seen it." This condition apparently results in part from the tendency to look only for what one is interested in and to ignore, for example, all the other blips that appear on a radar screen that one does not recognize or have a developed interest in.

REFERENCES

L.L. Bonham and L.V. Blake, "Radar echoes from birds and insects," Sci. Monthly, vol. 82, pp. 204-209, Apr. 1956.

E. Eastwood, Radar Ornithology. London, England: Methuen & Co., 1967.

W.L. Flock, "Radar observations of bird migration at Cape Prince of Wales," Arctic, vol. 25, pp. 83-98, June 1972.

_____, "Radar observations of bird movements along the Arctic Coast of Alaska," The Wilson Bull., vol. 85, pp. 259-275, Sept. 1973.

Flock, W.L. and J.L. Green, "The detection and identification of birds in flight, using coherent and noncoherent radars," Proc. IEEE, vol. 62, pp. 745-753, June, 1974.

J.L. Green and W.L. Flock, "Characteristics of radar echoes from birds in flight," presented at the 1972 USNC-URSI Spring Meet., Washington, D.C., Apr. 13-15, 1972.

E.W. Houghton, Use of Bird Activity Modulation Waveforms in Radar Identification, Malvern, Worcs., England: Royal Radar Establishment, June 6, 1972.

_____, "Highlights of the NATO-Gibralter bird migration radar study," Paper to Bird Strike Committee Europe, Paris Conference, May, 1973. Malvern, Worcs., England: Royal Radar Establishment, 1973.

P A R T   I I I

WORKSHOP REPORTS

STATISTICAL DESCRIPTION OF TARGETS AND ENVIRONMENT INCLUDING ATMOSPHERIC EFFECTS

W.L. Flock, H.S. Hayre, R.A. Helvey, A.G. Kjelaas,
A. Plaisant, D. Rauch, R.C. Runnels, K. von Schlachta,
G. Tacconi

1. INTRODUCTION

A random phenomenon can be described by its probability density function (PDF). The specification of the first moment (mean value), second moment (variance), and possibly the third moment is also useful and highly desirable (see for example [1, 2]). Other quantities of practical usefulness are correlation distance and correlation time. Random processes may be treated as stationary or nonstationary and it is desirable to specify the degree of stationarity and to state the ergodicity if the process is treated as stationary. In the case of nonstationary anisotropic phenomena, the four-dimensional structure function may be needed [3].

The radar cross section (RCS) of a target or clutter is a quantity of major importance to radar operations. Normally the cross section for monostatic operation is specified, but cross sections can also be specified for bistatic geometries. In order to adequately describe radar cross sections and to detect and identify targets in the presence of the environment and/or clutter, the specification of the probability density functions and/or the first, second, and possibly third moments is needed, as mentioned previously for random phenomena in general. Both static and dynamic characteristics and both temporal and spatial variations of targets must be taken into account.

2. TARGETS

The statistical description of targets is dependent upon the size of the

Jeske (ed.), Atmospheric Effects on Radar Target Identification and Imaging, 439–446.

volume element considered and the dwell time as well as upon the characteristics of the target itself.

2.1. Range Cells

Often the azimuthal and elevation angle beamwidths of a radar system are
fixed, and attention is concentrated on the location and length of the range
cells (or range gates or bins). If the extent of a range cell is greater than
the length of the target, clutter then also tends to occur over a distance
greater than the length of the target.

2.2. Dwell Time

A continuously rotating surveillance radar dwells on a particular target for
only a short time and thus scanning employs a very short time window for
sampling and processing (≈ 50 msec). One of two scattering models, fast
and slow [4] may apply, depending on the carrier frequency and the use of
frequency agility. In contrast, the dwell time, or time of observation, may
be several seconds or longer for a tracking radar or a radar that points in a
fixed direction for a period of time.

2.3. Radar-Cross-Section (RCS)

The backscatter characteristics of a target are strongly dependent on aspect
angle since the transmitter wavelength is smaller compared to its size.
Typical values characterizing the RCS should be the median value of RCS
(which is the same as the fiftieth percentile value), its second moment, and
perhaps the third moment. These should be calculated by averaging over
aspect angle windows. ($10°$ averaging cells are used in [5] for plotting
experimental data.)

2.4. Target Noise

Since the echo from a target is the vector sum of contributions from various
parts of the target, its magnitude and phase fluctuate with time and their
probability density functions (PDF) are dependent on scanning time, the
carrier frequency, and any use that is made of frequency agility. The PDF
of amplitude often approximates a chi square distribution with 1 to 10
degrees of freedom [6, 7, 8].

The fluctuation in amplitude of man-made targets may fall in either low or high frequency ranges. Yaw, pitch and roll motions of the target, as well as atmospheric effects contribute to the low frequency component, whereas random and periodic modulations due to rotating and vibrating parts contribute to the high frequency components.

Doppler frequency is the derivative of phase, and targets can be described either in terms of doppler frequency or phase. The use of doppler spectra (or the PDF's of doppler frequency) is common. Such spectra or PDF's may be normally distributed in ground-to-air observation of airplanes [9] or show an asymmetric distribution when derived from an airborne doppler velocity measuring system [10, 11, 12]. Using a tracking-radar the PDF's of long-term phase fluctuations tend to have a symmetrical exponential distribution [13].

2.5. Clutter Characteristics

What is clutter for one party may be the desired signal for another and vice versa. The PDF of clutter is a function of the resolution cell employed and the type of clutter. Particular types of clutter have been shown to have chi square, log normal, or Weibull PDF's. The power spectral density of an echo processed in a coherent system has its maximum at a mean doppler frequency f_{do}. Some general observations on the f_{do}'s for different types of clutter are given below.

- Ground clutter, in the case of ground-based observations, has zero f_{do}.

- The f_{do} of weather clutter represents the radial component of wind speed. With larger cells one integrates over windshear [14].

- Clutter due to chaff is similar to weather clutter [14]. (Chaff can be used to deduce wind speed when hydrometeors are not available.)

The power spectral density of the echo from a coherent source may have a Gaussian PDF with the second moment or variance depending upon the type of clutter [14, 15]. For considering the second moment, one obtains a

- low value for ground clutter from rocks or towers, with somewhat larger value for ground covered by vegetation;

- a value dependent on the resolution cell and the direction of the radar beam with respect to wind in the case of sea clutter;

- a larger value for sea and weather clutter than for ground clutter.

Radar echoes can be received from individual birds but are most commonly from flocks or aggregations of birds. The intensity of bird echoes (RCS) and its second moment depend on the type (size) of birds, the density of birds in a flock and the number of different flocks in one resolution cell. The wing-beat of birds is known to cause characteristic modulation of the amplitude and doppler frequency of echoes from birds [16], the amplitude modulation occurring mostly in the 2 to 20 Hz range. Birds have ground speeds ranging from about 20 to 70 knots and are most commonly encountered at heights of 7000 meters or more (larger birds such as waterfowl and cranes may consti-tute a serious hazard to aircraft).

Clutter due to insects may need to be considered at times.

3.   TARGETS AND THE ENVIRONMENT

Targets are generally complex sets of aspect-dependent scattering centres and, when turning as in the case of submarine [17] and aircraft, can be represented by a time-varying filter [18]. Similarly the environment con-sisting of the atmosphere for the radar case and the underwater medium for the sonar case, may also be represented by another time-varying filter. Since these two phenomena, the scattering by the target and the effect of the medium, are generally assumed to be statistically independent, the target echo may then be obtained by convolving the transmitted signal with the convolved joint transfer function of the target and the environment. The transfer-function approach assumes linearity of the processes involving the target and the media, and when it is invalid, one must then consider the non-linear interaction between these instead of employing simple con-volution.

4.   DESCRIPTION OF THE MEDIUM BETWEEN RADAR AND TARGET RELEVANT TO RADAR APPLICATIONS

The performance of a radar, whether we are dealing with an air-to-air, air-to-ground, or ground-to-air radar, will depend upon the characteristics of the medium (the atmosphere), between the target and the radar.

4.1. Refractive Index

The basic parameter describing the radar atmosphere is the refractive index n . This index is a function of pressure P , temperature T and water vapour pressure e , and it varies in both time and space, and accordingly can be treated as a statistical random variable with a mean value $\bar{n}(r)$ and a fluctuating component $n'(\bar{r}, t)$. Variations in both space and time of $\bar{n}$ and n' will cause phase errors and time delays, as well as amplitude errors. In order to describe these phase- and amplitude errors one needs to know the statistical characteristics of the refractive index over distance and time relevant to radar applications. For that no systematic information is available although some scattered information on the variability of both $\bar{n}$ and n' with height and time does exist for certain geographical areas for instance Ref.[19, 20, 21].

The variability of n' is well described using the structure function [3] which includes both the intensity of refractive index fluctuations C_n^2 and the rate at which the turbulent energy is decaying with the frequency and the scale of turbulence.

Recent work using the structure function and calculating phase variations for side-looking radar is encouraging. Models do exist for the variability of C_n^2 with height, and further information will certainly be gained by the application of different remote sensing techniques or aircraft measurements.

Based on conventional radiosondes, some information about the phase decorrelation distance can also be gained [23].

Data from conventional radiosondes can be used to estimate frequency-of-occurrence of refractive layer characteristics such as altitude, intensity and thickness. Statistics on occurrence of elevated ducts are of particular interest and have been compiled for various localities.

Significant errors result, however, from resolution and accuracy limitations in the radiosonde system and because of the methods in reducing and encoding the data. Substantially better information can be obtained by the use of airborne refractometers, but at greater expense and with less ready availability than from radiosondes.

4.2. Ducts

Forecasts of elevated ducting are possible with considerable success for periods of at least 24 hours and are being prepared routinely for such regions as Northern Europe and the southern California coastal area. (In Northern Europe they are available for a 10 years period.)

4.3. Turbulence

Theoretical models of phase and amplitude fluctuations, caused by turbulence, do exist but as it is not possible to separate them from the effects produced by layers and turbulence, a model describing the effects on the radar performance should include both.

4.4. Rain

In addition to turbulence the existence of rain introduces additional effects. Although rain has been subject to extensive theoretical and experimental analysis, very little information is available on such phenomena as the statistical behaviour of rain cells and depolarization models for both back- and forward scatter.

It is recommended that more work on three-dimensional drop shapes, drop-size distribution, as well as drop states be actively encouraged.

5.   CONCLUDING REMARKS

It is fair to say that much work on modelling and simulation remains to be done before a final conclusion regarding the atmospheric effects on radar target identification and imaging can be reached.

Although models of target and clutter behaviour seem to be known in literature, for any special application additional data are needed concerning the PDF's and/or spectra of the amplitude and phase or doppler frequency variations as well as depolarization properties of the echoes from various types of targets and clutter.

REFERENCES

1.   G.A. Sitton: Acoustic Segmentation of Speech; Rept. No.ORO-2572-18, AEC Contract AT-(40-1)-2572, June 1969

2.   C. Stewart: Design of Real Time Statistical Time-Series Analyzer for Speech Segmentation; M.S. Thesis, E.E. Dept. University of Houston, Texas, USA, 1970

3.   V.I. Tatarski: The Effects of the Turbulent Atmosphere on Wave Propagation; Israel Program for Scientific Translations, Jerusalem 1971, IPST Cat.No. 5319

4.   P. Swerling: Probability of Detection for Fluctuating Targets; IRE Trans. Information Theory, Vol. IT-6, pp. 269-308, April 1960

5. I.D. Olin, F.D. Queen: Dynamic Measurement of Radar Cross Section; Proc. IEEE Vol. 53, pp. 954-961, August 1965

6. J.D. Wilson: Probability of Detecting Aircraft Targets; IEEE Trans. Aerospace Electr.Syst., Vol. AES-8, No. 6, pp. 757-761, November 1972

7. J.P. Reilly: On the Statistical Representation of Targets for Detection Studies; IEEE Trans. Aerospace and Electr.Syst., Vol. AES-5, May 1969 (Correspondence)

8. P. Swerling: Recent Development in Target Models for Radar Detection Analysis; AGARD Conference Proceedings No. 66, Istanbul, Turkey, 1970 (No. 7)

9. K. von Schlachta: Remarks on Target Models for the Design of Radar Systems; IEEE International Radar Conference, Arlington, VA, 1975, IEEE Publication 75 CHO 938-1 AES, pp. 440-445

10. R.F. Broderick, H.S. Hayre: Doppler Return from a Random Rough Surface; IEEE Trans. Aerospace and Electr.Syst., Vol. AES-5, pp. 441-449, May 1969

11. H.S. Hayre: Doppler Tracking Radar Systems; J.Inst.Eng.(India), Vol. 51, No. 10, ET 3, pp. 169-172, May 1971

12. M.S. Sohel, H.S. Hayre: Doppler Radar Return from Two-Dimensional Random Rough Surfaces; IEEE Trans. Geoscience Electronics, Vol. GE-10, pp. 33-47, January 1972

13. J.H. Dunn, D.D. Howard: Radar Target Amplitude, Angle, and Doppler Scintillation from Analysis of the Echo Signal Propagating in Space; IEEE Trans. Microwave Theory Tech., Vol. MTT-16, pp. 715-728, September 1968

14. F.E. Nathanson: Radar Design Principles; McGraw-Hill, Inc., 1969

15. K. von Schlachta: The Use of Target and Clutter Data for Different Methods of Discrimination between Targets and Unwanted Clutter; AGARD Lecture Series No. 59, paper 2(d), 1973

16. W.L. Flock, J.L. Green: The Detection and Identification of Birds in Flight Using Coherent and Noncoherent Radars; Proc. IEEE, Vol. 62, No. 6, pp. 745-753, June 1974

17. K.A. Sostrand: Measurement of Coherence and Stability of Underwater Acoustic Transmission; NATO-ASI on Signal Processing, Enschede, Netherlands, 1968

18. P.A. Bello: Characterization of Randomly Time Variant Linear Channel; IEEE Trans. on Communication Systems, Vol.CS, No. 4, pp. 360-393, December 1963

19. H.A. Panofsky: Spectra of Atmospheric Variables in the Boundary Layer; Radio Science, Vol. 4, No. 12, pp. 1101-1109, 1969

20. N.K. Vinnichenko, J.A. Dutton: Empirical Studies of Atmosphere Structure and Spectra in the Free Atmosphere; Radio Science, Vol. 4, No. 12, pp. 1115-1126, 1969

21. D.T. Gjessing, A.G. Kjelaas, E. Golton: Small-Scale Atmospheric Structure Deduced from Measurements of Temperature, Humidity, and Refractive Index; Boundary-Layer Meteorology, No. 4, pp. 475-492, 1973

22. D.M. Willis: Phase Variation at Millimetric Wavelengths on an Earth-Space Path through Model Atmospheres; Electronics, Vol. 10, No. 14, July 1974

23. D.T. Gjessing: On the Relation between Atmospheric and Radar Signal Parameters; NATO ASI: Atmospheric Effects on Radar Target Identification and Imaging, Goslar, F.R. Germany, 1975

24. L.J. Porcello: Turbulence-Induced Phase Errors in Synthetic Aperture Radars; IEEE Trans. on Aerospace and Electr.Syst., Vol. AES-6, pp. 636-644, September 1970

25. M.C. Thompson Jr., H.B. Janes: Measurement of Phase Front Distortion on an Elevated Line-of-Sight Path; IEEE Trans. on Aerospace and Electr.Syst., Vol. AES-6, pp. 645-656, September 1970

26. W.M. Brown, J.F. Riordan: Resolution Limits with Propagation Phase Errors; IEEE Trans. on Aerospace and Electr.Syst., Vol. AES-6, pp. 657-662, September 1970

SHORT PULSE, FM/CW AND CHIRP RADARS

R. G. Taylor, A. Alongi, J. T. Hall, N. Klint Hansen,
D. Thomson, E. Topuz, A. T. Waterman

1.   As a starting point in its discussions on the effect of
atmospheric factors on short pulse, FM/CW and chirp radars the
working group assigned the same fractional bandwidth and mean
power to each type. This was necessary to make them equivalent
and permit a valid comparison to be made. In view of the amount
of experimental evidence below 10 GHz and the practical difficulties of system implementation beyond 90 GHz these two frequencies
were made the lower and upper limits of the frequency range. The
water vapour and oxygen absorption bands at 22 GHz and 60 GHz
were excluded since radar design would normally avoid these
regions. A bandwidth of one gigahertz within the range was taken
to be representative for the purposes of estimation.

2.   The main difference between the radar systems is the instantaneous spatial extent, or pulse length, of the transmission.
Fluctuations in the propagation medium on a timescale of a few
milliseconds or less were not considered to be significant and it
was concluded that the three systems would be comparably affected
by atmospheric disturbances. Further discussion was concerned
with the distortion to be expected in the generalised, equivalent
bandwidth, system.

3.   Across antenna apertures of the order of 10-100 wavelengths,
which are relevant to the radar types being considered, phase
perturbations are considered to be negligible. For larger
antenna apertures reference should be made to the report of
Working Group 'E' (Synthetic Aperture Radar).

4.   In the design of a high resolution radar it is essential to
produce and maintain a specified waveform with a high degree of

Jeske (ed.), Atmospheric Effects on Radar Target Identification and Imaging, 447–449.
All Rights Reserved. Copyright © 1976 by D. Reidel Publishing Company, Dordrecht-Holland.

accuracy. Any mechanism which alters the relative magnitudes and phases of the spectral components will degrade the resolution of the system. The refractive index of the clear atmosphere itself is not significantly dispersive, except near resonance absorption lines (which we have excluded above). The fluctuations caused by turbulence in the clear atmosphere have a weakly dispersive effect, but over a fractional bandwidth of no more than 10 per cent (a 1 GHz band at 10 GHz carrier frequency) this dispersion is expected to be minor. Frequency covariances greater than 0.9 and phase excursions less than 40 degrees should be typical. It appears that dispersive effects in the clear atmosphere are most likely to be significant in cases of duct propagation, or multi-path from layers. This is felt to be an important area where knowledge of the dispersive effect of these atmospheric donditions is insufficient and where future effort should be directed.

5.   During considerations on the nature of the atmospheric inhomogeneities it became apparent that meteorological measure-ments can now be made with a resolution which exceeds that commonly used for radar design and evaluation. These measurements have resulted in greatly improved models, some of which have the potential of reducing range and pointing errors obtained with the routines used to predict radar performance.

6.   The relative magnitude of the radar backscatter from atmos-pheric phenomena with respect to a target echo which has undergone attenuation has not been adequately measured in this frequency range. Precipitation, clouds and clear air turbulence, where it is significant, could conceivably mask a target and reduce the probability of obtaining adequate echo power for identification and imaging purposes. Quantitative estimates are needed of the effective radar cross-sections for these atmospheric features.

7.   In the light of these considerations the following recommen-dations are made.

7.1 Investigations should be made into the dispersion over giga-hertz bandwidths caused by multipath propagation and by ducts.

    a.   In this connection investigations should also be made of the scale of roughness, the horizontal extent and temporal variation of refractive layers, particularly in stable atmos-pheric conditions.

2.   Existing dynamical models and detailed observations of the atmosphere's planetary boundary layer should be evaluated for radar applications and then utilized to improve predictions of the space-time refractive index field and associated propagation phenomena.

7.3 In the range of frequencies 10 to 90 GHz the magnitude of backscatter from precipitation, clouds and, possibly, clear air turbulence should be investigated and related to attenuated target echoes.

SYNTHETIC APERTURE RADAR

MEMBERS: S. R. Brooks, P. Cabion, K. G. Corless (Chairman),
W. Gabsdil, D. B. Hodge, S. Marder, G. Nilsson, R. Olsen,
M. C. Thompson, R. Trotter

INTRODUCTION

A SAR is subject to phase errors across the aperture due to
fluctuations in atmospheric refractive index caused by turbulence
in the region between the aircraft position and the mapped
terrain. These errors will affect radar azimuth processing by
raising sidelobe level and degrading resolution and pointing
accuracy.

Some work has already been done to predict the extent to
which SAR will be affected (references 1-4 and the papers by
Marder and Gjessing at this ASI), and the indications are that,
while these effects may only occasionally degrade the performance
of some current radars, future systems will be increasingly vul-
nerable to propagation effects as resolution and/or ranges
increase.

This Workshop did not concern itself with range resolution
effects as they are independent of the azimuthal effects and were
the province of another Workshop anyway.

CURRENT STATE OF THE PROBLEM

Tatarski has studied the case of plane wave propagation
through turbulence and has found that under certain conditions
the process can be described by the phase structure function
$D_\phi(\mu)$. Porcello[1] has extended this analysis to the problem of
atmospheric phase errors on SAR and has found that in this case

Jeske (ed.), Atmospheric Effects on Radar Target Identification and Imaging, 451–454.
All Rights Reserved. Copyright © 1976 *by D. Reidel Publishing Company, Dordrecht-Holland.*

the phase structure function is of the form

$$D_\phi(\mu) = b^{5/3} \mu^{5/3}$$

where μ is the length of the synthetic aperture

$$\text{and} \quad b^{5/3} = 4a \, k^2 \, (\csc \psi)^{8/3} \, I/R^{5/3}$$

where a = constant given by Tatarski

$\quad\quad\quad$ $k = 2\pi/\lambda$ and λ = wavelength

$\quad\quad\quad$ R = range from aperture to target

$\quad\quad\quad$ ψ = elevation angle of radar

$$I = \int_0^{h_o} C_n^2(h) \, h^{5/3} \, dh$$

$\quad\quad\quad$ h_o = altitude of radar

$\quad\quad\quad$ C_n = refractive-index structure constant

Brown and Riordan[2] have obtained a closed form expression for the resolution limits for a SAR operating through turbulence when the phase structure is of the form $D_\phi \sim \mu^n$. Thus, the work of Porcello and of Brown and Riordan permit the determination of the limiting resolution of a SAR when $C_n(h)$ can be determined. Though selected values of C_n are available, the validity of existing estimates of the profile $C_n(h)$ appears to be highly controversial.

Microwave measurements, such as those of Thompson and Janes[3,4], permit the direct determination of D_ϕ and so by-passing the profile of C_n. Their measurements combined with the analysis of references 1, 2 lead to estimates of limiting synthetic aperture lengths of the order of 100 to 500 m. They incidentally lead to values of the integral I which are some three orders of magnitude larger than Hufnagel's model would give for the values of C_n in the optical region. The extrapolations of these measurements to SAR performance at other altitudes and locations are tenuous. There have been no attempts at this time to assess the effects of atmospheric propagation on SAR imagery, because these effects have been expected to be small compared to other error sources (notably uncompensated cross-track motions) for the synthetic aperture lengths which have been employed.

RECOMMENDATIONS FOR FURTHER WORK

As a result of a considerable amount of Workshop discussion, reading of relevant papers, personal experience of a number of the members and consultation with others in the ASI the following recommendations were agreed upon.

1   To carefully examine the imagery from current and future SAR for any evidence of propagation-induced distortion.

2   To develop improved models of the atmosphere.

3   To improve the analytic methods for interpreting the electro-magnetic consequences of these models (including amplitude and phase errors across synthetic apertures of varying lengths) leading to theoretical predictions of degradations in azimuth sidelobes, resolution, pointing errors, etc. (Some experimental data on amplitude fluctuations does exist. It had not been included in the analyses performed hitherto because it was then regarded as of less significance than the phase data. The report of the Holography Workshop also supports this view.)

4   To continue further microwave measurements of D_ϕ using fixed-path propagation, especially in diverse climatic regions and in association with measurements of the C_n profile. Gjessing's proposal to use available radiosonde data should be validated in this connection. An assessment should also be made of the value of using a water absorption radiometer, and/or local water vapour measurement, co-located with the receiver array in estimating the integrated water vapour content along the line of sight. (Distance measuring equipment used to measure ranges of up to 10 Km to a precision of a few mm employs range correction factors based on localised water vapour measurements made at each end of the range.)

5   To explore the use of a satellite-borne microwave beacon in a low orbit earth-stabilised satellite to derive phase and amplitude statistics as a function of elevation angle. Such an experiment would require a beacon oscillator at a frequency high enough to avoid ionospheric effects, and with a good short-term phase stability. Primary dependence should be placed on the use of existing ground terminals, each of which should have at least 2 antennas to provide spatial-variance data. The Space lab may be a convenient platform for the beacon launch.

6   To consider the use of a radiometer, as referred to in 4 above, as an atmospheric sensor in a joint installation with a SAR to assess the effects of atmospheric propagation on SAR

performance.   The radiometer line-of-sight would be coincident in azimuth with that of the SAR but would have to be elevated above the horizon.

In all the experimental work mentioned above it is essential to complement the radar measurements with simultaneous atmospheric soundings using such techniques as are available.

In conclusion it is strongly recommended that the contact between meteorologists and radar designers, which has been so well established by this ASI, be maintained in the future as it is now abundantly clear that as radar performance continues to improve, designers must increasingly reckon with the vagaries of the propagation medium.

REFERENCES

1   L J Porcello, "Turbulence-Induced Phase Errors in Synthetic Aperture Radars", IEEE Transactions Aerospace Electronic Systems, Vol AES-6, pp 636-644, Sept 1970.

2   W M Brown, J F Riordan, "Resolution Limits with Propagation Phase Errors", ibid, pp 657-662.

3   M C Thompson, Jr; H B Janes, "Measurements of Phase-Front Distortion on an Elevated Line-of-Sight Path", ibid, pp 645-656.

4   H B Janes, M C Thompson, Jr, "Comparison of Observed and Predicted Phase-Front Distortion in Line-of-Sight Microwave Signals", IEEE Trans Ant and Prop, Vol 21, pp 263-266, March 1973.

INFLUENCE OF ATMOSPHERIC PROPAGATION EFFECTS ON HOLOGRAPHIC IMAGING TECHNIQUES

M.E. Bechtel, D.T. Gjessing, G. Graf, O. Loevhaugen,
K. Magura, E. Mehlum, A. Nania, M. Vogel, B. Yazgan

1. INTRODUCTION

Several decades ago attempts of microwave imaging had already been made by using lenses or reflectors [e.g. 1, 2, 3, 4]. During the last ten years an increasing interest in the application of holographic techniques for microwave imaging has been noted [e.g. 5, 6, 7, 8, 9]. Regardless of the specific signal processing methods used, a holographic imaging system shall be understood as a device which measures in a first step the relative amplitudes and phases of the scattered object field over a real aperture in one or two dimensions. This definition is used here to exclude synthetic aperture systems, the subject of another workshop (E). Contrary to the optical case in which intensity patterns of superposed scattered and reference waves must be used to record phase information, the microwave frequencies permit a more direct measurement of the relative time variable phases as well as amplitudes. In a second step an image is reconstructed from these data, either via a scaled hologram by a coherent optical system (an analog processor), or by a digital computer and a display unit. Without going into further details it should be mentioned that nowadays it becomes possible to use holographic imaging techniques in real time [e.g. 10].

This workshop report deals with the influence of atmospheric propagation effects on holographic imaging with microwaves. To begin with, the inherent limitations of an ideal system in free space are given without consideration of atmospheric influences. Then the atmospheric influences on coherence degradation and attenuation due to the refractive index structure, including fog and precipitation, are discussed, and conclusions and recommandations are derived from the discussion.

Jeske (ed.), Atmospheric Effects on Radar Target Identification and Imaging, 455–463.

2. LIMITATIONS OF AN IDEAL HOLOGRAPHIC IMAGING SYSTEM

The imaging process is essentially the convolution of the reflectivity distribution on the object with the focused diffraction pattern (point source response) of the imaging aperture. The imaging quality is often expressed by the lateral and longitudinal resolutions δ_{Lat} and δ_{Long} (dimensions of the focal region) according to the Rayleigh criterion [e.g. 9]. Approximate equations are (for distances R > D = aperture dimension)

$$\delta_{Lat} \approx \frac{\lambda R}{D} ,   \tag{1}$$

$$\delta_{Long} \approx 8\lambda \left(\frac{R}{D}\right)^2 \quad \text{for} \quad D < R << \frac{D^2}{\lambda} ,   \tag{2}$$

where λ is the wavelength and R the distance between object and aperture. Equation (1) and (2) show in a simple and drastic manner the inherent limitations of a (monochromatic) imaging system. δ_{Lat} increases with the distance. If, for example, a lateral resolution of δ_{Lat} = 1 m is required with an aperture size of D = 3 m at a wavelength of λ = 3 mm, the operation of the system is limited to distances below R = 1 km. Very limited longitudinal resolution δ_{Long} by the focussing effect is available, but only near the aperture.

Good range resolution for all ranges can, however, be obtained by broadband waveforms of frequency bandwidth B, giving a radial resolution $\Delta R \approx c/2B$, where c is the velocity of light.

A time duration τ of the waveform, used in a doppler spectrum analysis of the scattered wave, limits the doppler resolution to a frequency bandwidth of $\Delta f_{doppler.} \approx 1/\tau$ with a corresponding resolution in radial velocity $\Delta R = \Delta f_{doppler} \cdot \lambda/2$.

Further it should be mentioned that rather severe limitations are introduced by the mirror-like reflection properties of typical objects in the microwave region. The images look very different from optical images of the objects in diffuse illumination; bright spots occur only on those parts of the smooth surface where the phase is stationary. An improvement of the recognizability may, however, be obtained by incoherent [9] or coherent [15] superposition of partial images recorded at several illuminating and/or receiving aspect angles, which is, of course, equivalent to some sort of synthetic aperture realized through a relative change of aspect angle.

3. ATMOSPHERIC INFLUENCES

The propagation properties can formally be described by a complex refractive index $\underline{n}$ (x, y, z, t, f) as a function of the space coordinates x, y and z, the time t, and the wave frequency f. The effects of fog and precipitation can probably be integrated into an equivalent refractive index. The local refractive index is determined by the local state of matter, molecule concentrations, and temperature. Water vapour molecules, having a permanent dipole moment and a variable concentration, are of special importance. The exact de facto spatial fine structure of the refractive index can not be determined and a smooth average function with a statistically described superimposed granular structrue must usually be used in a model for computations.

The qualitative effects of such a structure on a holographic radar imaging system are

> a systematic delay of the waves (range error) due to an integral refractive effect along the propagation path, plus a delay jitter;

> a systematic phase front distortion (ray bending) due to a transversal refractive gradient, plus fine-scale phase-front distortions;

> loss due to absorption in molecules and hydrometeors;

> scattering produced by fine-scale refractive granularity, esp. in rain.

All these effects result in displacement, distortion or blurring, and weakening of the image, with some background (sum of "sidelobes" of scattering center images) enhancement.

These effects may become important for a given system if they begin to degrade the resolution limits of the ideal free space system.

It appeared to the workshop team that the phase front distortion due to the refractive granularity of the atmosphere was the most serious effect and the team concentrated therefore in the available time on a discussion of this effect.

3.1 Effects of trophospheric turbulence

Spatial and temporal variations of the tropospheric index of refraction introduce temporal and spatial fluctuations in the phases and amplitudes of the waves recorded in the holographic aperture after propagation through the turbulent atmosphere. Only random effects on phase will be discussed.

If the rms value $\Delta\phi_{rms}$ of the phase variations and their correlation length l_c are known, their effect on the image quality can be estimated by the results of antenna tolerance theory [11]. The phase distribution "roughness" over the aperture is equivalent to the surface roughness of an imaging mirror of the same aperture size.

If l_c is very small relative to the aperture dimension D, the roughness enhances the sidelobes in all directions, whereas a larger l_c/D increases the sidelobes near to the main lobe, and a very large $l_c > D$ causes angular jitter of an otherwise little distorted pattern. Further effects are directivity degradation, and beam broadening.

An intuitive semiquantitative discussion of the problem [12] assumes that the spatial random variation of the real refractive index n is a stationary process with a given Δn_{rms}. Time variations are assumed to be much slower than the recording time constant of the system. For two points in space, 1 and 2, with the distance d_{12} between them, we have

$$\overline{\Delta n_1} = \overline{\Delta n_2} = 0 \tag{3}$$

and

$$\overline{\Delta n_1^2} = \overline{\Delta n_2^2} = \Delta n_{rms}^2 . \tag{4}$$

The correlation between Δn_1 and Δn_2 as a function of d_{12} shall be

$$\overline{\Delta n_1 \Delta n_2} = \Delta n_{rms}^2 \cdot e^{-(d_{12}/l_c)^2} . \tag{5}$$

To facilitate the calculation, the medium will be considered to consist of isotropic cells or blobs of size l_c, each with a refractive index slightly different from the mean by an independent Δn. A single layer of blobs with thickness l_c in front of the aperture will result in a random effect on the phase of

$$\Delta\phi_{1,rms} = \frac{2\pi}{\lambda} \cdot l_c \cdot \Delta n_{rms} . \tag{6}$$

Along each propagation path of length L are L/l_c uncorrelated blobs. The total phase fluctuation is then

$$\Delta\phi_{rms} = \sqrt{\frac{L}{l_c}} \cdot \Delta\phi_{1,rms} = \sqrt{L \cdot l_c} \cdot \frac{2\pi}{\lambda} \cdot \Delta n_{rms} . \tag{7}$$

The correlation between $\Delta\phi$'s across the aperture will approximately obey an equation similar to (5)

$$\overline{\Delta\phi_1 \cdot \Delta\phi_2} = \Delta\phi_{rms}^2 \cdot e^{-(d_{12}/l_c)^2} . \tag{8}$$

We will, for simplicity, neglect the fact that rays from a point source in the propagation medium will diverge and project from nearby blobs a contribution on the aperture with a correlation length $> l_c$.

If the maximum rms phase difference value across the aperture is to remain smaller than, say, π, the condition

$$\sqrt{\overline{(\Delta\phi_1 - \Delta\phi_2)^2}} < \pi \tag{9}$$

results. This requirement is, after squaring, with equation (8), and $d_{12} = D$, equivalent to

$$\overline{\Delta\phi_1^2} + \overline{\Delta\phi_2^2} - 2 \cdot \overline{\Delta\phi_1 \Delta\phi_2} = 2 \cdot \Delta\phi_{rms}^2 \cdot (1 - e^{-(D/l_c)^2}) < \pi^2. \tag{10}$$

For small $(D/l_c) < 1$ the approximation

$$e^{-(D/l_c)^2} \approx 1 - (D/l_c)^2 \tag{11}$$

can be used, giving, with (7), the condition

$$2 \cdot \Delta\phi_{rms}^2 \cdot (\frac{D}{l_c})^2 = 2 \cdot \frac{L}{l_c} \cdot \frac{4\pi^2 D^2}{\lambda^2} \cdot \Delta n_{rms}^2 < \pi^2 \tag{12a}$$

or

$$\sqrt{\frac{2L}{l_c}} \cdot \Delta n_{rms} < \frac{1}{2} \cdot \frac{\lambda}{D} . \tag{12b}$$

If the aperture diameter D is larger than the correlation length l_c, the maximum rms phase difference saturates (the exponential function in (10) approaching zero) to $\sqrt{2} \cdot \Delta\phi_{rms}$. If we relate this result to a distance between independent sampling points of l_c, we get an estimate for the rms phase front tilt (angle fluctuation) $\Delta\beta_{rms}$, caused by the refractive index blob structure, of

$$\Delta\beta_{rms} = \frac{\sqrt{2} \cdot \Delta\phi_{rms} \cdot \lambda/2\pi}{l_c} = \sqrt{\frac{2L}{l_c}} \cdot \Delta n_{rms} . \tag{13}$$

This value should not be larger than half the diffraction limited angular resolution (beamwidth) of the aperture in free space as given in (1)

$$\Delta\beta < \frac{1}{2} \cdot \frac{\delta_{Lat}}{R} = \frac{\lambda}{2D} . \tag{14}$$

The resulting condition

$$\sqrt{\frac{2L}{l_c}} \cdot \Delta n_{rms} < \frac{\lambda}{2D} \qquad (15)$$

is identical to equation (12b), but different independent beam
tilts occur simultaneously over the large imaging aperture and
blurring results. A similar condition was obtained in [13] with
a slightly different numerical factor.

For reasonable values of $\Delta n_{rms} = 10^{-6}$, $l_c = 10$ m [12], and
$L = 2$ R = 10 km, we get the result

$$D < \frac{\lambda}{2} \cdot \sqrt{\frac{l_c}{2L}} \cdot \frac{1}{\Delta n_{rms}} = \lambda \cdot \sqrt{\frac{10^3 \text{ cm}}{8 \cdot 10^6 \text{ cm}}} \cdot 10^6 \approx 10^4 \lambda \; . \quad (16)$$

At first sight the above condition could lead to the conclusion
that for feasible aperture sizes (we exclude interferometer type
systems using partial apertures spread over a very large incomple-
tely filled aperture) tropospheric turbulence effects on microwave
holographic imaging are of very little importance. This may, at
the present state of development, be true for systems operating
in the decimeter and centimeter wavelength range. We must, however,
keep in mind the uncertainty of our "reasonable" assumptions used
in the derivation of (16), and leave a final quantitative conclusion until more detailed knowledge is available about the atmospheric refractive index fine-structure under different weather conditions. There may be also reasons to reduce the tolerance limit for
the phase variation in (9) from π to one or even less.

3.2 Effects of fog and precipitation

Very often efforts toward short-range imaging with microwaves or
millimeter waves are motivated by the fact that for the longer
wavelengths (in relation to the water droplets) more favorable
propagation properties exist under bad weather conditions than in
the optical or infrared regions. High power CW and pulsed sources
with rather high spectral purity have been built. Curves in [14]
show that about 1 kW CW power and 100 kW pulse power at 100 GHz
are possible, although such tubes may not yet be available for
practical systems. According to [14] the available power decreases
with increasing frequency f with a f^{-5}-law. The attenuation by
rain and fog in the atmosphere increases also with frequency and
reaches at 100 GHz the order of 10 db/km for heavy fog (2.3 g/m^3)
or rain (16 mm/h) [8]. The point here is that the trend toward
shorter waves from the standpoint of image resolution is counterac-
ted by an increase in attenuation, which can be overcome to some
degree by higher power and receiver sensitivity. The optimum design

of a radar imaging system follows therefore from a joint consideration of atmospheric propagation and transmitter/receiver state
of the art.

To some extent, the effect of scattering by precipitation on imaging can be reduced by range-gating techniques. Range-gating re-
moves the direct signals from the drops but cannot remove the wavefront distortion caused by the index-of-refraction variations
along the wave-path, and some effects of multiple scattering.

Water droplets in fog and rain will also effect the phase of a
microwave and the question arose at the workshop how large the
refractive effect might be. The very general result was that the
random phase fluctuations would probably be worse in fog and still
more in rain than for a clear atmosphere, but no quantitative
value for an effective refractive index of rain or fog could be
found.

4. CONCLUSIONS AND RECOMMENDATIONS

Holographic imaging with real apertures, especially for millimeter
waves, could be of interest for short range applications under
conditions of bad visibility, where optical and infrared systems
fail. It seems that decorrelation effects due to the turbulent
clear atmosphere can, at the present state of development, be ne-
glected in most cases. Wether this conclusion is also true in the
presence of dense fog or rain is not certain and requires further
study. The atmospheric effects on microwave imaging systems are
very similar to the effects on very large radiotelescopes for high
resolution sky mapping and for communication with satellites and
space probes. Valuable information should be found in the relevant
literature [e.g. 16].

A severe hindrance for the trend to shorter wavelengths is the
lack of millimeter wave sources with high power output needed to
overcome atmospheric attenuation.

The present state of the art of imaging microwave systems appears
to be such that atmospheric decorrelation effects are not the
dominating difficulty. Nevertheless, because a strong trend to
higher resolution exists, it is recommended

> that the interaction between atmospheric/meteorological ex-
> perts and radar experts be continued and intensified to ex-
> change existing knowledge;
>
> that the small scale refractive index structure of the atmo-
> sphere as a function of weather and climate conditions be
> studied, especially the refractive effects of fog, clouds,

and rain; holographic systems, sensitive to this structure, could be used as a tool for this purpose;

that possible self correction techniques be investigated, based on the fact that holographic systems record also data relevant to the propagation medium in the field of view; many scattering centers on smooth targets can be considered as point targets and their reconstructed blurred image might be a reference for deblurring techniques; and finally

that, although only indirectly related to the ASI topic "atmospheric effects", the nature and behaviour of scattering centers on objects of interest be studied - especially for millimeter waves - by theory and experiment, to provide a basis for prediction and - later - interpretation of micro-wave imagery.

REFERENCES

1. F. Trenkle, Deutsche Ortungs- und Navigationsanlagen (Land und See 1939-1945). Deutsche Gesellschaft für Ortung und Navigation e.V., Düsseldorf, 1966, No. 1038 (CH. 4).
2. J.R. Patty, E.H. Hurlburt, Microwave Images. Proc. Nat. Electr. Conf. 10 (1954), p. 663-675.
3. H. Jacobs, R. Hofer, G. Morris, E. Horn, Conversion of Millime-ter-Wave Images Into Visible Displays. J.O.S.A. 58 (1968), p. 246-253.
4. B.J. Levin, B.R. Feingold, All-Weather Eye opens up with Milli-meter Wave Imaging. Electronics 43 (Aug. 1970), p. 82-87.
5. R.P. Dooley, X-Band Holography. Proc. IEEE 53 (1965), p. 1733-1735.
6. Y. Aoki, Microwave Holograms and Optical Reconstruction. App. Opt. 6 (1967), p. 1943-1946.
7. Y. Aoki, A. Boivin, Computer Reconstruction of Images from a Microwave Hologram. Proc. IEEE 58 (1970), p. 821-822.
8. N.H. Farhat, Microwave Holography and its Applications in Mo-dern Avionics. 1971 European Microwave Converence, Stockholm, Schweden.
9. K. Magura, Probleme bei der holographischen Abbildung im Mikro-wellenbereich. FHP-Report No. 6-72, Wachtberg-Werthhoven, W. Germany.
10. M. Wu, N.H. Farhat, Real-Time Optical Reconstruction of Micro-wave Holograms. Proc. IEEE 63 (1975), p. 1254-1255.
11. M. Skolnik (ed.), Radar Handbook (Ch. 9). McGraw-Hill Book Comp., New York, 1970.
12. H. Hodara, Laser Wave Propagation in the Atmosphere. 10th Sym-posium of the AGARD Ionospheric Research Committee on Propagation Factors in Space Communications, Rome, Italy, Sept. 1965.
13. J.P. Ruina, C.M. Angulo, Antenna Resolution as Limited by Atmospheric Turbulence. IEEE Trans. AP-11 (1963), p. 153-161.

14. Microwave Emgineers' Handbook Vol. 2, Artech House, Inc., Dedham, Mass., 1971.
15. G. Graf, Analyse des Dopplerfrequenzspektrums rotierender Körper, Deutsche Luft- und Raumfahrt, FB-72-46, 1972.
16. P.G. Smith, Atmospheric Distortion of Signals Originating from Space. IEEE Transactions on Aerospace and Electronic Systems, Vol. AES-3, No. 2, pp. 207-216, March 1967.